ACTIVE MINING

Frontiers in Artificial Intelligence and Applications

Series Editors: J. Breuker, R. López de Mántaras, M. Mohammadian, S. Ohsuga and W. Swartout

Volume 79

Volume 3 in the subseries
Knowledge-Based Intelligent Engineering Systems
Editor: L.C. Jain

Previously published in this series:

Vol. 78, T. Vidal and P. Liberatore (Eds.), STAIRS 2002
Vol. 77, F. van Harmelen (Ed.), ECAI 2002
Vol. 76, P. Sinčák et al. (Eds.), Intelligent Technologies – Theory and Applications
Vol. 75, I.F. Cruz et al. (Eds.), The Emerging Semantic Web
Vol. 74, M. Blay-Fornarino et al. (Eds.), Cooperative Systems Design
Vol. 73, H. Kangassalo et al. (Eds.), Information Modelling and Knowledge Bases XIII
Vol. 72, A. Namatame et al. (Eds.), Agent-Based Approaches in Economic and Social Complex Systems
Vol. 71, J.M. Abe and J.I. da Silva Filho (Eds.), Logic, Artificial Intelligence and Robotics
Vol. 70, B. Verheij et al. (Eds.), Legal Knowledge and Information Systems
Vol. 69, N. Baba et al. (Eds.), Knowledge-Based Intelligent Information Engineering Systems & Allied Technologies
Vol. 68, J.D. Moore et al. (Eds.), Artificial Intelligence in Education
Vol. 67, H. Jaakkola et al. (Eds.), Information Modelling and Knowledge Bases XII
Vol. 66, H.H. Lund et al. (Eds.), Seventh Scandinavian Conference on Artificial Intelligence
Vol. 65, In production
Vol. 64, J. Breuker et al. (Eds.), Legal Knowledge and Information Systems
Vol. 63, I. Gent et al. (Eds.), SAT2000
Vol. 62, T. Hruška and M. Hashimoto (Eds.), Knowledge-Based Software Engineering
Vol. 61, E. Kawaguchi et al. (Eds.), Information Modelling and Knowledge Bases XI
Vol. 60, P. Hoffman and D. Lemke (Eds.), Teaching and Learning in a Network World
Vol. 59, M. Mohammadian (Ed.), Advances in Intelligent Systems: Theory and Applications
Vol. 58, R. Dieng et al. (Eds.), Designing Cooperative Systems
Vol. 57, M. Mohammadian (Ed.), New Frontiers in Computational Intelligence and its Applications
Vol. 56, M.I. Torres and A. Sanfeliu (Eds.), Pattern Recognition and Applications
Vol. 55, G. Cumming et al. (Eds.), Advanced Research in Computers and Communications in Education
Vol. 54, W. Horn (Ed.), ECAI 2000
Vol. 53, E. Motta, Reusable Components for Knowledge Modelling
Vol. 52, In production
Vol. 51, H. Jaakkola et al. (Eds.), Information Modelling and Knowledge Bases X
Vol. 50, S.P. Lajoie and M. Vivet (Eds.), Artificial Intelligence in Education
Vol. 49, P. McNamara and H. Prakken (Eds.), Norms, Logics and Information Systems
Vol. 48, P. Návrat and H. Ueno (Eds.), Knowledge-Based Software Engineering
Vol. 47, M.T. Escrig and F. Toledo, Qualitative Spatial Reasoning: Theory and Practice
Vol. 46, N. Guarino (Ed.), Formal Ontology in Information Systems
Vol. 45, P.-J. Charrel et al. (Eds.), Information Modelling and Knowledge Bases IX

ISSN: 0922-6389

Active Mining

New Directions of Data Mining

Edited by

Hiroshi Motoda

Division of Intelligent Systems Science,
The Institute of Scientific and Industrial Research,
Osaka University, Osaka, Japan

IOS Press
Ohmsha

Amsterdam • Berlin • Oxford • Tokyo • Washington, DC

ISBN 1 58603 264 X (IOS Press)
ISBN 4 274 90521 7 C3055 (Ohmsha)
Library of Congress Control Number: 2002106944

Publisher
IOS Press
Nieuwe Hemweg 6B
1013 BG Amsterdam
The Netherlands
fax: +31 20 620 3419
e-mail: order@iospress.nl

Distributor in the UK and Ireland
IOS Press/Lavis Marketing
73 Lime Walk
Headington
Oxford OX3 7AD
England
fax: +44 1865 75 0079

Distributor in the USA and Canada
IOS Press, Inc.
5795-G Burke Centre Parkway
Burke, VA 22015
USA
fax: +1 703 323 3668
e-mail: iosbooks@iospress.com

Distributor in Germany, Austria and Switzerland
IOS Press/LSL.de
Gerichtsweg 28
D-04103 Leipzig
Germany
fax: +49 341 995 4255

Distributor in Japan
Ohmsha, Ltd.
3-1 Kanda Nishiki-cho
Chiyoda-ku, Tokyo 101-8460
Japan
fax: +81 3 3233 2426

PRINTED IN THE NETHERLANDS

Preface

Our ability to collect data, be it in business, government, science, and perhaps personal life has been increasing at a dramatic rate. However, our ability to analyze and understand massive data lags far behind our ability to collect them. The value of data is no longer in "how much of it we have". Rather, the value is in how quickly and how effectively can the data be reduced, explored, manipulated and managed.

Knowledge Discovery and Data mining (KDD) is an emerging technique that extracts implicit, previously unknown, and potentially useful information (or patters) from data. Recent advancement made through extensive studies and real world applications reveals that no matter how powerful computers are now or will be in the future, KDD researchers and practitioners must consider how to manage ever-growing data which is, ironically, due to the extensive use of computers and ease of data collection, ever-increasing forms of data which different applications require us to handle, and ever-changing requirements for new data and mining target as new evidences are collected and new findings are made. In short, the need for 1) identifying and collecting the relevant data from a huge information search space, 2) mining useful knowledge from different forms of massive data efficiently and effectively, and 3) promptly reacting to situation changes and giving necessary feedback to both data collection and mining steps, is ever increasing in this era of information overload.

Active mining is a collection of activities each solving a part of the above need, but collectively achieving the various mining objectives. By "collectively achieving" we mean that the total effect outperforms the simple add-sum effect that each individual effort can bring. Said differently, a spiral effect of these interleaving three steps is the target to be pursued. To achieve this goal the initial action is to explore mechanisms of 1) active information collection where necessary information is effectively searched and pre-processed, 2) user-centered active mining where various forms of information sources are effectively mined, and 3) active user reaction where the mined knowledge is easily assessed and prompt feedback is made possible.

This book is a joint effort from leading and active researchers in Japan with a theme about active mining. It provides a forum for a wide variety of research work to be presented ranging from theories, methodologies, algorithms, to their applications. It is a timely report on the forefront of data mining. It offers a contemporary overview of modern solutions with real-world applications, shares hard-learned experiences, and sheds light on future development of active mining.

This collection evolved from a project on active mining and the papers in this collection were selected from among over 40 submissions.

The book consists of 3 parts. Each part corresponds to one of the three mechanisms mentioned above. Namely, part I consists of chapters on Data Collection, part II on User-centered Mining, and part III on User Reaction and Interaction. Some of the chapters overlap each other but have to be placed in one of these three parts. The topics covered in 27 chapters include online text mining, clustering for information gathering, online monitoring of Web page updates, technical term classification, active information gathering, substructure mining from Web and graph structured data, web community discovery and classification, spatial data mining, automatic configuration of mining tools, worst case analysis of exceptional rule mining, data squashing applied to boosting, outlier detection, meta-learning for evidenced based medicine, knowledge acquisition from both human expert and data, data visualization, active mining in business application world, meta analysis and many more.

This book is intended for a wide audience, from graduate students who wish to learn basic concepts and principles of data mining to seasoned practitioners and researchers who want to take advantage of the state-of-the-art development for active mining. The book can be used as a reference to find recent techniques and their applications, as a starting point to find other related research topics on data collection, data mining and user interaction, or as a stepping stone to develop novel theories and techniques meeting the exciting challenges ahead of us.

Active mining is a new direction in the knowledge discovery process for real-world applications handling huge amounts of data with actual user need.

Hiroshi Motoda

Acknowledgments

As the field of data mining advances, the interest in as well as the need for integrating various components intensifies for effective and successful data mining. A lot of research ensues. This book project resulted from the active mining initiatives that started during 2001 as a grant-in-aid for scientific research on priority area by the Japanese Ministry of Education, Science, Culture, Sports and Technology. We received many suggestions and support from researchers in machine learning, data mining and database communities from the very beginning of this book project. The completion of this book is particularly due to the contributors from all areas of data mining research in Japan, their ardent and creative research work. The editorial members of this project have kindly provided their detailed and constructive comments and suggestions to help clarify terms, concepts, and writing in this truly multi-disciplinary collection. I wish to express my sincere thanks to the following members: Numao Masayuki, Yukio Ohsawa, Einoshin Suzuki, Takao Terano, Shusaku Tsumoto and Takahira Yamaguchi.

We are also grateful to the editorial staff of IOS Press, especially Carry Koolbergen and Anne Marie de Rover for their swift and timely help in bringing this book to a successful conclusion.

During the process of this book development, I was generously supported by our colleagues and friends at Osaka University.

Contents

III. User Reaction and Interaction

I

DATA COLLECTION

Toward Active Mining from On-line Scientific Text Abstracts Using Pre-existing Sources

TuanNam Tran and Masayuki Numao
tt-nam@nm.cs.titech.ac.jp, numao@cs.titech.ac.jp
Department of Computer Science,
Tokyo Institute of Technology
2-12-1 O-okayama, Meguro-ku, Tokyo 152-8552, JAPAN

Abstract. As biomedical research enters the post-genome era and most new information relevant to biology research is still recorded as free text, there is an extensively increasing needs of extracting information from biological literature databases such as MEDLINE. Different from other work so far, in this paper we presents a framework for mining MEDLINE by making use of a pre-existing biological database on a kind of *Yeast* called *S.cerevisiae*. Our framework is based on an active mining prospect and consists of two tasks: an information retrieval task of actively selecting articles in accordance with users' interest, and a text data mining task using association rule mining and term extraction techniques. The preliminary results indicate that the proposed method may be useful for consistency checking and error detection in annotation of MeSH terms in MEDLINE records. It is considered that the proposed approach of combining information retrieval making use of pre-existing databases and text data mining could be expanded for other fields such as Web mining.

1 Introduction

Because of the rapid growth of computer hardwares and network technologies, a vast amount of information could be accessed through a variety of databases and sources. Biology research inevitably plays an essential role in this century, producing a large number of papers and on-line databases on this field. However, even though the number and the size of sequence databases are growing rapidly, most new information relevant to biology research is still recorded as free text. As biomedical research enters the post-genome era, new kinds of databases that contain information beyond simple sequences are needed, for example, information on protein-protein interactions, gene regulation etc. Currently, most of early work on literature data mining for biology concentrated on analytical tasks such as identifying protein names [5], simple techniques such as word co-occurrence [12], pattern matching [8], or based on more general natural language parsers that could handle considerably more complex sentences [9], [15].

In this paper, a different approach is proposed for dealing with literature data mining from MEDLINE, a biomedical literature database which contains a vast amount of useful information on medicine and bioinformatics. Our approach is based on *active mining*, which focuses on *active information gathering* and data mining in accordance with the purposes and interests of the users. In detail, our current system contains two subtasks: the first task exploits existing databases and machine learning techniques for selecting useful articles, and the second one using association rule mining and term extraction techniques to conduct text data mining from the set of documents obtained by the first task.

The remainder of this paper is organized as follows. Section 2 gives a brief overview on literature data mining. Section 3 describes in detail the task of making use of existing databases to retrieve relevant documents (the information retrieval task). Given the results obtained from the Section 3, Section 4 introduces the text mining task by using association rule mining and term extraction. Section 5 describes some directions for future work. Finally Section 6 presents our conclusions.

2 Overview on literature data mining for biology

In this section we give a brief overview of current work on literature data ming for biology. As described above, even though the number and the size of sequence databases are growing rapidly, most new information relevant to biology research is still recorded as free text. As a result, biologists need information contained in text to integrate information across articles and update databases. Current automated natural language systems could be classified as *information retrieval* systems (which return documents relevant to a subject), *information extraction* systems (which identify entities or relations among entities in text) and *question answering* system (which answer factual questions using large document collections). However, it should be noted that most of these systems work on newswire, and text mining for biology is considered to be harder because the syntax is more complex, new terms are introduced constantly and there is a confusion between genes and proteins [6].

On the other hand, since natural language processing offers the tools to make information in text accessible, there are an increasing numbers of groups working on natural language processing for biology. Fukuda et. al. [5] attempt to identifying protein names from biological papers. Andrade and Valencia [2] also concentrate on extraction of keywords, not mining factual assertions. There have been many approaches to the extraction of factual assertions using natural language processing techniques such as syntactic parsing. Sekimizu et. al. [11] attempt to generate automatic database entries containing relations extracted from MEDLINE abstracts. Their approach is to parse, determine noun phrases, spot the frequently-occurring verbs and choose the most likely subject and object from the candidate NPs in the surrounding text. Rindflesch [10] uses a stochastic part-of-speech tagger to generate an underspecified syntactic parse and then uses semantic and pragmatic information to construct its assertions. This system can only extract mentions of well-characterized genes, drugs cell types, not the interactions among them. Thomas et. al. [13] use an existing information extraction system called SRI's Highlight for gathering data on protein interactions. Their work concentrates on finding relations directly between proteins. Blaschke et. al. [3] attempt to generate functional relationship maps from abstracts, however, it requires a pre-defined list of all named entities and cannot handle syntactically complex sentences.

3 Retrieving relevant documents by making use of existing database

We describe our information retrieval task, which can be considered as a specific task for retrieving relevant documents from MEDLINE. Current systems for accessing MEDLINE such as PubMed ([1]) accept keyword-based queries to text sources and return

[1]http://www.ncbi.nlm.nih.gov/PubMed/

documents that are hopefully relevant to the query. Since MEDLINE contains an enormous amount of papers and the current MEDLINE search engines is a keyword-base one, the number of returned documents is often large, and many of them in fact are non-relevant. The approach to solve this issue is to make use of existing databases of organisms such as *S.cerevisiae* using supervised machine learning techniques.

Figure 1 shows the illustration of the information retrieval task. In this Figure, YPD database (standing for Yeast Protein Database [2]) is a biological database which contains genetic functions and other characteristics of a kind of *Yeast* called *S.cerevisiae*. Given a certain organism X, the goal of this task is to retrieve its relevant documents, i.e. documents containing useful genetic information for biological research.

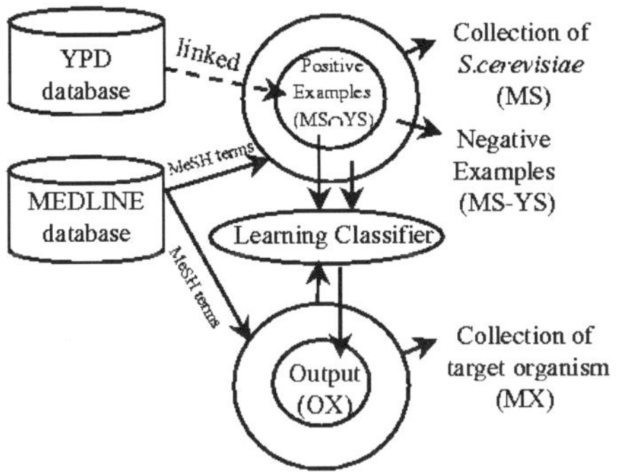

Figure 1: Outline of the information retrieval task

Let MX, MS be the sets of documents retrieved from MEDLINE by querying for the target organism X and *S.cerevisiae* respectively (without any machine learning filtering) and YS be the set of documents found by querying for the YPD terms for *S.cerevisiae* (YS is omitted in Figure 1 for the reason of simplification). The set of positive and negative examples then are collected as the intersection set and difference set of MS and YS respectively. Given the training examples, OX is the output set of documents obtained by applying Naive Bayes classifier on MX.

3.1 Naive Bayes classifier

Naive Bayes classifiers ([7]) are among the most successful known algorithms for learning to classify text documents. A naive Bayes classifier is constructed by using the training data to estimate the probability of each category given the document feature values of a new instance. The probability a instance d belongs to a class c_k is estimates by Bayes theorem as follows:

$$P(C = c_k|d) = \frac{P(d|C = c_k)P(C = c_k)}{P(d)}$$

Since $P(d|C = c_k)$ is often impractical to compute without simplifying assumptions, for the Naive Bayes classifier, it is assumed that the features X_1,X_2,...X_n are conditionally

[2]http://www.proteome.com/databases/index.html

independent, given the category variable C. As a result :

$$P(d|C = c_k) = \prod_i P(d_i|C = c_k)$$

3.2 Experimental results of information retrieval task

Our experiments use YPD as an existing database. From this database we obtain 14572 articles pertaining to *S.cerevisiae*. For the target organisms, initially we collect 3073 and 8945 articles for two kinds of *Yeast* called *Pombe* and *Candida* respectively. After conducting experiments as in Figure 1, we obtain the output containing 1764 and 285 articles for *Pombe* and *Candida* respectively.

A certain number of documents (50 in this experiment) in each of dataset is taken randomly, checked by hand whether they are relevant or not. Figure 2 shows the Recall-Precision curve for *Pombe* and *Candida*. It can be seen from this Figure that using machine learning approaches remarkably improved the precision. The reason the recall in the case of *Candida* is rather lower compared to the case of *Pombe* is that *Pombe* is a yeast which has many similar genetic characteristics than *Candida*.

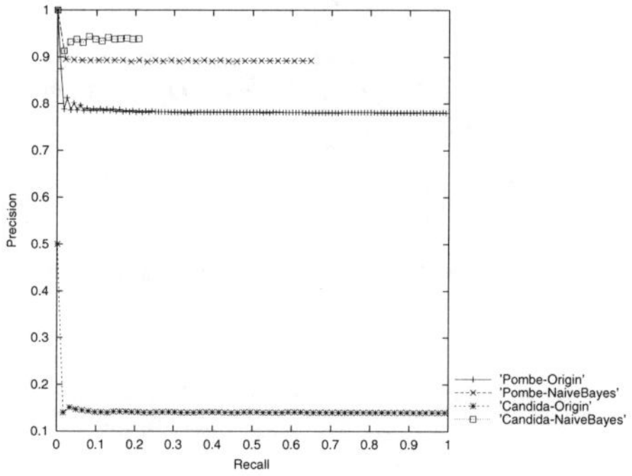

Figure 2: Recall-Precision curve for *Pombe* and *Candida*

4 Mining MEDLINE by combining term extraction and association rule mining

In this section, we attempt to mine the set of MEDLINE documents obtained in the previous section by combining term extraction and association rule mining.

The text mining task from the collected dataset consists of two main modules: the Term Extraction module and the Association-Rule Generation module. The Term Extraction module itself includes the following stages:

- **XML translation**: This stage translates the MEDLINE record from HTML form into a XML-like form, conducting some pre-processing dealing with punctuation.
- **Part-of-speech tagging**: Here, the rule-based Brill part-of-speech tagger [4] was used for tagging the title and the abstract part.

- **Term Generation**: sequences of tagged words are selected as potential term candidates on the basis of relevant morpho-syntactic patterns (such as "Noun Noun", "Noun Adjective Noun", "Adjective Noun", "Noun Preposition Noun" etc). For example, "in vivo", "saccharomyces cerevisiae" are terms extracted from this stage.

- **Stemming**: Stemming algorithm was used to find variations of the same word. Stemming transforms variations of the same word into a single one, reducing vocabulary size.

- **Term Filtering**: In order to decrease the number of "bad terms", in the abstract part, only sentences containing verbs listed in the "verbs related to biological events" Table in [14] have been used for Term Generation stage.

After necessary terms have been generated from the Term Extraction module, the Association-Rule Generation module then applies the Apriori algorithm[1] using the set of generated terms to produce association rules (each line of the input file of Apriori-based program consists every terms extracted from a certain MEDLINE record in the dataset).

Figure 3 and Figure 4 show the list of twenty rules among obtained rules demonstrating the relationships among extracted terms for *Pombe* and *Candida* respectively. For example, the 5th rule in Figure 4 implies that "the rule that in a MEDLINE record if aspartyl proteinases occurs then this MEDLINE document is published in the Journal of Bacteriology has the support of 1.3% and the confidence of 100.0%.". It can be seen that the relation between journal name and terms extracted from the title and the abstract has been discovered from this example. It can be seen from Figure 3 and 4 that making use of terms can produced interesting rules that cannot be obtained using only single-words.

5 Future Work

5.1 For the information retrieval task

Although using an existing database of *S.cerevisiae* is able to obtain a high precision for other yeasts and organisms, the recall value is still low, especially for the yeasts which are different remarkably from *S.cerevisiae*. Since yeasts such as *Candida* might have many unique attributes, we may improve the recall by feeding the documents checked by hand back to the classifier and conduct the learning process again. The negative training set has still contained many positive examples so we need to reduce this noise by making use of the learning results.

5.2 For the text mining task

By combining term extraction and association rule mining, it is able to obtain interesting rules such as the relations among journal names and terms, terms and terms. Particularly, the relations among MeSH terms and "Substances" may be useful for error detection in annotation of MeSH terms in MEDLINE records. However, the current algorithm treats extracted terms such as "cdc37_caryogamy_defect", "cdc37_in_mitosy",

```
1: fission_yeast_schizosaccharomyc_pomb <-
      transcript_control  (0.3%, 80.0%)
2: cell_cycle <- period  (0.6%, 77.8%)
3: mutant <- other_mutant  (0.4%, 83.3%)
4: essenty <- gene_disrupt_expery  (0.5%, 75.0%)
5: mitosy <- passag_through_start  (0.3%, 80.0%)
6: transcript <- mat2-mat3_interval  (0.3%, 80.0%)
7: embo_j <- p34cdc2_kinas_activity  (0.5%, 75.0%)
8: nucleu <- periphery  (0.3%, 80.0%)
9: structur <- function_similar  (0.3%, 80.0%)
10: meiosy <- premeiot_dna_synthesy  (0.5%, 75.0%)
11: meiosy <- pair  (0.3%, 80.0%)
12: s_phase <- complet_of_s_phase  (0.4%, 83.3%)
13: amino_acid_sequ <- alignment  (0.4%, 83.3%)
14: amino_acid_sequ <- _residu  (0.3%, 80.0%)
15: human <- mous_homolog  (0.3%, 80.0%)
16: open_read_frame <- uninterrupt  (0.4%, 83.3%)
17: subunit <- rpb2  (0.3%, 80.0%)
18: centromer <- central_core  (0.4%, 83.3%)
19: centromer <- centromer_function  (0.4%, 83.3%)
20: wee1 <- mik1  (0.5%, 85.7%)
```

Figure 3: First twenty rules obtained for the set of *Pombe* documents obtained in Section 3 (*minimum support = 0.003, minimum confidence = 0.75)*

"cdc37_mutat" to be mutually independent. It may be necessary to construct semi-automatically term taxonomy, for instance users are able to choose only interesting rules or terms then feedback to the system.

5.3 Mutual benefits between two tasks

Gaining mutual benefits between two tasks is also an important issue for future work. First, by applying text mining results, it should be noted that we can decrease the number of documents being "leaked" in the information retrieval task. As a result, it is possible to improve the recall. Conversely, since the current text mining algorithm create many unnecessary rules (from the viewpoint of biological research), it is also possible to apply the information retrieval task first for filtering relevant documents, then apply to the text mining task to decrease the number of unnecessary rules obtained and to improve the quality of the text mining task.

6 Conclusions

This paper has introduced a framework for mining MEDLINE by making use of existing biological databases. Two tasks concerning information extraction from MEDLINE have been presented. The first task is used for retrieving useful documents for biology research with high precision. Given the obtained set of documents, the second task attempts to apply association rule mining and term extraction for mining these documents. It can be seen from this paper that making use of the obtained results is useful for consistency checking and error detection in annotation of MeSH terms in MEDLINE records. In future work, combining these two tasks together may be essential to gain mutual benefits for both two tasks.

```
1: open_read_frame <- molecular_weight  (1.8%, 75.0%)
2: open_read_frame <- molecular_mass  (1.8%, 75.0%)
3: open_read_frame <- cdna_clone  (1.3%, 100.0%)
4: virul <- growth_rate  (1.8%, 75.0%)
5: j_bacteriol <- aspartyl_proteinas  (1.3%, 100.0%)
6: j_bacteriol <- gene_code  (1.3%, 100.0%)
7: j_bacteriol <- sucros  (1.3%, 100.0%)
8: organism <- immunoelectron_microscopy
      (1.3%, 100.0%)
9: resist <- transport  (1.8%, 75.0%)
10: similar <- hyphal_growth  (1.8%, 75.0%)
11: clone <- southern_blot  (1.3%, 100.0%)
12: white <- opaqu  (1.8%, 75.0%)
13: white <- opaqu_phase  (1.8%, 75.0%)
14: white <- opaqu_cell  (1.8%, 75.0%)
15: amino_acid_sequ <- comparison  (2.7%, 83.3%)
16: amino_acid_sequ <- escherichia_coly  (1.8%, 75.0%)
17: amino_acid_sequ <- alignment  (1.8%, 75.0%)
18: fragment <- molecular_mass  (1.8%, 75.0%)
19: cell_wall <- moiety  (1.3%, 100.0%)
20: cell_wall <- immunoelectron_microscopy
      (1.3%, 100.0%)
```

Figure 4: First twenty rules obtained for the set of *Candida* documents obtained in Section 3 (*minimum support = 0.01, minimum confidence = 0.75*)

References

[1] R. Agrawal and R. Srikant. Fast algorithms for mining association rules. In *Proceedings of the 20th International Conference on Very Large Databases*, 1994.

[2] M.A. Andrade and A. Valencia. Automatic annotation for biological sequences by extraction of keywords from medline abstracts. development of a prototype system. In *Proceedings of the 5th International Conference on Intelligent Systems for Molecular Biology*, 1997.

[3] C. Blaschke, M.A. Andrade, C. Ouzounis, and A. Valencia. Automatic extraction of biological information from scientific text: protein-protein interactions. In *Proceedings of the 7th International Conference on Intelligent Systems for Molecular Biology*, 1999.

[4] E. Brill. A simple rule-based part of speech tagger. In *Proceedings of the Third Conference on Applied Natural Language Processing*, 1992.

[5] K. Fukuda, A. Tamura, T. Tsunoda, and T. Takagi. Toward information extraction: identifying protein names from biological papers. In *Proceedings of the Pacific Symposium on Biocomputing*, 1998.

[6] L. Hirschman. Mining the biomedical literature: Creating a challenge evaluation. Technical report, The MITRE Corporation, 2001.

[7] D.D. Lewis and M. Ringuette. A comparison of two learning algorithms for text categorization. In *Third Annual Symposium on Document Analysis and Information Retrieval*, 1994.

[8] S. K. Ng and M. Wong. Toward routine automatic pathway discovery from on-line scientific text abstracts. *Genome Informatics*, 10:104–11, December 1999.

[9] J. C. Park, H. S. Kim, and J. J. Kim. Bidirectional incremental parsing for automatic pathway identification with combinatory categorial grammar. In *Proceedings of the Pacific Symposium on Biocomputing*, 2001.

[10] T.C. Rindflesch. Edgar: Extraction of drugs, genes and relations from the biomedical literature. In *Proceedings of the Pacific Symposium on Biocomputing*, 2000.

[11] T. Sekimizu, H.S. Park, and J. Tsujii. Identifying the interaction between genes and gene products based on frequently seen verbs in medline abstracts. *Genome Informatics*, pages 62–71, 1998.

[12] B. J. Stapley and G. Benoit. Biobibliometrics: Information retrieval and visualization from co-occurrences of gene names in medline abstracts. In *Proceedings of the Pacific Symposium on Biocomputing*, 2000.

[13] J. Thomas, D. Milward, C. Ouzounis, S. Pulman, and M. Carroll. Automatic extraction of protein interactions from scientific abstracts. In *Proceedings of the Pacific Symposium on Biocomputing*, 2000.

[14] J. Tsujii. Information extraction from scientific texts. In *Proceedings of the Pacific Symposium on Biocomputing*, 2001.

[15] A. Yakushiji, Y. Tateisi, Y. Miyao Y., and J. Tsujii. Event extraction from biomedical papers using a full parser. In *Proceedings of the Pacific Symposium on Biocomputing*, 2001.

Active Mining
H. Motoda (Ed.)
IOS Press, 2002

Data Mining on the WAVEs — Word-of-mouth-Assisting Virtual Environments

Masayuki Numao, Masashi Yoshida and Yusuke Ito
numao@cs.titech.ac.jp
http://www.nm.cs.titech.ac.jp
Department of Computer Science
Tokyo Institute of Technology
2-12-1 O-okayama, Meguro 152-8552, JAPAN

Abstract. Recently, computers play an important role not only in knowledge processing but also as communication media. However, they often cause troubles in communication, since it is hard for us to select only useful pieces of information. To overcome this difficulty, we propose a new tool, WAVE (Word-of-mouth-Assisting Virtual Environment), which helps us to communicate and spread information by relaying a message like *Chinese whispers*. This paper describes its concept, an implementation and its preliminary evaluation.

1 Introduction

Chinese whispers *a game in which a message is distorted by being passed around in a whisper (also called Russian scandal).*

word of mouth *(a) oral communication or publicity; (b) done, given, etc., by speaking; oral.*

— *New Shorter Oxford English Dictionary*

WWW and e-mail are very useful tools for communication. However, we sometimes feel uncomfortable because of flaming or mental barriers to participate in Computer-Mediated Communication (CMC). There are some important differences between CMC and direct communication[5].

Another problem is that computer networks deliver too many pieces of information, by which it is too hard to select useful pieces. Although search engines, such as *Yahoo*, *Goo* and *Google*, are very useful to find web pages, we need another type of tool without requiring a keyword for search. Good candidates are a mailing list and a network news system, where we need a filtering system to select only useful messages. Although content-based filtering[6] and collaborative filtering[8] are good solutions, the current methods have not achieved high precision and recall. This paper presents another approach by relaying a message like Chinese whispers to gather useful information, to alleviate mental barriers and to block flames.

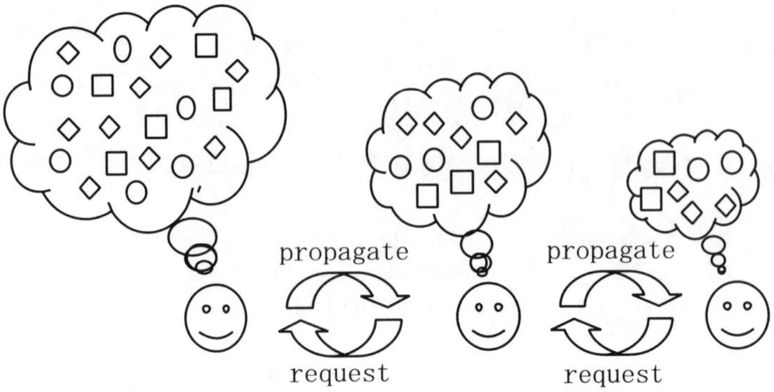

Figure 1: Spread of information

2 Spread of information by Chinese whispers

Fig. 1 shows spread of information by word of mouth, where each person relays a message like Chinese whispers. Although a message is distorted by being passed around in the game, in a computer-assisted environment we expect that a delivered message is the same as its original. In such a process, we even have a merit that, as a result of evaluation and selection by each person, this process delivers only useful information. Each person knows whom (s)he should ask on a current topic, and retrieve a small amount that can be handled, where only interesting information survives.

3 WAVE

To assist spread of information by Chinese whispers, we propose a system WAVE (Word-of-mouth-Assisting Virtual Environment) for smooth communication and information gathering. Compared to agent systems proposed to automate word of mouth [1, 9, 2, 7], WAVE is a simpler tool and works as directed by the user except for a separated recommendation module. The authors believe that, in most situations, a simple and intuitive tool is better than an automated complicated tool, since users construct a model of the tool easily.

Fig. 2 shows a diagram of WAVE. The user's operations are posting, opening and reviewing an article. In addition, in a recommendation window, the system shows some good articles based on the user's log.

3.1 Posting an article

The user can post an article as shown in Fig. 3, which may contain a text and URLs of web pages or photos. (S)he gives evaluation 1-5 (1 for the worst and 5 for the best) and a category to the article. The posted article is open to others as shown in Fig. 4 and referred by other users like WWW and a mailing list.

The user can browse articles posted by her/his friends. Fig. 5 shows a list of friends. Each person is identified by an address 'user_name@host:port'. If an article is interesting, (s)he can post its review, by which (s)he relays the article to his friends as shown in Fig. 2. Fig. 4 shows a list of articles the user has posted or reviewed.

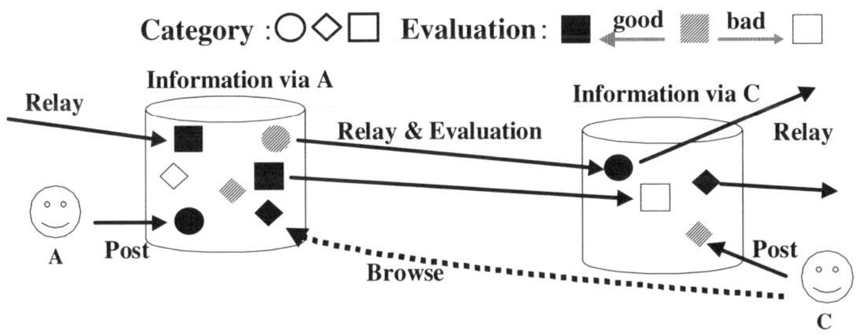

Figure 2: Word-of-mouth-Assisting Virtual Environment

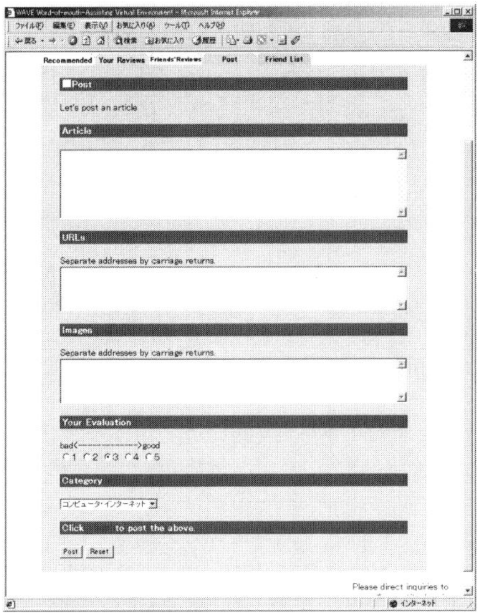

Figure 3: Posting an article

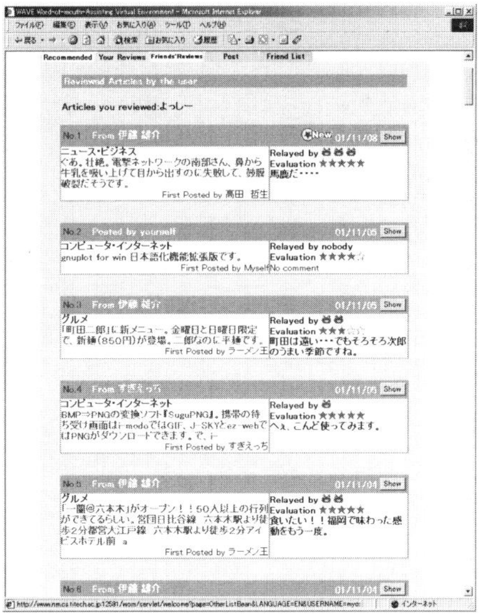

Figure 4: Articles posted or reviewed

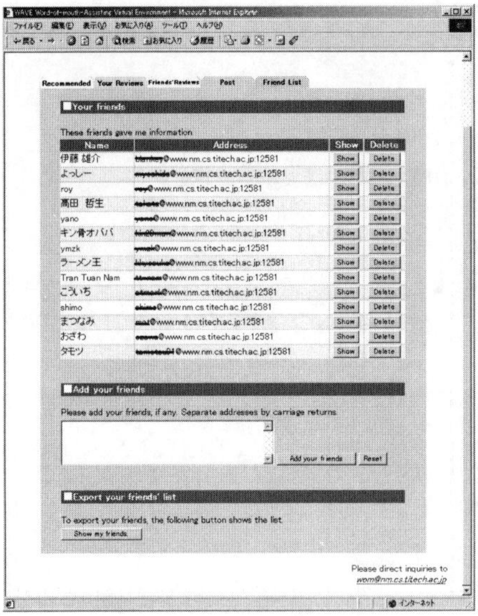

Figure 5: Your friends

Figure 6: Reviews by your friend

3.2 Open articles

Articles posted or reviewed by the user are stored in her/his database. It is open to people who registered her/him as a friend. The user can register an address of her/his friends, or notify her/his address to another user. For example, if C registered A and B to her/his friend's list, C can see the databases of A and B.

Since each user knows her/his friends, (s)he can judge their reliability, which is very useful to select information from them. In addition, it is comfortable to join the community because (s)he exchanges messages only with her/his friends.

3.3 Review an article

If C is interested in an article from A in Fig. 2, C can browse its body and give an evaluation and a comment as shown in Fig. 7. After this operation, the article is automatically retrieved and stored in C's database, which is open to C's friends. Chaining the operation propagates an article.

As such, WAVE seamlessly assists opening, browsing, evaluation, retrieve of an article. This saves us a lot of time and labor of uploading, advertisement, etc. In BBS and mailing lists, most participants feel mental barriers to post an article. In contrast, a user first posts an article only to his friends in WAVE. Mental barriers are alleviated in this fashion. ROMs (Read Only Members) often form a bridge between two communities. WAVE is useful to activate a bridge.

3.4 Automatic recommendation

When a user has many friends, it might be good to order articles based on her/his model. Modeling a person is difficult since we cannot directly measure a mental state. Even if it can by using MRI or other devices, it is still hard to clarify a relation between

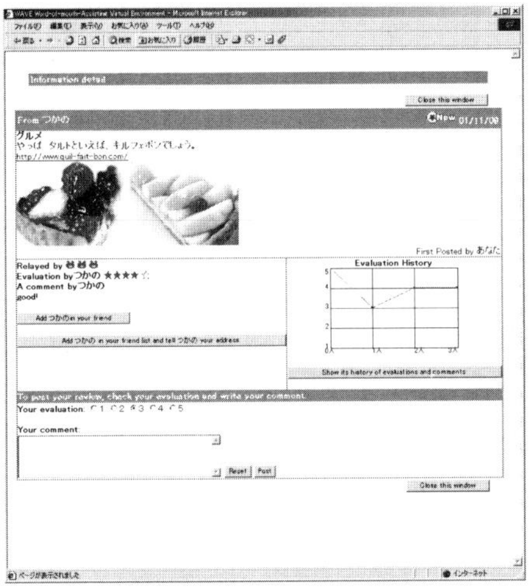

Figure 7: An article

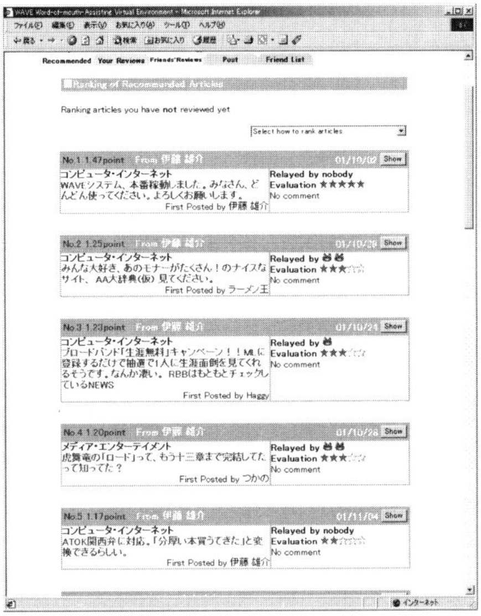

Figure 8: Recommendation

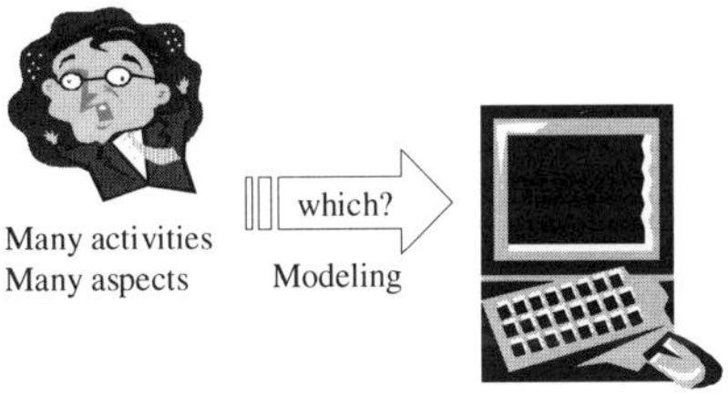

Figure 9: Modeling

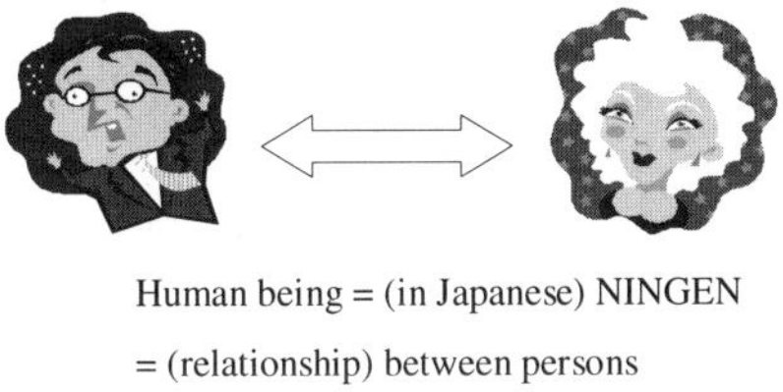

Figure 10: Modeling based on communication

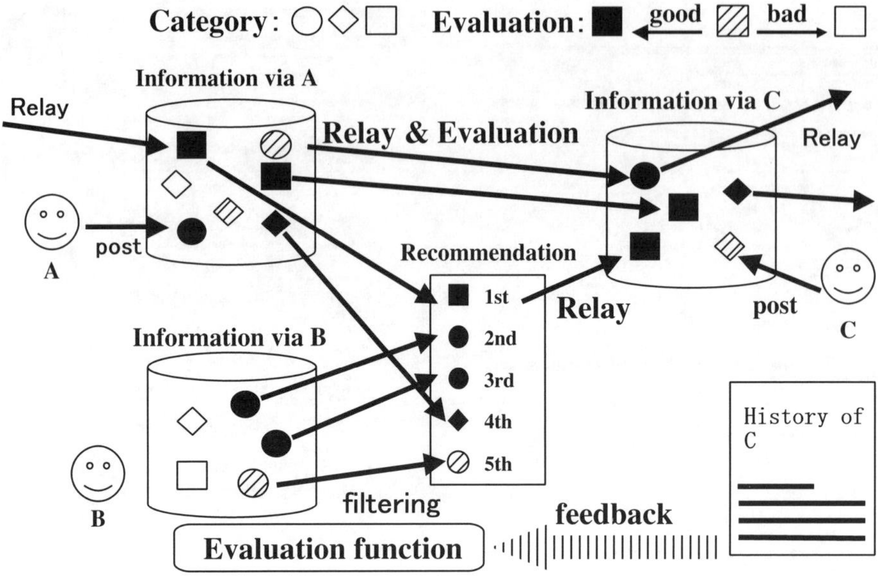

Figure 11: Recommending process

a brain state and its social effects, since a person has many activities and aspects (Fig. 9). Instead, we propose to model a relation between two persons by logging their communication.

To model a relation between two persons, we need a log of communications between all combinations of persons. This causes a trouble in analyzing WWW, a news system or a mailing list. In contrast, all communications are occurred only among friends in WAVE. We have no combinatorial problem in analyzing communications and modeling relations, since the number of friends of one person is not usually large.

Fig. 11 shows a process of ordering articles for recommendation, where C's history is analyzed based on an evaluation function to order articles in databases of A and B, and evaluation is based on the following factors:

- Evaluation of the article by the last reviewer.
- Evaluation of the last reviewer by the user.
- The user's preference for the category of article.
- How old is the article?
- How many people relay the article?

3.5 Distributed implementation

The system is implemented on Java servelet and works on a web server as shown in Fig. 12. The user first registers her/his name and password, and accesses the system by using a web browser.

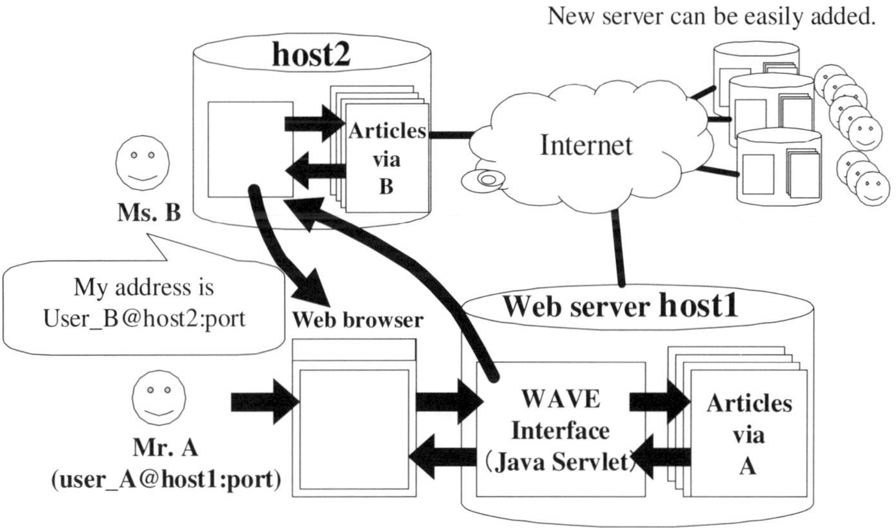

Figure 12: Distributed implementation

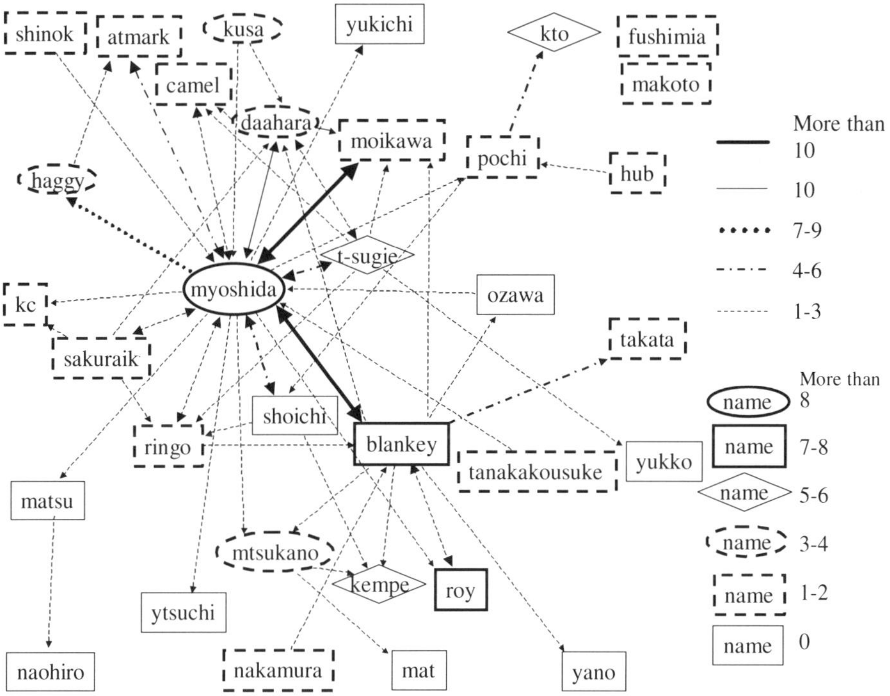

Figure 13: Social network analysis

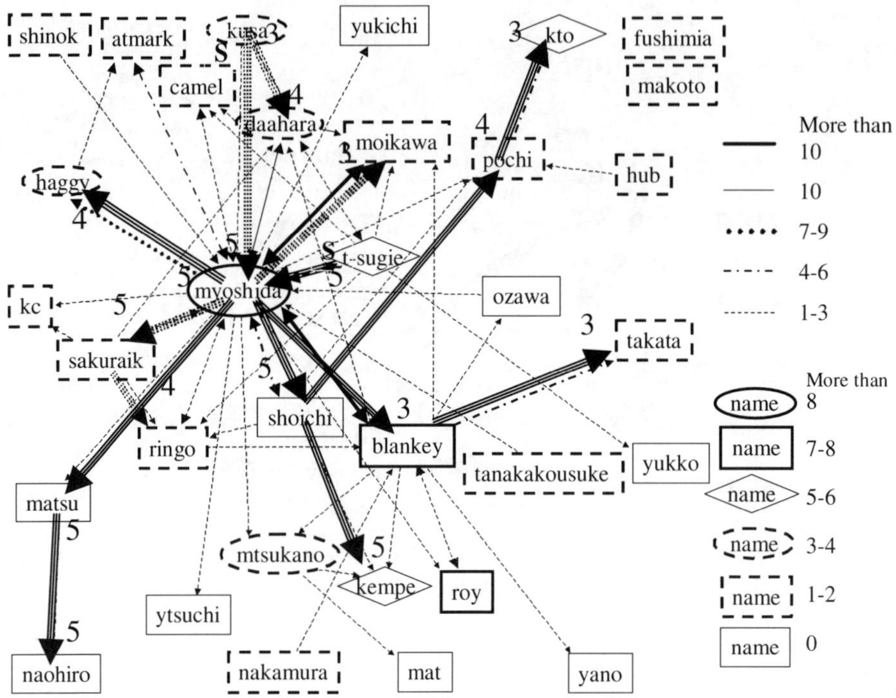

Figure 14: Two example flows of an article

The system is distributed easily to several hosts. In Fig. 12, Mr. A registered on host1 to use the system. Ms. B registered on host2. Mr. A can see Ms. B's article by specifying her address. As such, the system is scalable by being distributed over many hosts.

4 Preliminary evaluation

33 users test the system for 20 days. The result is visualized as shown in Fig. 13. This map is based on one by KrackPlot[4], which is a program for network visualization designed for social network analysts.

Each node denotes a user, whose shape denotes the number of articles (s)he posts. Here, myoshida, blankey, roy and t-sugie are opinion leaders that post many articles. A directed arc denotes that articles are retrieved and reviewed in that direction. Its thickness denotes the number of articles retrieved. In the network, we can see many triangles, each of which forms *triad* strongly connecting each other.

Two example flows of an article are shown in Fig. 14. One flow is in thick solid line. The other is in thick dotted line. *S* denotes their origin. Each attached number denotes evaluation by each person. In most cases, the evaluation degrades as people relay an article.

Each island circled in Fig. 15 shows a community the authors observed, where people know each other in their real life. An article moves mainly in a community. Some people appear in multiple communities, and play a role of *gatekeeper*[3], who bridges information between communities.

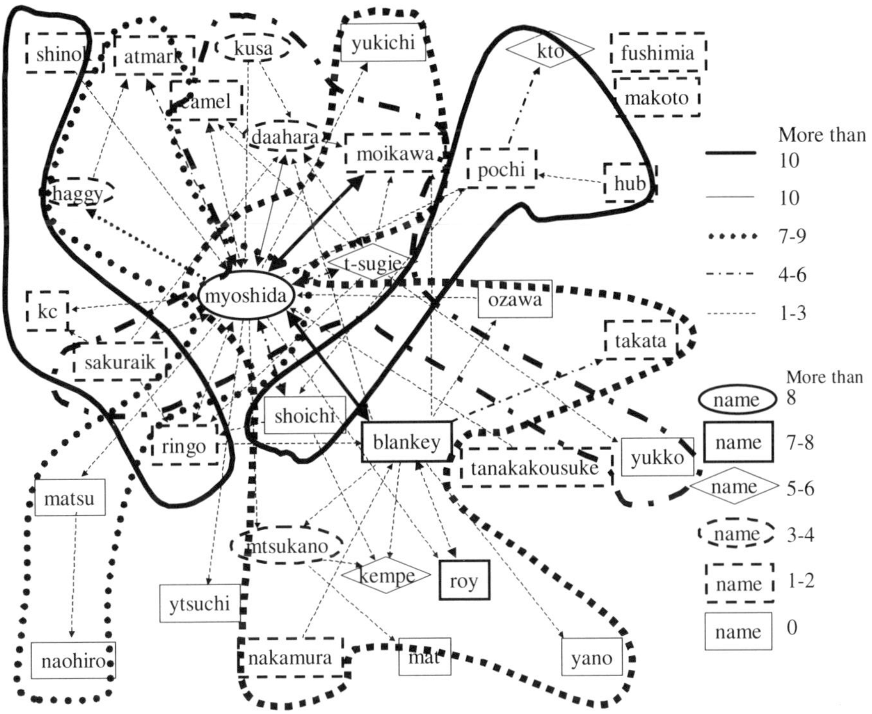

Figure 15: Communities in the real life

5 Conclusion

We have proposed a system for information propagation and gathering by relaying a message like Chinese whispers. The URL of the experimental system is:

http://www.nm.cs.titech.ac.jp:12581/wom/

The authors are preparing a distribution package of the system for experiments in the distributed manner shown in Fig. 12.

References

[1] L. N. Foner. A multi-agent referral system for matchmaking. In *Proceedings of the International Conference on the Practical Applications of Intelligent Agents and Multi-Agent Technology*, 1996.

[2] L. N. Foner. Yenta: a multi-agent, referral-based matchmaking system. In *AA-97*, pages 301–307, 1997.

[3] S. Goto and H. Nojima. Analysis of the three-layered structure of information flow in human societies. *Journal of Japanese Society for Artificial Intelligence (in Japanese)*, 8(3):348–356, 1993. This paper also appears in *Artifical Intelligence*.

[4] KrackPlot, URL: http://www.contrib.andrew.cmu.edu/~ krack/.

[5] M. Lea. Contexts of computer-mediated communication. *Harvester Wheatsheaf*, pages 30–65, 1992.

[6] Pattie Maes. Agents that reduce work and information. *CACM*, 37(7):30–40, 1994.

[7] Takeshi Otani and Toshiro Minami. Searching for information resources by word of mouth. In *MACC 97 (In Japanese)*, 1997. http://www.kecl.ntt.co.jp/csl/msrg/events/macc97-/ohtani.html.

[8] P. Resnick, N. Iacovou, M. Suchak, P. Bergstrom, and J. Riedl. Grouplens: An open architechture for collaborative filtering of netnews. In *CSCW '94*, pages 175–186, 1994.

[9] U. Shardanand and P. Maes. Social information filtering: Algorithms for automating "word of mouth". In *CHI*, pages 210–217, 1997.

Active Mining
H. Motoda (Ed.)
IOS Press, 2002

Immune Network-based Clustering for WWW Information Gathering/Visualization

Yasufumi Takama[1,2] and Kaoru Hirota[1]
{takama,hirota}@hrt.dis.titech.ac.jp
[1]Tokyo Institute of Technology
4259 Nagatsuta, Midori-ku, Yokohama 226-8502 JAPAN
[2]PREST, Japan Science and Technology Corporation, JAPAN

Abstract. A clustering method based on the immune network model is proposed to visualize the topic distribution over the document set that is found on the WWW. The method extracts the keywords that can be used as the landmarks of the major topics in a document set, while the document clustering is performed with the keywords. The proposed method employs the immune network model to calculate the activation values of keywords as well as to improve the understandability of the web information visualization system. The questionnaires are performed to compare the quality of clusters between the proposed method and k-means clustering method, of which the results show that the proposed method can get better results in terms of coherence as well as understandability than k-means clustering method.

1 Introduction

A WWW information visualization method to find topic distribution from document sets is proposed. When the WWW is considered as the information resource, it has several significant characteristics, such as hugeness, dynamic nature, and hyperlinked structure, among which we focus on the fact that the information on the WWW tends to be obtained by users as a set of documents. For example, there are so many online-news sites on the WWW, which constantly release a set of news articles of various topics day by day. As another example, a series of user's retrieval processes also provides the user with a sequence of document sets. Although the hugeness of the WWW as well as its dynamic nature is burden for the users, it will also bring them a chance for business and research if they can notice the trends or movement of the real world from the WWW, which cannot be found from a single document but from a set of documents.

Information visualization systems[6, 15, 16, 18] are promising approaches to help the user notice the trends of topics on the WWW. The Fish View system[15] extracts the user's viewpoint as a set of concepts, and the extracted concepts are used not only to construct the vector space that is sensitive to the user's viewpoint, but also to present the user's current viewpoint in an explicit manner.

In this paper, an information visualization method based on document set-wise processing is proposed to find the topic distribution over a set of documents. One of the characteristic features of the proposed method is the generation of *keyword map* as well as document clustering. That is, a *landmark* that is a representative keyword on a keyword map is found, while the documents containing the same landmark form a document cluster.

When landmark keywords are found based on the propagation of keywords' activation values over the keyword network, the keywords should be activated with related keywords, while the keywords relating to each other should not be highly activated at the same time. To achieve this kind of nonlinear activation, the *immune network model* [1, 5, 7, 8] is employed to calculate the activation values of keywords.

The understandability of the information visualization system for users can be improved by employing an appropriate *metaphor*. From this viewpoint, the method based on the immune network model is expected to improve the understandability of the keyword map, by incorporating the additional information, such as landmark and its suppressing keywords, into the ordinary keyword map, on which only the distance between keywords is a clue to understand the topic distribution over a document set.

The concept of the clustering method based on the immune network model as well as its algorithm are proposed in Section 2, followed by the experimental results that compare the quality of the clusters generated by the proposed method and that by k-means clustering method in Section 3. An application of the proposed method to information visualization / gathering systems is considered in Section 4.

2 Immune Network-based Clustering Method

2.1 Concept of Immune Network-based Clustering

Generally, the information visualization systems designed for handling documents are divided into 2 types, an information visualization system based on document clustering, and *a keyword map*. In this paper, the information visualization system that arranges the keywords extracted from documents on (usually) a 2-D space according to their similarities is called a *keyword map* [6, 9, 16]. A keyword map is often adopted to visualize the topic distribution over a document set.

The clustering method[11, 12, 13, 14] proposed in this paper aims to generate a keyword map, while performing a document clustering. On a keyword map, the keywords relating to the same topic are assumed to gather and form a cluster. The proposed method extracts a representative keyword, called *landmark*, from each cluster. As the border of keyword clusters on a keyword map is usually not obvious, another constraint for extracting a landmark is adopted from the viewpoint of document clustering. That is, when the documents containing the same landmark are classified into the same cluster, there should not exist overlapping among clusters. From the viewpoint of document clustering, a landmark is called as a *cluster identifier*, because it defines the member of a document cluster.

To extract a landmark (a cluster identifier) from a keyword map, the proposed method calculates an activation value of each keyword based on the interaction between the keywords that relate to each other. In this paper, the *immune network model* is employed to calculate a keyword's activation value, which is described in Section 2.2.

2.2 Immune Network Model

Th Immune network model has been proposed by Jerne[5] to explain the functionality of an immune system, such as variety and memory. The model assumes that an antibody can be active by recognizing the related antibody as well as the antigen of a specific type. As antibodies form a network by recognizing each other, the antibody that has once recognized an invading antigen can outlive after the antigen has been removed.

Concerning the immune network model, several models have been proposed in the field of computational biology[1, 7, 8], among which one of the simplest model is employed in this paper:

$$\frac{dX_i}{dt} = s + X_i(f(h_i^b) - k_b), \quad (1)$$

$$h_i^b = \sum_j J_{ij}^b X_j + \sum_j J_{ij}^g A_j, \quad (2)$$

$$\frac{dA_i}{dt} = (r - k_g h_i^g) X_i, \quad (3)$$

$$h_i^g = \sum_j J_{ji}^g X_j, \quad (4)$$

$$f(h) = p \frac{h}{(h+\theta_1)} \frac{\theta_2}{(h+\theta_2)}, \quad (5)$$

here X_i and A_i are the concentration (activation) values of antibody i and antigen i, respectively. The s is a source term modeling a constant cell flux from the bone marrow and r is a reproduction rate of the antigen, while k_b and k_g are the decay terms of the antibody and antigen, respectively. The J_{ij}^b and $J_{ij}^g (\in \{0, WC, SC\})$ indicate the strength of the connectivity between the antibodies i and j, and that between antibody i and antigen j, respectively. The influence on antibody i by other connected antibodies and antigens is calculated by the proliferation function (5), which has a log-bell form with the maximum proliferation rate p.

Using Eq. (5) does not only activate the antibody by recognizing other antibodies or antigens, but also suppresses the antibody if the influence by other objects is too strong. The characteristics of immune systems such as immune response and tolerance[1] can be explained by the model[1, 7, 10].

The dynamics and the stability of the immune network model have been analyzed by fixing the structure or the topology of the network[1, 7, 10]. As the structure of the keyword network that is generated in the proposed method is defined based on the occurrence of keywords in a set of documents, the analysis noted above cannot be applicable. However, the consideration about the combination of the activation states between the connected antibodies leads to the following constraints[13]:

- An antibody can take one of 4 states in terms of activation value; virgin state, suppressed state, weakly-activated state, and highly-activated state.
- It is unstable that both of the antibodies connected to each other take highly-activated state at the same time.
- When there are several antibodies that connect to the same antibody of highly-activated state, the antibodies with strong connection[2] are suppressed, while those with weak connection become weakly-activated.

Applying such a nonlinear activation mechanism of immune network model enables to satisfy the following contradictory conditions for a landmark.

[1]A tolerance indicates the fact that the immune system of a body does not attack the cells of oneself.

[2]As noted in Section2.3, there are two types of connections in terms of strength.

- A landmark should form a keyword cluster with a certain number of connected keywords.
- There should not exist any connection between landmarks.

2.3 Algorithm of Immune Network-based Clustering

In this paper, the immune network model(Eq. (1)–(5)) is applied to the calculation of activation values of keywords, by considering a keyword as an antibody and a document as an antigen. The algorithm is as follows:

1. Extraction of keywords (nouns) from a document set with using the morphological analyzer[3] and the stopword list. In this paper, only the keywords contained in more than 2 documents are extracted.
2. Construction of the keyword network by connecting the extracted keywords k_i to other keywords k_j or documents d_j:
   (a) Connection between k_i and k_j: (D_{ij} indicates the number of documents containing both keywords.)
   Strong connection (SC): $D_{ij} \geq T_k$.
   Weak connection (WC): $0 < D_{ij} < T_k$
   (b) Connection between k_i and d_j: (TF_{ij} indicates the term frequency of k_i in d_j.)
   SC: $TF_{ij} \geq T_d$
   WC: $0 < TF_{ij} < T_d$
3. Calculation of keywords' activation values on the constructed network, based on the immune network model (Eq. (1)–(5)).
4. Extraction of the keywords that activate much higher than others as landmarks after the convergence.
5. Generation of document clusters according to the landmarks

In Step 4, a convergence means that the same set of keywords always becomes active. It is observed through most of the experiments that the same set of keywords have much (about 100 times) higher activation values than others[11], after 1,000 times calculation.

[3]As the current system is implemented to handle Japanese documents, Japanese morphological analyzer *Chasen*(http://chasen.aist-nara.ac.jp/) is used to extract nouns.

Table 1: Parameter Settings Used in the Experiments

Parameter	s	r	k_g	k_b	θ_1	θ_2	p
Value	10	0.01	10^{-4}	0.4	10^3	10^6	1.0
Parameter	$X_i(0)$	$A_i(0)$	T_k	T_d	SC	WC	
Value	10	10^5	3	3	1.0	10^{-3}	

3 Experimental Results

The quality of clusters generated by the proposed clustering method is compared with that by k-means clustering[3], of which the applicability is widely demonstrated in many applications.

While k-means generates the clusters so that each data (documents) in a set can be covered by one of the generated clusters, the proposed method does not intend to cover all the documents. It is observed through many experiments that 60–80% of a document set is covered by the generated clusters. Therefore, it is meaningless to compare both methods in terms of coverage. In this paper, questionnaires are performed to compare the clusters generated by the proposed method and that by k-means, from the following viewpoints.

- Coherence: how closely the documents within a cluster relate to each other.
- Understandability: how easily the topic of a cluster can be understood by users.

The sets of documents used for the experiments are collected from the following online news sites.

Set1 Documents in entertainment category of Yahoo! Japan News site[4] , released on September 18, 2001. The 75 keywords are extracted from 25 documents.

Set2 Documents in entertainment category of Yahoo! Japan News site, released on September 21, 2001. The 62 keywords are extracted from 24 documents

Set3 Documents in local news category of Lycos Japan[5] , released on September 28, 2001. The 22 keywords are extracted from 23 documents.

The parameter values used in the experiments are shown in Table 1. These values are empirically determined based on the values used in the field of computational biology[1, 7, 8].

The STATISTICA2000 (Statistica Soft, Inc.) is used to perform k-means clustering. The number of clusters generated by k-means, which has to be determined in advance, is specified as much as the number of clusters generated by the proposed clustering method. The naive k-means clustering tends to generate the clusters of various sizes, and sometimes the cluster containing only one document is generated, which is removed from questionnaires.

The questionnaires are answered by 9 subjects, consisting of researchers and students. Each subject is asked to evaluate the clustering results of 2 document sets, one

[4] http://news.yahoo.co.jp/
[5] http://www.lycos.co.jp/

Table 2: Comparison of Clustering Results between Proposed Method and K-means Clustering

Data	Item	Proposed	K-means
	Number of clusters	5	4
	Variance of Cluster Size	**0.48**	3.6
Set1	Average score	**4.33**	3.90
	Score≥3.5	**5**	2
	2.5≤Score<3.5	0	1
	Score<2.5	0	1
	Number of clusters	5	4
	Variance of Cluster Size	**0.32**	4.625
Set2	Average score	**3.82**	3.13
	Score≥3.5	**4**	1
	2.5≤Score<3.5	1	2
	Score<2.5	0	1
	Number of clusters	5	5
	Variance of Cluster Size	**0.48**	4.25
Set3	Average score	2.3	**4.00**
	Score≥3.5	1	**4**
	2.5≤Score<3.5	1	0
	Score<2.5	3	1

generated by the proposed method and another by k-means. Of course, subjects do not know by which method each result is generated.

In the questionnaires, the documents in a cluster and the *related keywords* are presented for each cluster. The related keywords of the proposed method are landmarks as well as their suppressing keywords. As for the k-means clustering method, the keywords of which the weight in the cluster center is higher than others are used as the related keywords. The number of related keywords of the proposed method is not fixed, while 5 related keywords are presented in the case of k-means for each cluster.

Subjects rate the coherence of each cluster with 5 grades, from score 5 as closely related to 1 as not related. As for the understandability, Subjects are asked to mark the related keyword that seems to represent the topic of a cluster[6] .

Table 2 shows the number of clusters, the variance of cluster size, average score of clusters, and the score distribution of the clustering results generated by both method from 3 document sets.

From this table, it is shown that the proposed method (Proposed) can obtain better results than k-means clustering (K-means) for Set1 and Set2. The reason why the proposed method cannot obtain good result for Set3 seems to relate with the fact that the number of keywords extracted from Set3 is much less than those from Set1 and Set2. That is, it seems that there are less topical keywords in the local news category than in the entertainment category. Extracting not only keywords but also phrases will be required to handle this problem.

It is observed that some clusters are generated by both of the proposed method and k-means clustering method. As k-means clustering tends to generate one large clusters, which leads to large variance of cluster size as shown in Table 2, it is also observed that some clusters generated by the proposed method are subset of the cluster generated by k-means. Table 3 and Table 4 shows the distribution of scores of the clusters, dividing the case when the clusters are generated by both methods (SAME),

[6]Multiple keyword selection for a cluster is allowed.

Table 3: Score Distribution of Clusters Generated by Plastic Clustering Method

Type	1	2	3	4	5	Total
SAME	0(0%)	2(14%)	0(0%)	7(50%)	5(36%)	14(100%)
SUBSET	1(8%)	2(15%)	0(0%)	8(62%)	2(15%)	13(100%)
DIFFERENT	4(22%)	1(6%)	0(0%)	10(55%)	3(17%)	18(100%)
TOTAL	5(11%)	5(11%)	0(0%)	25(56%)	10(22%)	45(100%)

Table 4: Score Distribution of Clusters Generated by K-means Clustering Method

Type	1	2	3	4	5	Total
SAME	1(7%)	1(7%)	0(0%)	6(43%)	6(43%)	14(100%)
SUBSET	1(10%)	2(20%)	0(0%)	4(40%)	3(30%)	10(100%)
DIFFERENT	2(20%)	2(20%)	0(0%)	2(20%)	4(40%)	10(100%)
TOTAL	4(12%)	5(15%)	0(0%)	12(35%)	13(38%)	34(100%)

the clusters generated by the proposed method is a subset of a cluster of k-means (SUBSET), and others (DIFFERENT). From these tables, it can be seen that the clusters generated by both methods can obtain higher scores than others. Although the scores of clusters in SUBSET and DIFFERENT are lower than those in SAME, the proposed method can obtain good score (4 and 5) compared with k-means clustering.

As for the understandability, Table 5 shows the ratio of the related keywords that are marked by more than one subjects among the related keywords presented to them. It is shown i Table 5 that the ratio becomes high when the clustering results obtain high scores in terms of coherence, i.e., the results of Set1 and Set2 by the proposed method, and the results of Set1 and Set3 by k-means clustering method. That is, the cluster with high score relates to a certain, obvious topic, which can be understood by several subjects from the same viewpoint.

4 WWW Information Visualization System with Immune Network Metaphor

An information visualization system is one of the promising approaches for handling the growing WWW information resource. The information visualization system that aims to support browsing process often tries to make it easy to understand a link structure by using 3D graphics as well as by introducing the interaction with the user[16]. When a information visualization system is designed to support the information retrieval process with using WWW search engines, it often employs the document clustering method for improving the efficiency of browsing retrieval results[4, 18, 19].

On the other hand, a keyword map[6, 9, 12, 16], which has not been so famous in

Table 5: Ratio of Keywords Extracted More Than Once

Document Set	Proposed	K-means
Set1	**0.286**	**0.304**
Set2	**0.368**	0.095
Set3	0.167	**0.241**

the field of WWW information visualization, is useful to visualize the topic distribution over a set of documents. Visualizing topic distribution is expected to be also suitable for supporting interactive information gathering process.

In the proposed method, as a landmark suppresses the related keywords on the constructed keyword network, this relationship among keywords is also useful as the metaphor to improve the understandability of a keyword map, as shown in Fig. 1. While the ordinary keyword map uses only the distance information, the immune network metaphor is used to improve the keyword map by emphasizing the keyword cluster of which the representative is a landmark. In Fig. 1, the immune network metaphor is incorporated into the spring model[16], so that the spring constant of the spring connected to a landmark can be set to be stronger than others, and the length of the spring between landmarks can be set to be longer than others. A landmark is indicated in white color, while dark-colored one is the keyword suppressed by a landmark. From Fig. 1, five distinct topics represented with landmarks and their related keywords can be shown clearly, while the suppressed keywords "Terrorism" and "Simultaneous" are arranged near the center of the map, because the topic about N. Y. tragedy is contained in many documents.

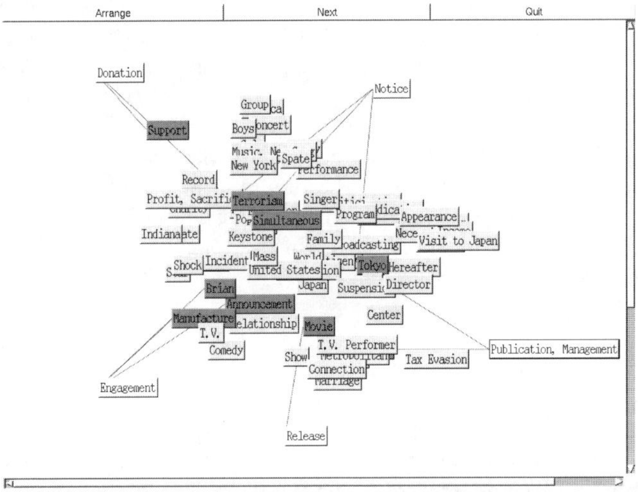

Figure 1: keyword Map Generated from Set1

5 Conclusion

A clustering method based on the immune network model is proposed to visualize the topic distribution over the document set found on the WWW. The method extracts the keywords that can be used as the landmarks of the major topics in a document set, while the document clustering is performed with the keywords. The proposed method employs the immune network model to calculate the activation values of keywords.

The questionnaires are performed to compare the clusters generated by the proposed method and those generated by k-means clustering method, of which the results show that the proposed method can get better results in terms of the coherence than k-means, in two of three document sets. From the viewpoint of understandability, it is shown that the landmark and their related keywords can represent the topic of the cluster.

Furthermore, the immune network metaphor is incorporated into an ordinary keyword map to improve its understandability. As the future work, the ways of incorporating the immune network model into a keyword map will be considered to further improve the understandability of a keyword map.

References

[1] Anderson, R. W., Neumann, A. U.,, Perelson, A. S., "A Cayley Tree Immune Network Model with Antibody Dynamics," *Bulletin of Mathematical Biology*, 55, 6, pp. 1091–1131, 1993.

[2] Cole, C., "Interaction with an Enabling Information Retrieval System: Modeling the User's Decoding and Encoding Operations," *Journal of the American Society for Information Science* , 51, 5, pp. 417–426, 2000.

[3] Duda, R. O., Hart, P. E., Stork, D. G., "10. Unsupervised Learning and Clustering," *in Pattern Classification (2nd Ed.)*, Wiley, New York, 2000.

[4] Hearst, M. A. and Pedersen, J. O., "Reexamining the Cluster Hypothesis: Scatter/Gather on Retrieval Results," *SIGIR'96*, pp. 76–84, 1996.

[5] Jerne, N. K., "The Immune System," *Sci. Am.*, 229, pp. 52–60, 1973.

[6] Lagus, K., Honkela, T., Kaski, S., Kohonen, T., "Self-Organizing Maps of Document Collection: A New Approach to Interactive Exploration," *2nd Int'l Conf. on Knowledge Discovery and Data Mining*, pp.238–243, 1996.

[7] Neumann, A. U. and Weisbuch, G., "Dynamics and Topology of Idiotypic Networks," *Bulletin of Mathematical Biology*, 54, 5, pp. 699–726, 1992.

[8] Smith, D. J., Forrest, S., Perelson, A. S., "Immunological Memory is Associative," *Int'l Workshop on the Immunity-Based Systems (IBMS'96)*, 1996.

[9] Sumi, Y., Nishimoto, K., Mase, K., "Facilitating Human Communication in Personalized Information Spaces," *AAAI-96 Workshop on Internet-Based Information Systems*, pp. 123–129, 1996.

[10] Sulzer, B. et al., "Memory in Idiotypic Networks Due to Competition Between Proliferation and Differentiation," *Bulletin of Mathematical Biology*, 55, 6, pp. 1133–1182, 1993.

[11] Takama, Y. and Hirota, K., "Application of Immune Network Model to Keyword Set Extraction with Variety," *6th Int'l Conf. on Soft Computing (IIZUKA2000)*, pp. 825–830, 2000.

[12] Takama, Y. and Hirota, K., "Development of Visualization Systems for Topic Distribution based on Query network", *SIG-FAI-A003*, pp. 13–18, 2000.

[13] Takama, Y. and Hirota, K., "Employing Immune Network Model for Clustering with Plastic Structure," *2001 IEEE Int'l Symp. on Computational Intelligence in Robotics and Automation (CIRA2001)*, pp. 178–183, 2001.

[14] Takama, Y. and Hirota, K., "Consideration of Memory Cell for Immune Network-based Plastic Clustering method," *InTech'2001*, pp. 233–239, 2001.

[15] Takama, Y. and Ishizuka, "FISH VIEW System: A Document Ordering Support System Employing Concept-structure-based Viewpoint Extraction," *J. of Information Processing Society of Japan (IPSJ)*, 42, 7, 2000 (written in Japanese).

[16] Takasugi, K. and Kunifuji, S., "A Thinking Support System for Idea Inspiration Using Spring Model," *J. of Japanese Society for Artificial Intelligence*, 14, 3, pp. 495–503, 1999 (written in Japanese).

[17] Watanabe, I., "Visual Text Mining," *J. of Japanese Society for Artificial Intelligence*, 16, 2, pp. 226–232, 2001 (written in Japanese).

[18] Zamir, O. and Etzioni, O., "Grouper: A Dynamic Clustering Interface to Web Search Results," *Proc. 8th Int'l WWW Conference*, 1999.

[19] Zamir, O. and Etzioni, O., "Web Document Clustering: A Feasibility Demonstration," *Proc. SIGIR'98*, pp. 46–54, 1998.

Active Mining
H. Motoda (Ed.)
IOS Press, 2002

Interactive Web page Retrieval with Relational Learning based Filtering Rules

Masayuki Okabe
okabe@mm.media.kyoto-u.ac.jp
Japan Science and Technology CREST
Yoshida-Nihonmatsu-Cho, Sakyo-ku, Kyoto 606-8501, JAPAN

Seiji Yamada
yamada@ymd.dis.titech.ac.jp
CISS, IGSSE, Tokyo Institute of Technology
4259 Nagatuta-Cho, Midori-ku, Yokohama 226-8502, JAPAN

Abstract. WWW Search Engines usually return a hit-list including many irrelevant pages because most of the users just input a few words as a query which is not enough to specify their information needs. In this paper we propose a system which applies relevance feedback to the interactive process between users and Web Search Engines, and accelerates the effectiveness of the process by using a query specific filter. This filter is a set of rules which represents the characteristics of Web pages that a user marked as relevant, and is used to find new relevant Web pages from unidentified pages in a hit-list. Each of the rules is made of logical and proximity relationships among keywords which exist in a certain range of a Web page. That range is one of the areas partitioned by four kinds of HTML tags. The filter is made by a learning algorithm which adopts separate-and-conquer strategy and top-down heuristic search with limited backtracking. In experiments with 20 different kinds of retrieval tests, we demonstrate that our proposed system makes it possible to get more relevant pages than the case not using the system as the number of feedback increases. We also analyze how the filters work.

1 Introduction

With the rapid growth of WWW, there are various information sources on the Internet today. Search engines are indispensable tools to access useful information which might exist somewhere on the Internet. While they have been getting higher capability to meet various information needs and large amounts of transactions, they are still insufficient in the ability to support the users who want to collect a certain number of Web pages which are relevant to their requirements.

When a user inputs a query, which is usually composed of a few words[1], search engines return a "hit-list" in which so many Web pages are presented in a certain order. However it does not often reflect the user's intent, and thus the user would waste much time and energy on judging Web pages in the hit-list.

To resolve this problem and to provide efficient retrieval process, we propose a system which mediates between users and search engines in order to select only relevant Web pages out of a hit-list through the interactive process called "relevance feedback"[8]. Given some Web pages marked with their relevancy(relevant or non-relevant) by a user, this system generates a set of filtering rules, each of which is a rule to decide whether

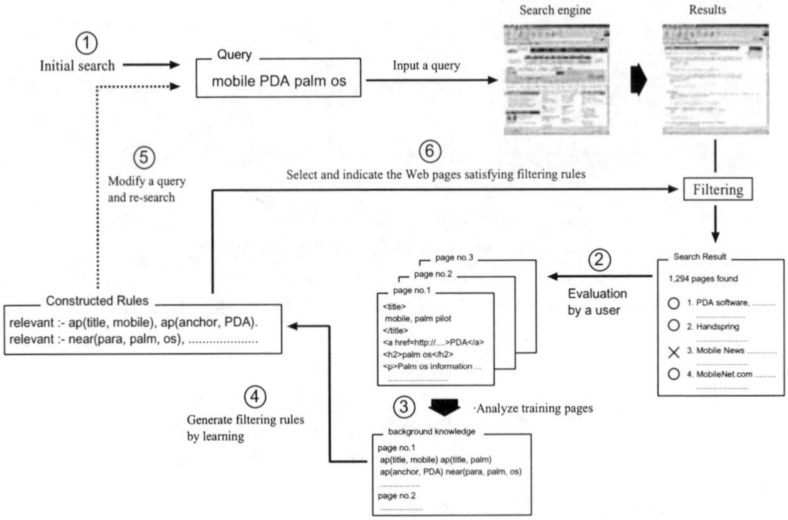

Figure 1: Interactive Web search

the user should look a Web page or not. The system constructs filtering rules from the combinations of keywords, relational operators and tags by a learning algorithm which is superior to learn structural patterns. We have developed this basic framework in document retrieval[6] and found our approach was promising. In this paper, we applied this method to the intelligent interface which coordinates the hit-lists of search engines in order for individual user to find their wanted information easily.

The remainder of the paper is organized as follows. Section 2 describes the interactive process and the way how to apply filtering rules. Section 3 describes the representation and the learning algorithm of filtering rules. Section 4 shows the results of retrieval experiments to evaluate our system.

2 Interactive Web search with relevance feedback

Figure 1 shows the overview of interactive Web search with relevance feedback. In this section, we explain the procedures of each step in this search process. The number assigned to them correspond to the numbers in circles of Figure 1.

1. **Initial search:** A user inputs a query (a set of terms) to our Web search system. Then the system puts the query through to a search engine and obtains a hit-list.

2. **Evaluation of results by a user:** After getting a hit-list from a search engine, the system asks the user to evaluate and mark the relevancy(relevant or non-relevant) of a small part of Web pages in the hit-list (usually upper 10 pages), and stores those pages as *training pages*, especially the relevant pages as *positive training pages* and the non-relevant pages as *negative training pages*.

3. **Analyzing training pages:** Then the system breaks up each positive training page into the minimal elements which can be a part of filtering rules. The concrete procedures are the followings.

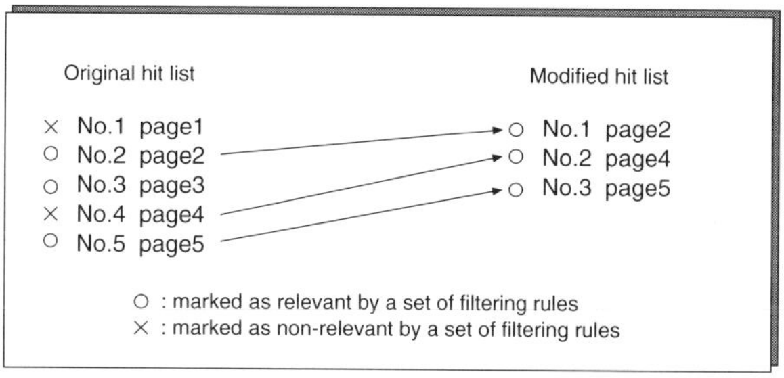

Figure 2: Filtering Web Pages

- *Generating candidates for additional keywords*: The extended keywords mean the terms which can be substituted to the arguments of a predicate. It is often said that users usually input only a few terms which are quite insufficient not only for specifying Web pages but for making effective filtering rules, thus this procedures is very important to widen the variations of rule representation. Our system uses TFIDF method[4] to extract additional keywords.

- *Generating literals for constructing bodies of filtering rules*: Using the extended keywords, the system generates literals which can be one of the elements which compose the body of each filtering rule. These literals are called a *condition candidate set* and used to construct a body of a filtering rule.

4. **Generating filtering rules by learning:** Using the condition candidate set, the system generates filtering rules by relational learning. The detail procedures will be developed in the next section.

5. **Modify a query and re-searching:** The system expands the query using terms which have been extracted through the analysis of training pages. Then the modified query is inputed into a search engine and the new results are obtained.

6. **Select and indicate the Web pages satisfying filtering rules:** As shown in Figure 2, the system selects the Web pages satisfying the filtering rules from the hit-list returned by search engine, and indicates them to the user. The pages which the user has already evaluated are eliminated from the indication.

The information retrieval is done using the above procedures, and the steps from 2 to 6 are repeated until the user collects enough relevant pages.

This system provides the two following functions which are used for filtering the results of simple relevance feedback.

- Modify a query and re-searching. (corresponding to Step5)

- Select and indicate the Web pages satisfying filtering rules. (corresponding to Step6)

The search engine usually selects the candidates of relevant Web pages and ranks them before returning a hit-list. By modifying a query and re-searching, a system is able to modify the ranking. Also by selecting and indicating the Web pages satisfying filtering rules, the filter is modified.

The modification of a query is done by using the query expansion techniques which have been studied so well in information retrieval[9, 10]. Thus we omit the discussion on the modification of a query in this paper. We develop representation and generation of filtering rules using the structure of HTML file in the next section.

3 Filtering rules

This section explains the representation and the generation of filtering rules in detail. We deal with the construction of filtering rules as *inductive learning* of *machine learningd*, in which relevant and non-relevant pages indicated by the user are used as *training examples*.

3.1 Rule representation

We use horn clause to represent filtering rules. The body of a rule consists of the following predicates standing for relations between terms and tags.

- $ap(region\_type, word)$: This predicate is true *iff* a word *word* appears within a region of *region_type* in a Web page.
- $near(region\_type, word1, word2)$: This predicate is true *iff* both of words w_i and w_j appear within a sequence of 10 words somewhere in a region of *region_type* of a Web page. The ordering of the two words is not considered.

The predicates *ap* and *near* represent basic relations between keyword(s) and the position of the keyword(s). Several types of relations among keywords can be assumed, however, we use only neighbor relation because it has been proven to be very useful in several researches.[2, 5].

Furthermore we can easily consider that the importance of words significantly depends on tags of HTML. For example, the words within `<TITLE>` seem to have significant meaning because they indicate the theme of the Web page. Hence we use the *region_type* to restrict a tag with which words are surrounded. We prepare the *region_type* in the followings.

- *title* : The region surrounded with title tags `<TITLE>`.
- *anchor* : The region surrounded with anchor tags `<A>`. For example, the `<A HREF=...>`.
- *head* : The region surrounded with heading tags `<H1∼4>`.
- *para* : The region surrounded with paragraph tags `<P>`. This means the region of the same paragraph.

We can represent various features of pages by combining these relations. Here is an example set of rules.

$$\begin{cases} relevant \text{ :- } ap(title, \text{mobile}), ap(anchor, \text{PDA}). \\ relevant \text{ :- } near(para, \text{palm}, \text{os}). \end{cases}$$

Filtering rules are interpreted disjunction. Thus if any rule is satisfied in a Web page, the page will be considered relevant and otherwise non-relevant. The above filtering rules means that a Web page is relevant if "mobile" appears in the title and "PDA" appears in an anchor text, or "palm" and "OS" appear near in the same paragraph.

Input: E^+ : a set of positive training pages, E^- : a set of negative training pages
C : a condition candidate set, K : a set of extended keywords
Output: R : a set of filtering rules.
Variables: *rule* : a filtering rule, S : a set of exception literals,
l_1 : an exception literal
Initialize: $K \leftarrow$ a set of words in a query. R, S, $l_1 \leftarrow$ *empty*. *rule* $\leftarrow$ *relevant:-*.
Repeat
1: · Investigate the number p of positive training pages satisfying the *rule*
and the number n of negative training pages satisfying the *rule*.
2: **if** $n = 0$ **then**
3: · Add *rule* to R.
4: · Remove a positive training page satisfying the *rule* from E^+.
5: **if** E^+ is empty **then Finish**
6: **else** Initialize *rule*, S, l_1.
7: **else**
8: · For all literals in $C \cap \overline{S}$, compute the information gain G.
9: **if** No literal with $G > 0$ **then**
10: **if** the body of the *rule* is empty **then**
11: · Add a keyword to K.
12: · Update C.
13: **else**
14: · Initialize S and *rule*.
15: · Add l_1 to S, and initialize l_1.
16: **else**
17: · Select l_{max} having the maximum G.
18: **if** the body of the *rule* is empty. **then** $l_1 := l_{max}$
19: · Add l_{max} to *rule* and S.

Figure 3: Learning Algorithm

3.2 Learning algorithm

Figure 3 shows the learning algorithm for making filtering rules. This algorithm is based on the first order learning system FOIL[7] which adopts a greedy separate-and-conquer strategy[3]. This algorithm generates a filtering rule one by one, and adds the generated rule to R. When a *rule* is generated, the pages covered with the rule are removed from the set of positive training pages E^+. Thus, as the number of generated filtering rules increases, E^+ decreases, and the algorithm finishes if the E^+ becomes empty (step3~5).

In the generation of a single filtering rule, a literal is added into the body one by one (step19), and the rule is established if it includes no negative training page (step2). The added literal is selected from a condition candidate set C. This C consists of the literals having all of the *region_types* and keywords in K as its arguments and being satisfied in training pages. Concretely the following two types of literals are used.

- The *ap* literals having all of the *region_types* and keywords in K as its arguments and being satisfied in training pages.

- The *near* literals having all of the *region_types* and keywords in K as its arguments and being satisfied in training pages.

The criteria for selecting a literal which should be added to the body is based on the *information gain*(step8). It is computed by the following equations, and popular in learning of filtering tree.

$$G = e^{\oplus}_{new}\{I(e^{\oplus}_{old}, e^{\ominus}_{old}) - I(e^{\oplus}_{new}, e^{\ominus}_{new})\}$$

$$I(e^{\oplus}, e^{\ominus}) = -\log_2 \frac{e^{\oplus}}{e^{\oplus} + e^{\ominus}}$$

$e^{\oplus}_{old}$, $e^{\ominus}_{old}$, $e^{\oplus}_{new}$, $e^{\ominus}_{new}$ mean the numbers of positive/negative training pages before/after the addition of a literal. Using the information gain, a system is able to select a literal which obtains not only much information for a training page but also many positive training pages satisfying it (step17).

This rule construction using information gain is efficient because it is greedy. However it sometimes selects bad literal and stops before completion. In such a case, if a current rule has some literals in its body, this algorithm eliminates all the literals in its body and restarts a rule making process. This backtracking is done for literals in C except for a literal l_1 which was first added to the body (step14, 15).

If the body of a current rule has no literal, a new keyword is added to K and C is updated (step11,12). The added keyword is selected from terms in positive training pages E^+ by the following procedures.

1. Extract paragraphs from E^+ using `<P>` tags.

2. Investigate a subset of the paragraphs including any word in a query, and the subset is called T.

3. Compute the importance for every word w_i in T by the following equation.

   Importance of w_i = (average occurrence in T)×(the number of texts in which w_i occu

4. Select the literal which has the maximum importance and is not included in a query.

Backtracking and iterative literal making process are main difference from the algorithm in FOIL. They are very specific and empirical procedure. Without these extensions, however, many useless rules would be generated.

4 Experiments and Results

To evaluate the effectiveness of filtering rules, we conducted retrieval experiments. The question here is how many relevant pages we can find more with our proposed system in the condition we look over a certain number of Web pages.

<num> Number: 401

<title> foreign minorities, Germany

<desc> Description:

What language and cultural differences impede the integration of foreign minorities in Germany?

<narr> Narrative:

A relevant document will focus on the causes of the lack of integration in a significant way; that is, the mere mention of immigration difficulties is not relevant. Documents that discuss immigration problems unrelated to Germany are also not relevant.

Figure 4: An example of topic

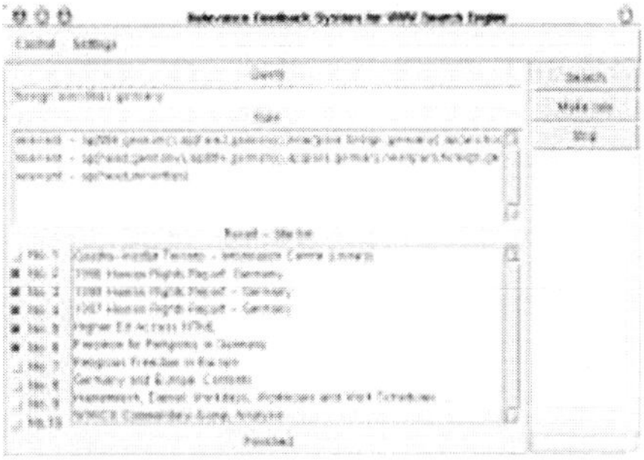

Figure 5: System Interface

4.1 Settings

We conducted two series of retrieval. The one is a retrieval from an original hit-list returned by a search engine (*retrieval1*). In this retrieval, we judged 50 pages from the top of the hit-list. The other is a retrieval using our system (*retrieval2*). In this retrieval, we made feedbacks every after judging 10 pages according to the procedure described in Section 2. We made total four feedbacks. 10 pages after each feedback are collected from the top of the hit-list (excluding the pages we've already judged and filtering rules don't satisfy). In both retrieval, total 50 pages from the same hit-list were evaluated.

We used the Google [1] as a test WWW search engine, which is recognized as one of the most powerful search engines. For test questions, we used 20 *topics*(No. 401~420) provided by the small web track in TREC-8[2] . This test collection is often used for evaluating the performance of retrieval systems in Information Retrieval community. Figure 4 is an example of topic which is composed of four parts. Title part consists of 1~3 words. We used these title words as a query for search engine. Relevance judgment of each page is conducted by the same searcher according to the account written in the *description* and the *narrative* part of each topic.

4.2 Interface

Figure 5 shows the system interface which consists of *query input*, *rule view*, *title view* and several buttons. When users put the *make rule* button, filtering rules are constructed and displayed in rule view. We can see the rules directly, thus we find useful patterns or keywords to retrieve relevant pages. Once rules are constructed, the system starts to collect new relevant pages, and display their titles in title view. If the user clicks a title, a browser rises and shows the clicked page.

4.3 Results

Figure 6 shows the relation between judged pages and relevant pages found in the judged pages. The number of relevant pages is average value of 20 topics. About first 10 pages, there is no difference because both retrieval returns the same pages. The

[1] http://www.google.com

[2] http://trec.nist.gov

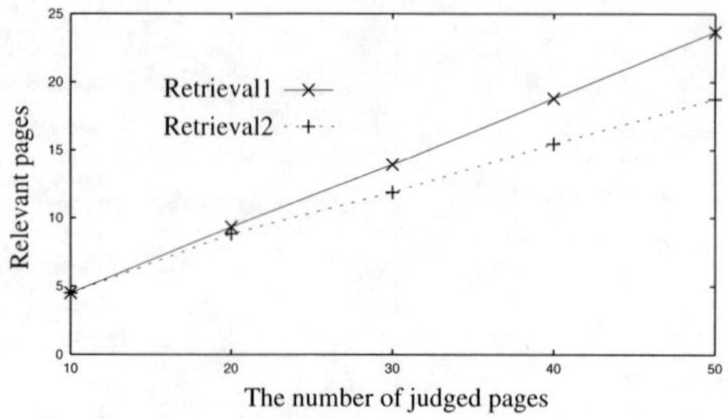

Figure 6: The average number of relevant pages

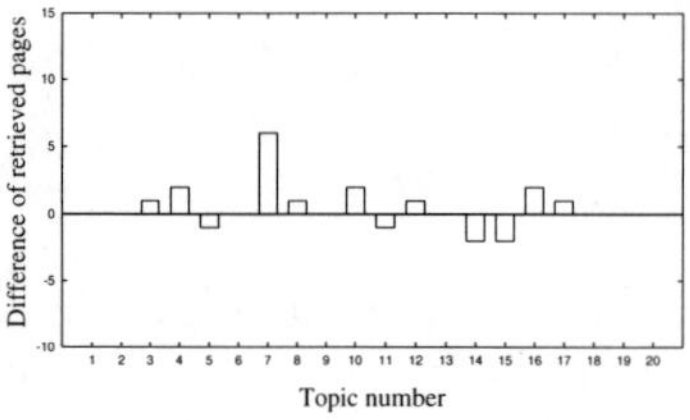

Figure 7: Difference after the first feedback (total 20 pages judged)

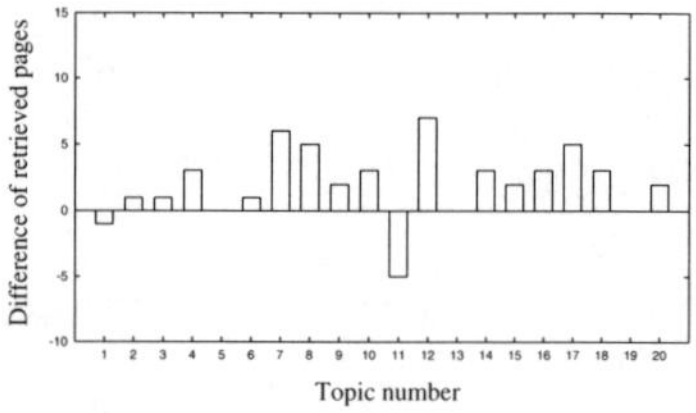

Figure 8: Difference after the second feedback (total 30 pages judged)

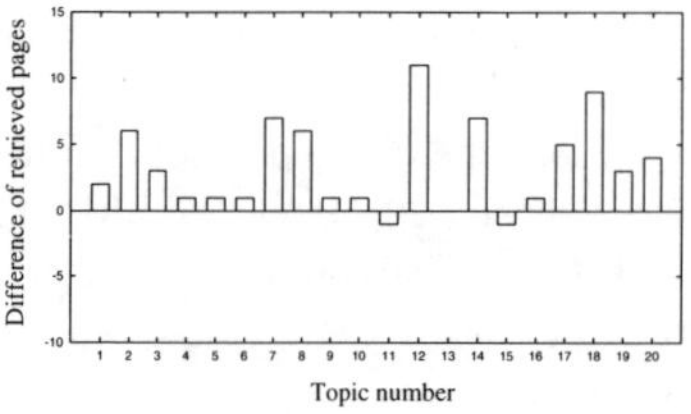

Figure 9: Difference after the third feedback (total 40 pages judged)

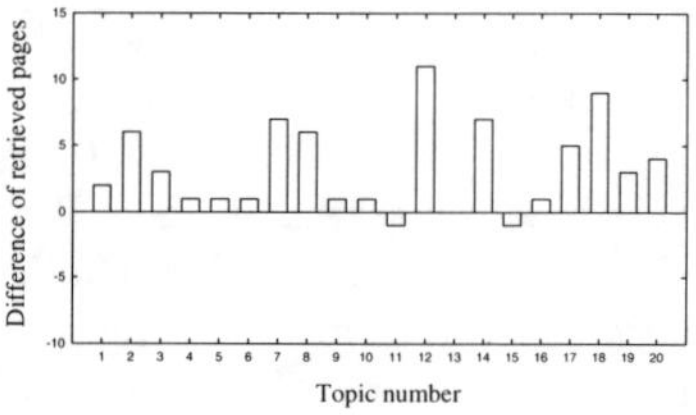

Figure 10: Difference after the fourth feedback (total 50 pages judged)

difference of the number of relevant pages increases after the first feedback. As a result, retrieval2 got about 5 relevant pages more than retrieval1 after four feedbacks. However the difference varies in each topic.

Figure 7 $\sim$ 10 shows the difference of relevant pages between retrieval1 and retrieval2 after each feedback. Let A be the number of relevant pages found in retrieval1 and B be the one in retrieval2, the difference D is calculated by $D = B - A$. In Figure 7, there is little effect of our system because we only judge small number of pages. In Figure 8 and 9, the effect gradually increases. In Figure 10, we can see the effect clearly. Our system produces good results for most of topics except a few topics such as no.4 and no.11.

4.4 Effective and Ineffective filtering rules

As seen in the results, the retrieval which uses our system enhanced the effectiveness for most topics. We show two types of examples, a good one that our system effectively worked, and a bad one that our system didn't work well.

Table 1: Filtering rules generated for topic no.12

relevant :- ap(anchor,screening).
relevant :- near(para,security,system), ap(title,airport).
relevant :- near(para,security,airports), near(para,security,access).
relevant :- near(para,security,airports), near(para,faa,system).

Table 2: Filtering rules generated for topic no.11

relevant :- ap(anchor,shipwreck).
relevant :- ap(anchor,shipwreck), ap(anchor,salvaging).

Topic 12 is an example that filtering rules worked most effectively. The objective of topic 12 is "to identify a specific airport and describe the security measures already in effect or proposed for use at that airport". Search engine returns many non-relevant pages which introduce "the security which travelers must prepare". Removing such pages by filtering rules, our system could provide proper results. Table 1 shows the filtering rules generated for this topic. These rules represent the pages which introduce specific security systems by using the words "faa" and "screening".

Topic 11 is an example that filtering rules didn't work well. The objective of this topic is "To find information on shipwreck salvaging: the recovery or attempted recovery of treasure from sunken ships". Relevant pages for this topic include various types of pages such as links, bulletin board, news and individual home pages. The filtering rules generated for this topic are too general or too specific, thus they could not select appropriate pages and it leads to the bad results. Table 2 shows the filtering rules generated for this topic. These rules uses only two keywords and they are insufficient to restrict relevant pages.

5 Conclusion

We described a system which enhances the effectiveness of WWW Search Engine by using relevance feedback and relational learning. The main function of our system is the application of filtering rules which is constructed by relational learning technique. We presented its representation and learning algorithm. Then we evaluated their effectiveness through retrieval experiments. The results showed that our system enables us to find more relevant pages though the effect differs in every questions.

Our system need quick response and moderate machine power. Thus it should be a user side application because search engines cannot afford to attach such a function. One of the future problem is to reduce the cost which users need to judge pages. We plan to apply clustering methods for this problem.

References

[1] Baeza-Yates, R. and Ribeiro-Neto, B.: Modern Information Retrieval: Addison-Wesley, Wokingham, UK, (1999)

[2] Cohen, W.W.: Text categorization and relational learning, In *Proceedings of the Twelfth International Conference on Machine Learning*, pp.124-132 (1995)

[3] Furnkranz, J.: Separate-and-Conquer Rule Learning, *Artificial Intelligence Review*, Vol.13, No.1 (1999)

[4] Joachims, T.: A Probabilistic Analysis of the Rocchio Algorithm with TFIDF for Text Categorization, *Proc. of the 14th International Conference on Machine Learning ICML97*, pp.143-151 (1997)

[5] Keen, E.M.: Some aspects of proximity searching in text retrieval system, *Journal of Information Science*, Vol.18, No.2, pp.89-98 (1992)

[6] M. Okabe and S. Yamada: Interactive Document Retrieval with Relational Learning, *Proc. of the 16th ACM Symposium on Applied Computing*, pp.27-31 (2001)

[7] Quinlan, J.R., and Cameron-Jones, R.M.: Induction of Logic Programs: FOIL and Related Systems, *New Generation Computing*, Vol.13, Nos.3,4, pp.287-312 (1995)

[8] Salton, G. and Buckley, C.: Improving Retrieval Performance by Relevance Feedback, *Journal of the American Society for Information Science*, Vol.41, No.4, pp.288-297 (1990)

[9] Mitra, M., Singhal, A., and Buckley, C.: Improving automatic query expansion, In *Proceedings of the 21st annual international ACM SIGIR*, pp.206-214 (1998)

[10] Xu, J. and Croft, W.B.: Query expansion using local and global document analysis, In *Proceedings of the 19th annual international ACM SIGIR*, pp.4-11 (1996)

Active Mining
H. Motoda (Ed.)
IOS Press, 2002

Monitoring Partial Update of Web Pages by Interactive Relational Learning

Seiji Yamada and Yuki Nakai
yamada@ymd.dis.titech.ac.jp
CISS, IGSSE
Tokyo Institute of Technology
4259 Nagatsuta, Midori, Yokohama 226-8502, JAPAN

Abstract. This paper describes an automatic monitoring system that constantly checks partial updates in Web pages and notifies them to a user. While one of the most important advantages of the WWW is frequent updates of Web pages, we need to constantly check them out and this task takes much cognitive load. Thus applications to automatically check updates of Web pages have been developed, however they can not deal with partial updates like updates in a particular cell in a table in a Web page. Hence we developed a automatic monitoring system that checks such partial updates. A user can give a system regions in which he/she wants to know the updates in a Web page as training examples, and it is able to learn rules to identify the partial updates by relational learning. By this learning, a user do not need to directly describe the rules. We fully implemented our system and presented executed results.

1 Introduction

We currently obtain various information from the WWW and utilize them for many purposes like business, education, personal use and so on. Since we can easily make, delete and modify Web pages, the WWW is growing as a huge and dynamic information resource. While one of the most important advantages of the WWW is its frequent updates of Web pages, we needs to constantly check them for acquiring the latest information and this task obviously forces much cognitive load on us. Thus a number of applications and services to automatically check and notify updates of Web pages have been developed[10][2][8]. Unfortunately almost all of them notify updates to a user whenever any part of a Web page is updated, and most of such updates may not useful to him/her.

For example, see a weather report Web page like Fig. 1. Consider a user who has a plan to go to a picnic on Sunday and is interested in the Sunday's weather. He/she needs to frequently check the Sunday's weather in the Web page of Fig. 1 because it is updated everyday. If a user employs a Web update checking application, it notifies him/her all of updates including other day's weather changes except Sunday thought such notifications are meaningless. Thus a *partial update* is defined as an update of a particular region in which a user is interested, not of any part of a Web page. We consider this partial update monitoring is widely necessary in a lot of fields like stock market Web pages, the exchange rate Web pages and so on. Hence a Web update checking application should deal with a partial update in a particular region like a cell in a table. Furthermore a user should be able to easily describe such a particular region by help of an application.

Weekly Weather
2001/1/1 17:00 JST

	Weather	Rain	High Temp			Low Temp		
			(F)	(C)	Diff	(F)	(C)	Diff
2002/01/03 (Thu)	cloudy, occasionally clear	30(%)	48	9.0	−2	32	0.0	−3
2002/01/04 (Fri)	cloudy	30(%)	51	11.0	0	33	1.0	−2
2002/01/05 (Sat)	clear, occasionally cloudy	20(%)	48	9.0	−2	37	3.0	0
2001/01/06 (Sun)	clear, occasionally cloudy	20(%)	50	10.0	−1	32	0.0	−3
2002/01/07 (Mon)	cloudy, passing rain	50(%)	46	8.0	−3	33	1.0	−2
2002/01/08 (Tue)	cloudy, occasionally clear	30(%)	48	10.0	0	34	0.0	−1

Figure 1: A table in a weather report Web page.

We developed an automatic monitoring system PUM(Partial Updates Monitoring) that constantly checks partial updates in Web pages and notifies them to a user. A user can give PUM regions in which he/she wants to know the updates in a Web page as training examples, and it is able to learn rules to identify the partial updates by relational learning. By this learning, a user do not need to directly describe the rules. Since describing such rules is significantly hard to a general user, this learning of PUM releases a user from much cognitive load. We implemented our system and made experiments to evaluate effectiveness. Finally we discuss on limitation and open problems.

WebBeholder[8] is a pioneer system which checks Web page updates and notifies only the difference between a old Web page and a new one. WebBeholder evaluates the differences using HTML tags and notifies important updates. Unfortunately a user can not indicate a partial update on which he/she wants to know in WebBeholder. The primary function of our PUM is to deal with such a task.

A lot of studies on information extraction from semi-structured text have been done[5][1][9] and inductive learning methods were applied to the systems. Another approach is to extract relational knowledge from Web pages as hyper texts[4]. Several ways including a traditional statistic method, machine learning were applied to extract knowledge from Web pages, and the comparison was discussed. However there is few research on interactive learning system in such a field, and all of them need a large number of training examples in advance.

A few works on interactive relational learning has been done in information retrieval[7]. Their system can learn rules to distinguish relevant documents from non-relevant ones by interactive relational learning and relevance feedback. Such an interactive approach is concerned with our approach, however the purpose is quite different.

2 PUM: partial updates monitoring in a Web page

2.1 System overview

Fig. 2 shows overview of PUM. PUM is a system that identifies a region indicated by a user in a Web page, checks updates in the region and notifies a user the updates which

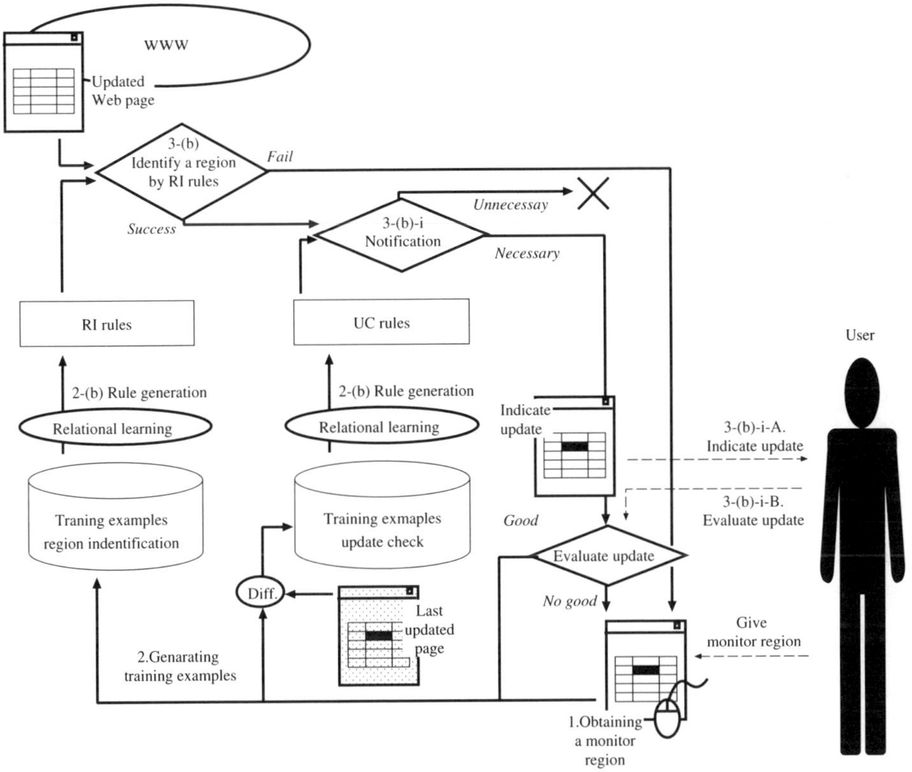

Figure 2: System overview.

he/she wants to know. A broken line indicates interaction between a user and PUM.

First PUM extracts training examples for both of *region identification* and *update check* from a region indicated by a user. Then a relational learning system automatically acquires two kinds of rules for region identification and update check. After such rules were generated, PUM becomes able to identify partial updates and determine whether it is one which a user wants to know or not by using two kinds of rules.

If PUM decides an update is useful to a user, it notifies the update to a user. Otherwise PUM indicates the updated Web page to a user and obtains his/her evaluation. PUM was implemented using Visual C++ and Ruby on Windows2000.

2.2 Region identification and update check

PUM learns *update monitoring rules* to check partial updates in a Web page. The update monitoring rules consists of two kinds of rules: *region identification rules* and *update check rules*. Region identification (RI) rules are used to identify and extract a region in which a user wants to know its updates. Update check (UC) rules are utilized to determine whether the update is one which a user wants to know or not.

Note that a meaningful update in a region can not be detected by using only RI

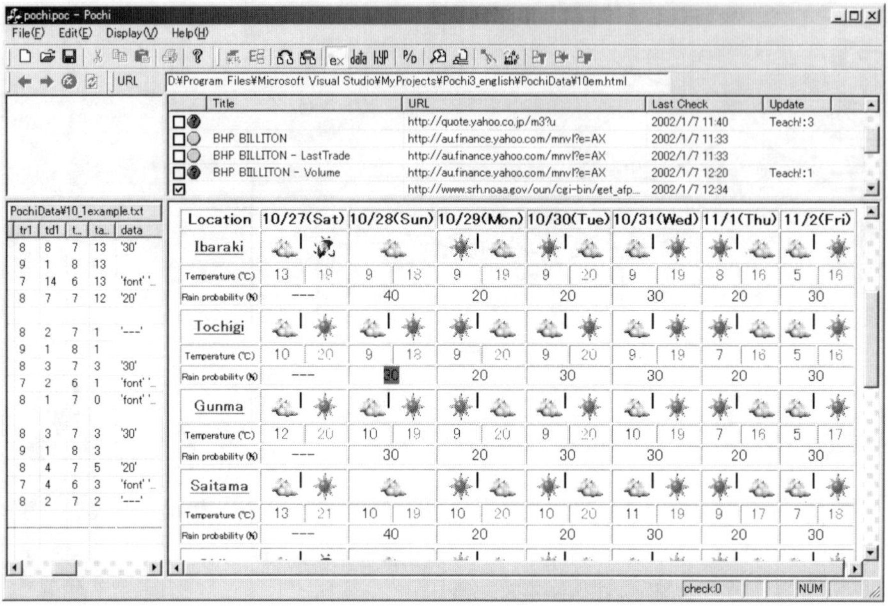

Figure 3: Interface of PUM.

rules. For example, we often wants to know updates only when a numeric value in a region increased or decreased in Web pages of stock value, exchange rate, temperature and so on. By using only RI rules, PUM can identify a region, however can not check such update's property. UC rules are necessary to check the property.

2.3 Procedure to monitor partial updates

In the following, we describe detail procedures of PUM to monitor partial updates. The number of a procedure corresponds to the number in Fig. 2.

1. Obtaining a monitor region.
   - (a) A user indicates a region in which he/she wants to know the update by mouse highlight operation on interface of PUM(Fig. 3).
   - (b) PUM obtains the region and analyses it.
2. Acquiring updates monitoring rules.
   - (a) Generating two kinds of training examples.
     - Training examples for RI consisting of the following properties are generated.
       - HTML source code of a region indicated by a user.
       - A sequence of ancestors of the region in a HTML tree.
       - Index of raw and column when the region is a cell of a table.

- Training examples for UC are generated with the difference between old values and updated values in an indicated region.

(b) The two kinds of generated examples are independently given to a relational learning system and it learns RI rules and UC rules.

3. Monitoring a Web page by update monitoring rules.
   (a) Monitor an update (not a partial update) in a Web page. If an update is detected, go to (b). Otherwise, repeat this monitoring.
   (b) Identify a region by RI rules.
      i. If the identification is success, determine whether the update is important or not by UC rules.
         A. If notification is necessary, indicate a user a Web page in which a region is highlighted and he/she evaluates it. If the update is correct, go to 3. Otherwise go to 1.
         B. If notification is not necessary, do nothing and go to 1.
      ii. If plural regions are matched, indicate a user a Web page in which the regions are highlighted and he/she evaluates it. If the update is correct, go to 2. Otherwise go to 1.
      iii. If no matched region, go to 1.

Fig. 3 shows interface of PUM. The window consists of three sub-windows: a Web browser window, an URL window and a training example window. A Web browser window (lower right in Fig. 3) shows a Web page in the same way to a Web browser and a user can easily indicate a region by highlighting it using a mouse. An URL window(upper in Fig. 3) stands for URLs of updated pages. A training example window(lower left in Fig. 3) indicates a table of attribute and value of stored training examples.

2.4 Relational learning

Relational learning[6] is a machine learning method that acquires rule for classify given examples into classes. Inductive learning approach is utilized to construct rules from a lot of positive/negative training examples.

PUM utilizes RIPPER[3] as a relational learning system. RIPPER acquires rules to classify examples into two classes, and the learned rule is described with symbolic representation, not weight distribution of neurons in neural network learning. Thus a user can easily understand rules and modify them. The another advantage of RIPPER is that it efficiently learns rules. For interactive system like PUM, fast learning is necessary.

RIPPER is given training examples consisting of attributes and their values. It is able to deal with a nominal value, a set value and a continuous value as an attribute value.

2.5 Generating training examples

At step 2a in procedures of the last subsection, PUM generates two kinds of training examples for learning RI rules and UC rules. In the following, we explain representation of such training examples.

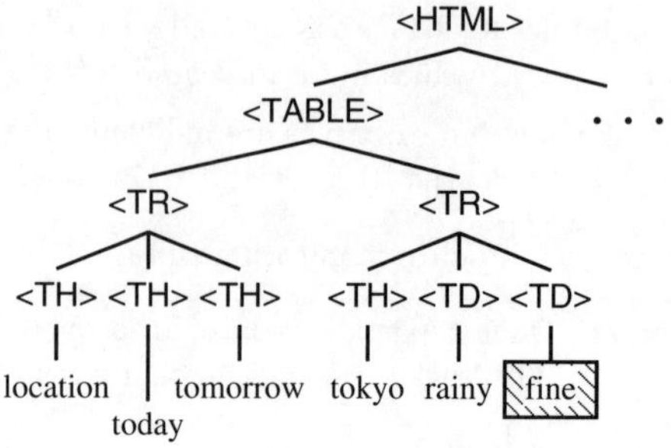

Figure 4: A HTML tree.

2.5.1 *Training examples for region identification*

A training example for RI is described with the following attributes.

$$(tag1, tag2, tag3, \cdots, cNo, rNo, data, cIndex, rIndex, class)$$

$tag1, tag2, tag3, ...$ are a sequence of ancestors of an indicated region in a HTML tree (Fig. 4). A value of a *tag* attribute is order of the tag when the other same tags are in the same depth. *data* is a set of `<image>` tag's `src`, `alt` attribute values and words included in the indicated region, and this attribute *data* is dealt as a set value in RIPPER.

When an indicated region is a cell in a table, a column number cNo and a row number rNo are also described in a training example. $cIndex$ and $rIndex$ stand for indexes for column and row respectively. They are obtained from the values of a cell with `<TH>` or the first cell in the same column or row. Last attribute *class* stands for whether an example is positive or negative, and has "good" or "nogood" as its value. An training example obtained from the shadowed region "fine" in Fig. 4 is described like the following.

html	table	tr	td	cNo	rNo	data	cIndex	rIndex	class
1	1	2	2	2	1	fine	tomorrow	tokyo	good

2.5.2 *Negative examples for region identification*

Since relational learning is a kind of inductive learning, negative examples play a important role to avoid over-generalization. However, in such an interactive system like PUM, Obtaining negative many negative examples may force much cognitive load to a user. Thus PUM automatically generates negative example for region identification to improve learning efficiency.

We consider the neighborhood of an indicated region was a near miss example. Hence PUM generates negative examples from four regions: left, right, upper and lower regions to an indicated region. We experimentally found out this strategy is effective to make learning more efficient.

2.5.3 Training examples for update check

Training examples for UC are generated from the last Web page and an updated Web page, and described with the following attributes.

$$(dataNew, relation, class)$$

dataNew means a value in an updated region. The *relation* means numerical relation between the last value and *dataNew*, and has <-> (decrease), <+>(increase), <=>(no change). If no relation between two values, the two values are assigned into *relation*. *class* stands for whether an example is positive or negative, and takes 'good' or 'nogood'.

3 Executed examples

3.1 Case-1: Weather report Web page

First example is on updates in a weather report Web page shown in Fig. 3. This page shows weekly weather report and weather report of next seven days is indicated by scrolling a table vertically.

In this example, a user wants PUM to notify updates when rain probability of Tochigi on the next Sunday (a highlighted cell on a Web browser window in Fig. 3) decreases less than 40%. Thus PUM needs to learn RI rules to identify a cell indicating weather probability of Tochigi on Sunday and UC rules to check that the value of weather probability is less than 40. Since the Web page is updated everyday, PUM detects an update once a day and notifies to a user if necessary. When an update is notified to a user, he/she evaluates it.

3.1.1 Learning RI rules

A part of training examples for RI is shown in Table 1. The first example was generated from a cell indicated by a user, and the remaining examples were automatically generated from neighbor cells by a method described in 2.5.2.

Table 2 stands for the number of user's evaluations and learned RI rules at that time. A RI rule learned from the first evaluation can identify a cell which is in 7th-row and has '10/28(Sun)' as its column index. This rule succeeded in identifying a region on the next day, however it failed after two days. Because the two rows of '10/27(Sat)' and '10/28(Sun)' disappeared by scrolling horizontally the table and a new target region included '11/3(Sun)' instead of '10/28(Sun)'. Then PUM requires user's evaluation and learned the second rule shown in Table 2. This rule identifies a correct cell using a more general condition 'Sun' as a column index, not '10/28(Sun)'.

3.1.2 Learning UC rules

Table 3 shows examples for update check and Table 4 indicates the number of user's evaluations and learned RI rules at that time. At seventh evaluation, the negative examples were given and rules with a nogood class and conditions '50' $\in$ dataNew and '—' $\in$ dataNew, where '—' means no value. These UC rules is not sufficient because if rain probability becomes 60~100 they notify the update. As progress of evaluation, the sufficient number of negative examples are given to PUM and correct conditions are learned.

Table 1: Training examples for RI.

html	body	center	table	tr	td	rNo	cNo	data	cIndex	rIndex	class
1	1	1	2	8	8	7	3	'30'	'10/28(Sun)' '10' '28' 'Sun'	'rain probability%'	good.
1	1	1	2	8	7	7	1	'—'	'10/27(Sat)' '10' '27' 'Sat'	'rain probability%'	nogood.
1	1	1	2	8	9	7	5	'20'	'10/29(Mon)' '10' '29' 'Mon'	'rain probability%'	nogood.

Table 2: RI rules.

Eval. No.	rule
1	good $\leftarrow$ '10/28(Sun)' $\in$ cIndex $\wedge$ '7' $\in$ rNo. nogood $\leftarrow$.
2	good $\leftarrow$ '7' $\in$ rNo $\wedge$ 'Sun' $\in$ cIndex. nogood $\leftarrow$.

Table 3: Training examples for UC.

dataNew	relation	class
'30'	'<->'	good
'20'	'<->'	good
'30'	'<+>'	good
'30'	'<=>'	good
'20'	'<+>'	good
'—'	'20' '—'	nogood
'50'	'—' '50'	nogood
'40'	'<->'	nogood

Table 4: UC rules.

Eval. No.	rule
1	good $\leftarrow$.
2	good $\leftarrow$.
⋮	⋮
6	nogood $\leftarrow$ '—' $\in$ dataNew. good $\leftarrow$.
7	nogood $\leftarrow$ '50' $\in$ dataNew. nogood $\leftarrow$ '—' $\in$ dataNew. good $\leftarrow$.

3.2 Case-2: stock market Web page

Another example is executed in a stock market Web page shown in Fig. 5. In this example, after several evaluation of a user, PUM became able to correctly notify partial updates of a cell indicating a stock value of a particular company. Also PUM is available to more complicated Web page with two tables shown in Fig. 6.

4 Discussion

4.1 Direct modification of rules by a user

Though RIPPER is a fast learning system, the learning is not sufficiently rapid for an interactive learning system like PUM. A way to improve performance is that a user directly modifies learned rules. Fortunately symbolic rule representation is far more suitable for a user to modify learned knowledge directly than weight distribution learning like neural network learning, regression and so on. Thus we are developing a human-computer interaction framework to deal with such user's help.

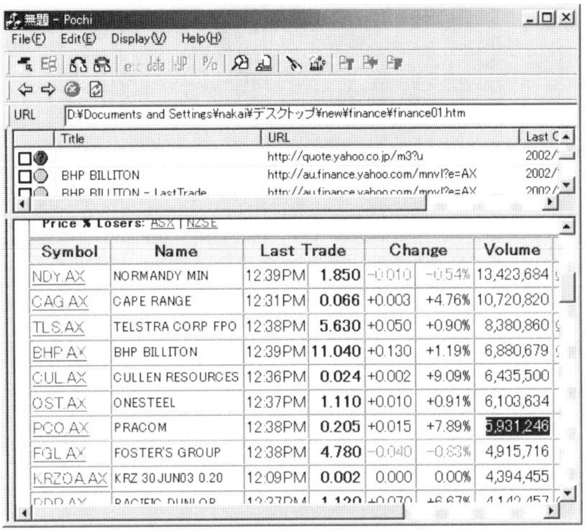

Figure 5: A stock market Web page.

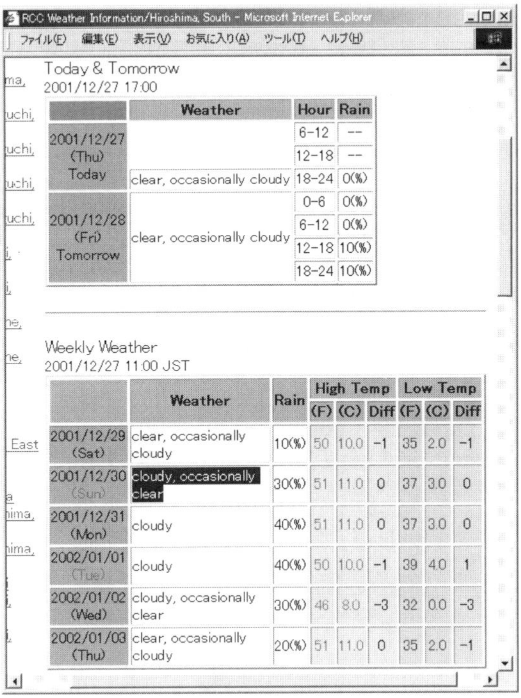

Figure 6: A Web page with two tables.

4.2 *Coverage of* PUM

In this paper, PUM dealt with a cell in a table as a region and we consider PUM is applicable to any table in Web pages of any field. However PUM has limitation on its coverage.

PUM can not deal with a region in plan text surrounded by <PR> tags. We are trying to utilize neighbor words of a region as context to identify it, however such approach may not sufficient. Also PUM may not be successfully applied to a list using <UL>, <OL> tags. Because effective constrains like index, column number are not available. To cope with this problem, we need to give much background knowledge like additional attributes to PUM.

There is no guarantee that a user always indicates the correct regions in a Web page in his/her evaluation. Thus PUM needs to deal with noisy training examples. We need to investigate the robustness of PUM against noise experimentally.

5 Conclusion

We proposed an automatic monitoring system PUM that constantly checks partial updates in Web pages and notifies them to a user. A user can give a system regions which he/she wants to know the update in a Web page as training examples, and it is able to learn rules to identify the partial update by relational learning. By this learning, a user do not need to describe the rules. We implemented our system and some executed examples were presented.

References

[1] M. E. Califf and R. J. Mooney. Relational learning of pattern-match rules for information extraction. In *Working Notes of AAAI Spring Symposium on Applying Machine Learning to Discourse Processing*, pages 6–11, 1998.

[2] ChangeDetection.com. http://www.changedetection.com/monitor.html.

[3] W. W. Cohen. Fast effective rule induction. In *Proceedings of the Twelfth International Conference on Machine Learning*, pages 115–123, 1995.

[4] M. Craven, D. DiPasquo, D. Freitag, A. McCallum, T. Mitchell, K. Nigam, and Seán Slattery. Learning to construct knowledge bases from the World Wide Web. *Artificial Intelligence*, 118(1-2):69–113, 2000.

[5] Nickolas Kushmerick, Daniel S. Weld, and Robert B. Doorenbos. Wrapper induction for information extraction. In *Intl. Joint Conference on Artificial Intelligence*, pages 729–737, 1997.

[6] S. Muggleton and L. De Raedt. Inductive logic programming: Theory and methods. *Journal of Logic Programming*, 19(20):629–679, 1994.

[7] M. Okabe and S. Yamada. Interactive web page filtering with relational learning. In *Proceedings of the First Asia-Pacific Conference on Web Intelligence*, pages 443–447, 2001.

[8] S. Saeyor and M. Ishizuka. WebBeholder: A revolution in tracking and viewing changes on the web by agent community. In *WebNet 1998*, 1998.

[9] S. Soderland. Learning information extraction rules for semi-structured and free text. *Machine Learning*, 34(1-3):233–272, 1999.

[10] Web Secretary. http://homemade.hypermart.net/websec/.

Active Mining
H. Motoda (Ed.)
IOS Press, 2002

Context-Based Classification of Technical Terms Using Support Vector Machines

Masashi Shimbo, Hiroyasu Yamada, and Yuji Matsumoto
{*shimbo,hiroya-y,matsu*}@*is.aist-nara.ac.jp*
Graduate School of Information Science
Nara Institute of Science and Technology
8916-5 Takayama, Ikoma, Nara 630-0101, Japan

Abstract. We investigate a practical method of classifying technical terms from the abstracts of medical and biological papers. The main objective is to identify a set of features relevant to the classification of technical terms. The features considered are: (1) spelling of a term, (2) words around the occurrence of a term, and (3) syntactic dependency of a term with surrounding words. We evaluated the effectiveness of these features in a task of classifying terms in the abstracts from MEDLINE database, in which target classes were determined in accordance with the first five top-level nodes of the MeSH tree. We first listed all the terms in the MeSH thesaurus which fall in exactly one class, and then retrieved MEDLINE abstracts containing the terms. Using these abstracts as training sets, we applied a support vector learning method to discriminate each class with various combination of the features. The result proved the effectiveness of dependency as features for classification, especially when complemented with spelling features.

1 Introduction

The ability to cope with technical terms is essential for natural language processing (NLP) systems dealing with scientific and technical documents. Since a majority of these terms are not in general-purpose dictionaries[1], domain-specific lexicons are often used in combination. However, it is still unrealistic to expect all technical terms to be enumerated in the lexicons, because compound terms, which share a high proportion of technical terminology, are quite productive. Hence, in the active fields of research such as biology and medicine, new terms are produced on a daily basis; in year 2000 and 2001 versions of NLM Medical Subject Headings (MeSH) [14], the numbers of new concepts introduced were 552 and 184, respectively. Furthermore, even if such terms are recognized with the help of lexicons, it may still be necessary to identify the meaning of each occurrence of the terms, because technical terms are often polysemous.

Robust techniques are thus required for (1) recognizing technical terms, and for (2) identifying the semantic class of those terms. The first task was tackled by several researchers, and some useful linguistic properties common to technical terminology have been identified. Moreover, recent advance in statistical NLP techniques allows the extraction of compound

[1]Gouhara et al. [5] reported that more than 15% of the words occurring in MEDLINE [13] abstracts were not in the Oxford Advanced Learner's Dictionary of Contemporary English [9].

terms at a practical level of accuracy. By contrast, semantic categorization of technical terms have attracted less researchers, mainly because the task is more involved and requires extensive expert knowledge to correctly evaluate the results.

As a step towards robust classification of technical terms, we construct a system for identifying semantic class of biological and medical terms using state-of-the-art NLP and machine learning techniques. Our objective with this system is to examine the effectiveness of new features extracted from syntactic dependency structure within sentences. Several other features are considered as well, such as spelling of the terms and words occurring around the terms. We also exploit publicly available resources as much as possible to avoid costly annotation of corpora by human experts.

Specifically, we apply Support Vector Machines (SVMs) [15] for classification of technical terms occurring in MEDLINE abstracts [13], using features obtained with natural language processing of the abstracts. Being a supervised learning method, SVM requires a volume of labeled examples for training. In our task, however, since MEDLINE holds the abstracts in plain text format, they first need to be annotated with the labels specifying the class of each technical terms. We take advantage of the MeSH Tree, also available publicly, to automatically annotate the abstracts.

The rest of the paper is organized as follows. We first review previous work dealing with technical terminology (Section 2) and describe the problems in machine learning approach to classification of terms (Section 3). We then discuss why syntactic dependency information is useful for the task and how it should be extracted (Section 4), and evaluate the effectiveness of the feature with some experiments (Section 5). Finally, we summarize the results and discuss possible future topics (Section 6).

2 Background

In their paper addressing identification of technical terms, Justeson and Katz [6] reported that 92.5–99% of the occurrences of technical terms were noun phrases. They also pointed out that technical terms share some common linguistic properties regardless of the domain of text, and presented a terminology identification algorithm motivated by these properties. Other work in this area includes researches by Su et al. [11] and by Maynard and Ananiadou [8], both exploiting domain-independent linguistic and statistical properties which discriminate technical terms from non-technical terms.

Unfortunately, such common properties of technical terms are useless in classifying them, as its objective is to discriminate among the terms. Classification task is more involved than identification because the properties that characterize a class is specific to the class; they are substantially different from one class to the other, and also from one domain of text to the other.

Approaches to classifying technical terms can be further divided into two. The first relies on handcrafted patterns. This approach was taken by Fukuda et al. [4], who achieved precision and recall rates of approximately 95% in identifying protein names, by using regular expression patterns and a series of heuristic rules created by human experts. A shortcoming of this approach, however, is non-negligible cost of constructing and maintaining such patterns and rules; because patterns and rules vary among semantic categories, this method substantially lacks portability.

By contrast, the second approach automatically acquires classification rules from large annotated corpora using supervised machine learning methods. It assumes that most of the relevant features are domain-independent, yet class- and domain-specific characteristics can be automatically extracted from these features. The work along this approach includes Collier et al. [2], Gouhara et al. [5] and Yamada et al. [16], as well as the present paper.

3 Supervised Learning Approach to Technical Terminology Classification

There are three factors dominating the performance of terminological classifiers in supervised learning approach: (1) the size and quality of training corpora, (2) the choice of the leaning algorithm, and (3) the choice of the features used for classification. Below, we review how these factors have been addressed in the literature, as well as our own approach.

3.1 The Size and the Quality of Training Corpora

Generally speaking, previous work in the area used relatively small number of examples because of the difficulty in constructing a large corpus of text with high-quality annotations. For instance, the corpora used by Collier et al. [2], Gouhara et al. [5], and Yamada et al. [16] consisted merely of 3312 technical terms (in 100 abstracts), 1526 terms (in 35 abstracts) and 2268 terms (in 77 abstracts), respectively. Moreover, Yamada et al., who employed two human experts to annotate the same set of abstracts in MEDLINE database, observed about 20% disagreement rate of annotated tags between the two annotators. A similar disagreement rate was also reported by Tateisi et al. [12]. Part of this disagreement comes from large cross-over in vocabulary of each semantic classes, yet these facts imply that classification task is non-trivial even for human experts.

Gouhara et al. utilized co-training technique [1] to augment small amount of labeled data with large amount of unlabeled data in order to reduce the cost of corpus annotation (labeling) by human experts. For that purpose, they used two sources of information, which are arguably mutually independent; namely, phrase-internal information such as character types and part-of-speech of words on the one hand, and contextual information such as syntactic dependency on the other hand. Unfortunately, the effect of co-training on performance was not apparent in their experiment.

The size and the quality of annotated corpora are thus non-negligible factors for supervised learning approach. In the present work, the difficulty of constructing a training corpus is alleviated by the use of existing thesaurus.

3.2 Learning Algorithms for Terminology Classification

Several machine learning algorithms have been applied for terminology classification. Collier et al. [2] used Hidden Markov Models; Gouhara et al. [5] used decision trees with co-training; and most recently, Yamada et al. [16] used Support Vector Machines (SVMs) to deal with high-dimensional feature space incurred by the use of abundant information on spelling, parts-of-speech, and substrings. Compared with Yamada et al., the former two researchers used smaller number of features, due to the limited scalability of the learning algorithms used.

Following Yamada et al., the present paper uses SVMs, which are known to perform well in the presence of many features as in our formulation of the problem.

3.3 Choice of Features

Both phrase-internal information and extra-phrase, or *contextual*, information has been used for classification of technical terminology. Phrase-internal information includes features such as character types and parts-of-speech of constituent words. The effectiveness of these features has been demonstrated in [2] and [16]. As to the contextual features, use of bigram or trigram sequence of words surrounding the terms is popular. However, fixed-length sequences are problematic in that how far we should look beyond its surroundings actually situation dependent. For instance, Sentences (1) and (2) below, both retrieved from MEDLINE database, are the examples in which the bigram feature fails to capture words that could possibly help in determining the class of terms.

Both the azide-insensitive and azide-sensitive components of F1-ATPase activity are equally inhibited by *labelling the enzyme with 7-chloro-4-nitrobenzofurazan*, by binding the natural inhibitor protein, or by cold denaturation of the enzyme.	(1)

Results suggest that E. chaffeensis infections are common in free-ranging coyotes in Oklahoma and that these wild canids could play a role in the *epidemiology of human monocytotropic ehrlichiosis*.	(2)

In Sentence (1), the bigram feature conveys only the information on two words preceding the term "7-chloro-4-nitrobenzofurazan," namely, "enzyme" and "with." They hardly provide us with a clue on the relation between the term and "enzyme" because the information on verb "labeling" is missing. Similarly, in Sentence (2), there are three words between the term "ehrlichiosis" and the key word "epidemiology" which strongly suggests that the term is the name of a disease.

Making sequence length larger (e.g., 4) solves the problem in the above examples, but it does not come without cost; it would indeed provide the classifier with richer information, but it would also result in data sparseness in a high dimensional feature space, making learning with a small number of examples extremely difficult. It is hence desirable to use a context feature more adaptive and flexible than fixed-length sequences.

4 Syntactic Dependency Structure as a Feature for Classification

One way to overcome the inflexibility of fixed-length context features, is to utilize the dependency structure of words within a sentence. It allows us to make selective use of information on distant words, without making the feature space too sparse. Such a structure can be detected in multiple ways, but in this paper we extract it from the parse trees of sentences. We will sketch how this is done with an illustration in Figure 1, which depicts a partial parse tree near the occurrence of "7-chloro-4-nitrobenzofurazan" in Sentence (1).

In the parse tree of Figure 1, each parent-child relation signifies an application of a context-free production rule of the form $X \rightarrow Y_1, \ldots, Y_n$, where X is a non-terminal symbol (denoting its syntactic categories such as NP, VP and PP) of the parent node, and $Y_1, \ldots, Y_n$ are the symbols of the children. A node is labeled not only with a symbol, but also with a *head word*. For a terminal node, it is the lexical entry of the node (shown in italics in the figure); for a non-terminal node, it is inherited from one of its children (shown in parentheses). If a node X has two or more children $Y_1, \ldots, Y_n$, $n \geq 2$, the so-called "head rule[2]" associated with

[2]We used a slightly modified version of the head rules used by Collins [3].

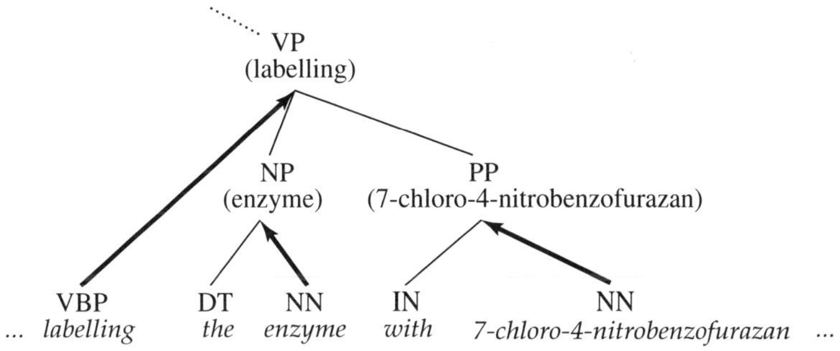

Figure 1: Parse tree of a verb phrase containing "7-chloro-4-nitrobenzofurazen." Bold arrows signify head word inheritance, and parenthesized words the inherited head words.

the production rule $X \longrightarrow Y_1, \ldots, Y_n$ determines a child (called *head constituent*) Y_i from which X inherits the head word. In the figure, bold arrows depict how head words are inherited; e.g., the bold arrow from NN to PP shows that NN is the head constituent[3] of production rule $\mathrm{PP} \longrightarrow \mathrm{IN}, \mathrm{NN}$.

When a parse tree is available, dependency structure can be extracted by recursively merging every head constituent node with its parent (i.e., by merging every parent-child pair connected with bold arrow in the figure), and marking the merged node with the same label as the head constituent. Then, in the resulting tree, a parent-child pair denotes a dependency of the head word of the child on that of the parent.

Applying this procedure to the tree in Figure 1, we can see that the preposition "with" depends on "7-chrolo-4-nitrobenzofurazan," the determinant "the" depends on "enzyme," and both "7-chrolo-4-nitrobenzofurazan" and "enzyme" depend on "labelling." Hence, by collecting the words that depend on and those depended by the term of interest, we can extract dependency information relevant to the term. This allows us, for instance, to detect a verb that collocates with a technical term regardless of the numbers of words inserted in between. For instance, it allows using the verb "labelling" as a feature for "7-chrolo-4-nitrobenzofurazan" in Sentence (1); and in Sentence (2), since "epimiology" is the head word of the noun phrase containing the term "ehrlichiosis," the dependency of the term on "epimiology" can be extracted with this procedure as well.

The experiments in Section 5 use this method of extracting dependency information from parse trees.

5 Experiments

To evaluate the effectiveness of the dependency information extracted from parse trees, we constructed a system for identifying the semantic class of technical terms in MEDLINE abstracts. The following assumptions were made in designing the system:

- As described in Section 2, existing NLP systems can extract a high proportion of noun

[3]This is one of the major modification we made to the head rules found in [3], in which IN instead of NN is the head consistuent of the production $\mathrm{PP} \longrightarrow \mathrm{IN}, \mathrm{NN}$.

Table 1: Number of examples for each class

Class	# of examples
A	4571
B	2811
C	5004
D	7335
E	4101
Total	23822

phrases from texts. Hence, we deal solely with classification of the terms, as opposed to Yamada et al. [16] who also took identification into account.

- It is not realistic to assume the availability of a large corpus of texts annotated by human experts. As described in Section 3.1, construction of a corpus is a major problem in the task. We explore the MeSH thesaurus, publicly available online, to automatically annotate corpus.

5.1 Experimental Setting

The experimental setting is described below.

Classes The target semantic classes were determined in accordance with the first five top-level nodes of the 2002 MeSH Tree. They are (A) Anatomy, (B) Organisms, (C) Diseases, (D) Chemicals and Drugs, and (E) Analytical, Diagnostic and Therapeutic Techniques and Equipments.

Data Sets A corpus of abstracts was obtained in the following way. First, 15000 terms from the above classes in the MeSH Tree were randomly sampled. Next, using these terms as query keywords, we retrieved 216404 abstracts from MEDLINE, and then resampled 1200 abstracts for each class from the set. Removing duplicates from the resampled collection resulted in a corpus of 5842 distinct abstracts.

In the corpus, 7531 terms belonging to exactly one of the classes (A) to (E) were identified and used as the target terms. This yielded a total of 23822 distinct examples. The number of examples for each class is shown in Table 1.

Features The types of features used by the classifiers were as follows. In addition to the ones using only one of these feature sets, we constructed the classifiers with various combinations of the feature sets as well.

- *Suffix features* — the suffix strings of the head words of target terms. A head word of a target term is determined by the same head rule as described in Section 4 We used the the suffixes of lengths 3 and 4.

- *Bigram features* — the surface and the parts-of-speech (POS) of words in the bigram sequences preceding and succeeding target terms. To obtain the POS, every sentence in

Table 2: Number of terms and examples in each cross-validation set.

Set ID	# of terms	# of examples
1	1507	5236 (22.0%)
2	1506	4361 (18.3%)
3	1506	4844 (20.3%)
4	1506	4715 (19.8%)
5	1506	4666 (19.6%)
Mean	1506.2	4764.4 (20.0%)
Total	7531	23822 (100.0%)

the corpus containing one or more technical terms was fed to Nakagawa et al.'s POS tagger [10][4].

- *Dependency features* — the words on which a target term depends, and the words which the term is depended on, together with their corresponding POS. To obtain these features, the output of the POS tagger was further fed to Yamada and Matsumoto's bottom-up parser [17], under the constraint that technical terms occuring in the sentence should be labeled as either NN (noun) or NP (noun phrase)[5]. The output parse trees were then used for extracting above features with the method of Section 4.

Algorithm Given a set of examples and a combination of features, we constructed an SVMs for each class (A) to (E). The examples whose target terms fall into other four classes were used as negative instances. In all cases, the SVMs used a linear kernel with a fixed soft margin parameter of $C = 1$.

Evaluation We conducted five-fold cross validation with the data set. The examples were partitioned into five sets so that no target terms appear in two sets, and so that each set contains a nearly equal number of distinct target terms. This partitioning scheme avoids a term to appear both in training and test sets during cross validation; since we make use of spelling (suffix) information as features, simply partitioning examples into five sets of equal size at random would make the problem much easier. As a result, the number of examples (occurrences of terms) in each set is not uniform, because some of these terms occur more than once in the abstracts. Table 2 shows the numbers of terms in each set.

5.2 Results

Under the setting described above, two experiments were conducted.

The first experiment compares the performance of the types of contextual features. Table 3 shows the performance of two classifiers, each using only one of the dependency or bigram features. The result clearly shows the superiority of dependency information over bigrams.

In the next experiment, we combined the contextual features with the phrase-internal suffix features. The performance of classifiers with various feature combinations is listed in

[4]The POS tagger performs well even in the presence of unknown words, with the accuracy of 87% for unknown words, and 96% overall in the Penn TreeBank [7]

[5]This constraint reflects the observation by Justeson and Katz [6] that a vast majority of the occurrences of technical terms are noun phrases.

Table 3: Performance with different contextual information. All the numbers are means over five cross validation trials.

	Dependency			Bigram		
Class	Precision	Recall	F-score	Precision	Recall	F-score
A	**0.802**	**0.464**	**0.587**	0.553	0.088	0.152
B	**0.792**	**0.324**	**0.459**	0.696	0.197	0.306
C	**0.829**	**0.522**	**0.640**	0.662	0.199	0.306
D	**0.767**	**0.456**	**0.571**	0.644	0.253	0.362
E	**0.739**	**0.369**	**0.491**	0.564	0.165	0.254

Table 4: Performance with various feature combinations. P: precision, R: recall, F: F-score

	Dependency + Bigram + Suffix			Dependency + Suffix			Bigram + Suffix			Suffix only		
Class	P	R	F	P	R	F	P	R	F	P	R	F
A	0.914	**0.707**	**0.794**	0.913	0.703	0.791	0.912	0.703	0.790	**0.916**	0.693	0.786
B	**0.878**	**0.548**	**0.674**	0.787	0.546	0.640	0.842	0.545	0.658	0.792	0.516	0.619
C	0.916	0.841	0.876	**0.920**	0.839	**0.877**	0.906	**0.842**	0.872	0.906	0.829	0.864
D	**0.857**	0.848	**0.851**	0.837	**0.861**	0.848	0.849	0.850	0.848	0.819	0.838	0.827
E	**0.895**	0.685	0.771	0.867	**0.693**	0.766	0.887	0.690	**0.772**	0.853	0.700	0.765

Table 4. As a base line, the performance of the classifier using only the suffix features is also included in the table.

The classifier using all of the dependency, bigram and suffix features performed best, but was only slightly ahead of the one with dependency and suffixes. Both of these outperformed the classifiers not using dependency information in most of the classes. Even in a few cases in which the latter surpassed the former, the difference was not significant at all. However, the performance advantage of dependency over bigrams was much smaller than the one observed in the previous experiment in which these features were used alone.

6 Summary and Future Directions

We have constructed a system for terminological classification in biological and medical papers. Motivated by practical considerations, the system takes advantage of state-of-the-art natural language processing and machine learning techniques, as well as publicly available resources. We have further evaluated the performance of the system over different set of features. Although more thorough experiments are desirable, the experimental results of Section 5 suggest the effectiveness of syntactic dependency information as features for classification, especially when they are used in combination with information on phrase-internal information.

Topics for future research include:

- Training of NLP pre-processing tools with a corpus of texts in similar domains. The part-of-speech tagger and parser programs used in the experiment were themselves based upon supervised learning techniques. They were trained on the Penn TreeBank [7] corpus consisting of newspaper articles from Wall Street Journal. It is likely that the difference of the corpora have given ill effects on the accuracy of part-of-speech tags and parse trees. There should be some room for improvement if we had a tagged corpus of papers available for training these programs.

- Evaluation of the robustness of the approach under various scales of training data. The experiment in this paper used about 6000 terms (and 20000 examples) for training classifiers. These numbers are much larger than those used in previous work mentioned in Section 3, but since tagging can be done automatically using the MeSH Tree in our setting, the system should be evaluated with more examples. It is also desirable to evaluate the generalization performance when the number of example is much smaller, because not all fields of research have an extensive thesaurus like the MeSH Tree.
- Classification into more detailed sub-categories. We used only the descriptors on the top-level nodes of the MeSH Tree Structure as semantic categories. It should be necessary to evaluate the performance of our methods in the tasks of classification into more detailed sub-categories.
- Measurement of performance in disambiguating multi-class terms. We trained classifiers only with terms whose class could be uniquely determined according to the MeSH Tree, and excluded multi-class terms from consideration. It would be interesting to apply the classifier trained this way to disambiguate the meaning of each occurrence of the multi-class terms in the corpus.
- Elaboration of the rules for extracting contextual information from parse trees. The current scheme for extracting dependency is based on the head rules of [3]. We needed to make modifictions to some of the rules involved, because the original rules were determined to be effective with respect to parsing, and not with respect to detecting dependency. For instance, for the production rule PP $\longrightarrow$ IN, NN, the original head rule specifies IN (preposition) as the head instead of NN (noun). We therefore modified the head rule and made NN the head constituent[6]. This is the only major modification made to the original head rule in our experiments, but there may be more rules which are not adequete for extracting contextual/dependency information from parse trees.
- Utility of information on multiple occurrences of terms. Justeson and Katz argued that when an entity is referred to by a terminological noun phrase and is rementioned subsequently, it is more likely that the full noun phrase is used intact. This property suggests that when a term is used more than once within an abstract, it is likely that the referent entity and hence its semantic class is unique in the abstract. It should be worth utilizing this clue to uniquely determine the semantic class of a technical terms within an abstract, for instance, by voting. Collier et al. [2] report that an improvement of 2.3% in F-score was achieved by a similar post-processing.

Acknowledgments. We are grateful to Taku Kudo and Tetsuji Nakagawa for providing us with their part-of-speech tagger and machine learning programs. This research was supported by the Ministry of Education, Culture, Sports, Science and Technology of Japan under Grant-in-Aid for Scientific Research on Priority Areas (B) no. 759.

[6]See the example in Figure 1.

References

[1] A. Blum and T. Mitchel. Combining labeled and unlabeled data with co-training. In *Proceedings of the Eleventh Annual Conference on Compuational Learning Theory (COLT'98)*, pages 92–100. Morgan Kaufmann, 1998.

[2] N. Collier, C. Nobata, and J. Tsujii. Extracting the names of gens and gene products with a Hidden Markov Model. In *Proceedings of COLING'2000*, pages 201–207, 2000.

[3] M. Collins. *Head-driven statistical models for natural language processing*. Phd dissertation, University of Pennsylvania, 1999.

[4] K. Fukuda, T. Tsunoda, A. Tamura, and T. Takagi. Towards information extraction: identifying protein names from biological papers. In *Proceedings of the Third Pacific Symposium on Biocomputing (PSB'98),*, pages 707–718, Maui, Hawaii, USA, 1998.

[5] H. Gouhara, T. Miyata, and Y. Matsumoto. Extraction and classification of medical and biological technical terms. IPSJ SIG Note 2000-NL-135-6, Information Processing Society of Japan, 2000. In Japanese.

[6] J. S. Justeson and S. M. Katz. Technical terminology: some linguistic properties and an algorithm for identification in text. *Natural Language Engineering*, 1(1):9–27, 1995.

[7] M. P. Marcus, B. Santorini, and M. A. Marcinkiewicz. Building a large annotated corpus of English: The Penn Treebank. *Computational Linguistics*, 19(2):313–330, 1994.

[8] D. Maynard and S. Ananiadou. TRUCKS: a model for automatic multi-word term recognition. *Journal of Natural Language Processing*, 8(1):101–125, 2001.

[9] R. Mitton. *A description of a computer-usable dictionary file based on the Oxford Advanced Learner's Dictionary of Current English*, 1992. ftp://ota.ox.ac.uk/pub/ota/public/dicts/710/.

[10] T. Nakagawa, T. Kudo, and Y. Matsumoto. Unknown word guessing with support vector machines. IPSJ SIG Note NL-141-13, Information Processing Society of Japan, 2001. In Japanese.

[11] K.-Y. Su, M.-W. Wu, and J.-S. Chang. A corpus-based approach to automatic compound extraction. In *Proceedings of the 32nd Annual Meetings of the Association for Computational Linguistics*, 1994.

[12] Y. Tateisi, T. Ohta, N. Collier, C. Nobata, and J. Tsujii. Building annotated corpus from biomedical research papers. In *Proceedings of the COLING'2000 Workshop on Semantic Annotation and Intelligent Content*, pages 28–34, 2000.

[13] U.S. National Library of Medicine. MEDLINE. http://www.ncbi.nlm.nih.gov/PubMed/.

[14] U.S. National Library of Medicine. MeSH: Medical Subject Headings. http://www.ncbi.nlm.nih.gov/mesh/.

[15] V. Vapnik. *Statistical Learning Theory*. John Wiley & Sons, 1998.

[16] H. Yamada, T. Kudo, and Y. Matsumoto. Using substrings for technical term extraction and classification. IPSJ SIG Note NL-140-11, Information Processing Society of Japan, 2000. In Japanese.

[17] H. Yamada and Y. Matsumoto. Deterministic bottom-up parsing with support vector machines. IPSJ SIG Notes 2002-NL-149, Information Processing Society of Japan, 2002. In Japanese (To appear).

Active Mining
H. Motoda (Ed.)
IOS Press, 2002

Intelligent Tickers: An Information Integration Scheme for Active Information Gathering

Yasuhiko Kitamura
kitamura@info.eng.osaka-cu.ac.jp
Department of Information and Communication Engineering
Graduate School of Engineering, Osaka City University
3-3-138 Sugimoto, Sumiyoshi-ku, Osaka 558-8585, JAPAN

Abstract.
An active information gathering system efficiently collects information from a number of frequently updating information sources on the Internet, considering the quality and cost of information gathering, to meet demands from human users or active mining systems. In this paper, we summarize functions required for active information gathering systems and show related works. We then propose a system called Intelligent Ticker. Intelligent Ticker consists of multiple information gathering modules and an information integration module. An information gathering module produces Tickers based on the difference between an updated Web page and the original one. The information integration module integrates multiple Tickers by using an ITT (Integration Template for Tickers) to assist the user in his decision making or problem solving.

1 Introduction

The Internet rapidly spreads into our society as one of infrastructures that support our daily life. Among a number of Internet based information services, the WWW (World Wide Web) is most popular and widely used for various purposes such as sharing information among researcher to advance research and development, disseminating sales information for electronic commerce, creating virtual communities that share a common interest, and so on. Considering the vast amount of various information stored on the Web, we may be able to regard the Web as a world wide knowledge base system on the Internet.

The most important feature of the Web is that it is built up in a bottom up manner, which is contrasting with conventional distributed database systems. Once an information provider connects his/her computer to the Internet and starts a Web server, he or she can immediately disseminate information toward the world. The Web can be viewed as a federated system where a huge number of distributed information sources are running autonomously and cooperating with each other without any centralized control mechanism.

On the other hand, from a viewpoint of information users, the Web has a drawback that it is not easy to locate required information in the huge amount of data widely distributed on the Internet. As remedies to deal with this drawback, various search engines have been developed. To a query with input keywords, however, a search engine sometimes just returns thousands of URLs, which often include ones unrelated to the user's request, and the user has to filter the output.

To make the Web more useful, we further continue to study technologies for not only improving outputs of Web search engines, but also developing new systems, which we call active mining systems, that can automatically discover useful information from a huge number of Web information sources by employing various techniques such as machine learning, information agents, information retrieval, database systems, computer human interaction, and so on.

To develop active mining systems, we need to consider the following features of Web information.

- Web information is widely distributed over a huge number of Web sites in the world.
- Web pages are normally described in HTML (Hyper Text Mark-up Language). HTML is suitable for representing the visual structure of Web page, which displays on a Web browser, but not for representing the semantic structure. To deal with this drawback, XML [1], which enables information providers to represent the semantic structure by inserting semantic tags into Web pages, has been standardized. Moreover, research on Semantic Web [2] aims at enabling computers to process the Web information without human interventions based on the XML standardization.
- The amount of Web information increases very rapidly day by day and a large number of Web sites are updated frequently. Especially, ones that deal with stock market or Internet auction are updated almost every minute. Even an active mining system succeeds to discover some information from such sites, if the information is obsolete, it is useless for the user. The active mining system should keep monitoring the Web sites and modify the discovered information depending on the updates of the original sites.

This paper discusses active mining systems that discover useful information for the user through gathering information from a number of Web information sources that may be updated frequently. An active mining system is a data mining system that mainly mines data gathered through the Internet. The activeness of active mining system comes from the dynamics of information sources (updates of information sources) and the user (changes of user's interests or requests), and an active mining system consists of three modules as shown in Figure 1.

- The active information gathering module monitors a number of Web sites, which is dynamically updated, on the Internet and gathers Web pages from them to provide them to the data mining module.
- The data mining module analyzes data gathered by the active information gathering module and discovers information useful to the user.
- The active reaction module plays a role of the user interface. It shows information discovered by the data mining module. By monitoring the user's response to the output, this module notifies changes of the user's interest or request to the data mining module.

The process of active mining is performed by three modules cooperating with each other. This paper focuses the discussion on the active information gathering module.

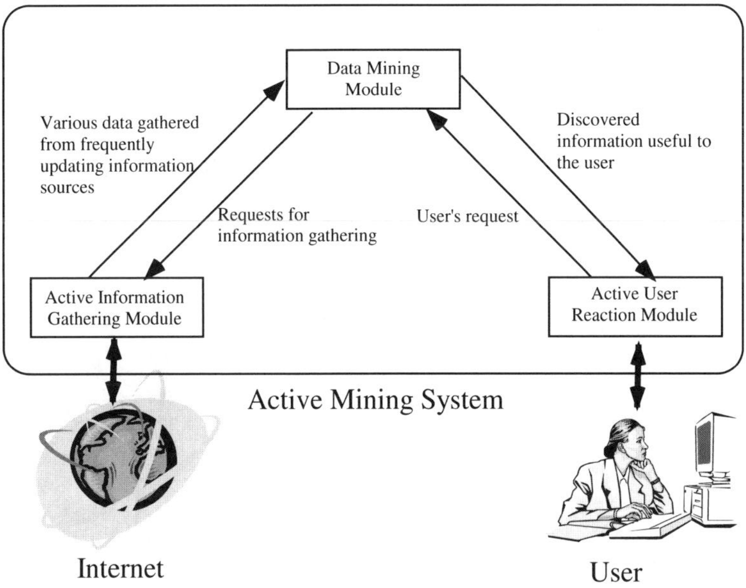

Figure 1: An Overview of Active Mining System

In Section 2, we discuss functionalities required for the active information gathering module and introduce some related works. In section 3, we propose Intelligent Ticker, which is an active information gathering system based on Ticker, which is a small piece of information extracted based on the difference between the updated Web page and the original one. We conclude this paper in Section 4.

2 Active Information Gathering Systems

Active information gathering system monitors Web information sources that are frequently updated and efficiently gathers information that meets requests given by the user. The functionalities required for active information gathering systems are summarized as follows.

2.1 Monitoring Information Sources

An important feature of Web information sources is that they are frequently updated. The frequency of updates depends on the type of information source. For example, top pages of many university Web sites are updated about once a week. Those of news sites are updated more frequently at several times an hour. Moreover, Web pages that carry information about stock market, sports, auction, highway traffic are updated more frequently at once a few minutes. An active information gathering system is expected to gather information efficiently from such dynamically updating information sources.

AIDE (AT&T Internet Difference Engine) [3] is a tool, which has been developed at AT&T, to track changes of Web pages. As a component of the tool, HtmlDiff[1] has been developed as a publicly available software that compares two Web pages and displays

[1] http://www.research.att.com/ douglis/aide/

the differences. It highlights the difference by using deleted text for data struck out and italic font for data added as shown in Figure 2.

Figure 2: An Output of HtmlDiff

TopBlend [4] is a revision of HtmlDiff and is implemented in Java, though it is not publicly available when we write this paper.

DataFoundry[2] [5, 6] is developed at Lawrence Livermore National Laboratory to maintain a data warehouse by detecting changes of information sources. In this project, the database schema of scientific data sources is represented as a graph that can be used to detect changes of the data and the schema itself. In scientific database systems, demands of database schema changes occur frequently, and to meet demands by manual operations costs much. This project aims at updating the schema automatically by tracking changes of information sources.

INRIA is also developing a data tracking system called Xyleme[3] [7] for maintaining XML-based data warehouses.

CONQUER [8] at Oregon Graduate Institute and NiagaraCQ [9] at University of Wisconsin are database approaches for monitoring information sources by formalizing the task as continuous queries.

2.2 Integrating Information

There are a number of Web information sources that deal with a similar topic. For example, there are a number of news sites on the Internet. The contents of each site is slightly different because of the difference of editors and/or news sources. The timing of updating contents is also different.

On the other hand, some readers may wish to read their favorite articles as soon as possible as they appear on the site and others may wish to read articles from a wide range of viewpoints even collecting articles takes time. The preference on reading

[2]http://www.llnl.gov/casc/datafoundry/index.html
[3]http://www.xyleme.com/

news articles depends on the reader. To provide a better service to each reader, the system needs to appropriately collect articles from multiple news sites considering the collecting time and the redundancy of articles.

Integrating information sources of different type adds more value to each of the information sources [10]. For example, there are a number of movie related sites on the Web. A movie site provides information about directors and actors, a theater site provides information about movies currently showing, and a critique site provides information about reviews on movie. By integrating information from these sites, a system can reply to a query such as "show me a movie directed by Steven Spielberg with three stars currently showing in Tokyo," which cannot be replied by information from any sole one of three sites.

To achieve an information integration task like above, we need to consider the quality and the cost of information gathering [11]. Generally speaking, there is a tradeoff between the quality of information and the cost of gathering the information. For example, if we approximate the quality of information by counting the number of information sources from which the information is gathered, obtaining information with high quality takes much time.

A planning mechanism would be required to make a good balance between the quality and the cost. To this end, an information gathering agent called BIG (resource-Bounded Information Gathering) [12] is developed at University of Massachusetts.

3 Intelligent Tickers: Toward An Active Information Gathering System

When we observe frequently updating information sources, we notice that the updates occur bit by bit. For example, the top page of some news site may be updated every ten minutes or so, but the amount of an update is only a few lines in the Web page, though the total amount of updates becomes quite large because such a small update occurs many times.

In this paper, we call such an object that carries a small piece of information "a ticker," and propose the Intelligent Ticker system that collects tickers from a number of Web information sources and integrates them to support the user's decision making or problem solving.

The Intelligent Ticker consists of an information gathering module and an information integration module as shown in Figure 3.

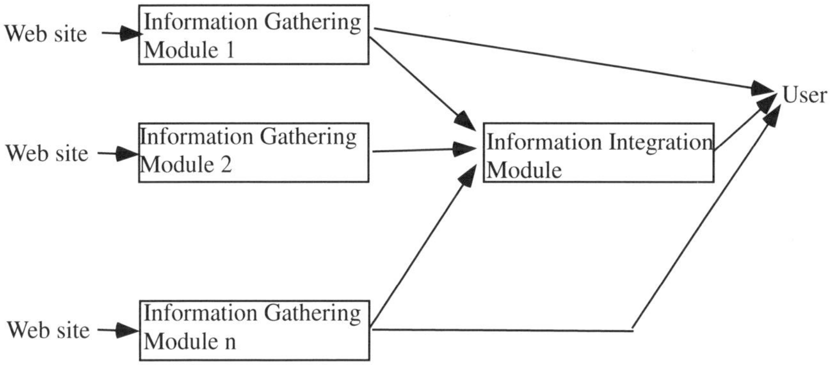

Figure 3: Overview of Intelligent Tickers

An information gathering module monitors a designated Web site and produces a ticker whenever an update occurs at the site. We assume a ticker is a part of Web page, and the user can show it directly on a Web browser.

The information integration module selects tickers sent from information gathering modules and integrates them to support the user's decision making and problem solving.

3.1 Information Gathering Module

Components of an information gathering module are shown in Figure 4. The Web access submodule fetches a Web page periodically from the designated Web site on the Internet and stores the sequence of the pages in the WebBase. The difference extraction module extracts the difference between two continuous pages in the sequence. This module uses the HtmlDiff mentioned in Section 2. The HtmlDiff shows differences by inserting tags into the original Web page, and the difference extraction module produces tickers from the output of HtmlDiff.

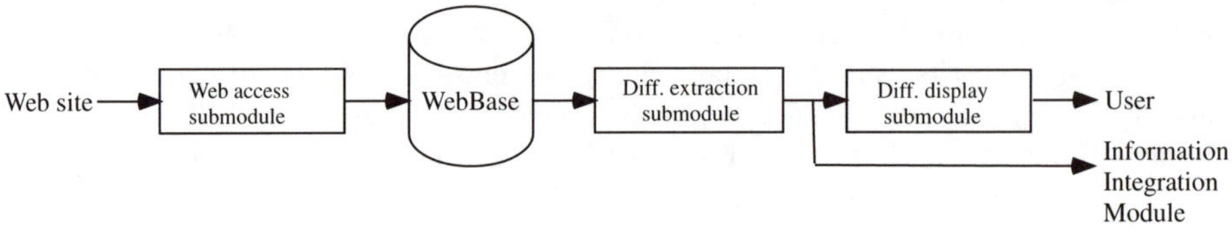

Figure 4: Information Gathering Module

The output of HtmlDiff can be analyzed as a tree structure as shown in Figure 5, for example, which is extracted from a news page shown in Figure 2. Each node represents a fragment of Web page consisting of text with the surrounding tags. Tags in a HTML document can be nested, so the document forms a tree structure. In this figure, a node depicted as a white thick circle represents a fragment which is actually updated, and should be left in the ticker. A node depicted as a white thin circle represents a fragment which is highly related to the updated fragment, such as the context or the category of updated article. We should leave such fragments also in the ticker to keep the updated fragment in the context. A node depicted as a gray circle is a fragment which should be left when the ticker is directly shown on a Web browser. Finally, a node depicted as a black circle is one which should be deleted.

How to build an analyzer that automatically categorizes the output of HtmlDiff into above 4 classes is an important and main research issue for developing the information gathering module.

A ticker, produced by an information gathering module, consists of the following elements.

- Object: An updated fragment of Web page. It is represented as text with surrounding tags.
- Time stamp: The time of update.
- Location: The URL of updated Web page.
- Context: The context of this ticker object.

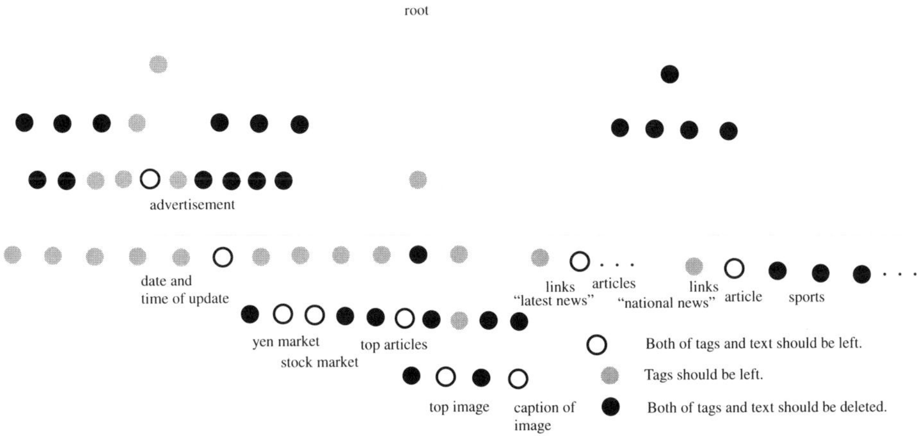

Figure 5: Analizing the output of HtmlDiff.

Each of information gathering modules needs to get access to the designated information source considering the frequency of updates. For example, it is useless to get access to an information source every hour when it is actually updated only once a day. Hence, we need to develop a Web access submodule with adaptability such that it learns the frequency of information source and dynamically change the interval of access.

3.2 Information Integration Module

The information integration module collects tickers produced by information gathering modules and integrates them to support the user's decision making or problem solving.

As the number of information gathering modules increases, the number of incoming tickers produced by the modules increases. Without selecting tickers, the information integration module may be overwhelmed by the incoming tickers, so we need to employ a ticker selection mechanism.

We use the ITT (Integration Template for Tickers), as shown in Figure 6, to integrate tickers. In this figure, choices of action that achieves a goal of traveling from Tokyo to Osaka are shown. We assume these choices are based on the user's preference. The user's first choice for traveling from Tokyo to Osaka are to take a bullet train or an airplane. The second choice is to take a night bus, and the last choice is to stay in Tokyo and to travel on the next day. By using this scheme, we know we need not collect tickers about the second and the last choices when the the first choice is satisfied.

For example, normally the information integration module collects tickers about bullet train and about airplane. We here assume that bullet trains are not available but airlines are available. Now let us assume that the module receives a ticker that airlines are not available, then the module begins to collect tickers about night bus. When it comes to know night buses are not available, the module begins to collect tickers about hotel and transportation of the next day. During such a process of information gathering, if it receives a ticker that bullet trains become available again, it stops collecting tickers about night bus and hotel. The information integration module dynamically changes the way of information gathering depending on the message of incoming tickers.

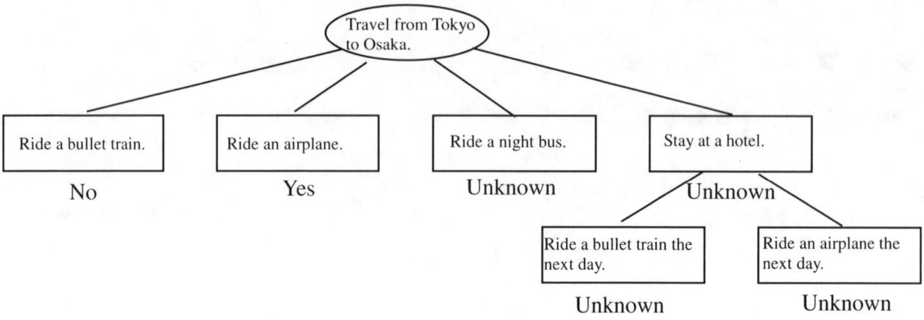

Figure 6: An example of Integration Template for Tickers.

When we consider an information gathering task, we can classify the information into the static one and the dynamic one. The dynamic information is frequently updated such as the availability of bullet train or airplane and is the main target of gathering. On the other hand, the static information is rather stable such as travel planning knowledge depicted in Figure 6 and is used to gather the dynamic information as mentioned above. Using the static information improves the performance of information gathering as discussed in [11].

Generally speaking, it is difficult to clearly define what is the static information and what is the dynamic information. The static information can be dynamic in a long run. For example, in the travel plan depicted in Figure 6, a night bus service may be abandoned or a cruise service between Tokyo and Osaka may start in the future. If so, the static information needs to be updated. Some static information may be updated directly by the user and others may be updated automatically by using collected information from the Web.

4 Conclusion

Active information gathering system gathers pieces of information from frequently updated information sources on the Internet and integrates them to assist the user in his/her decision making and problem solving task. It also works as an information gathering module of active mining system.

In this paper, we summarized required functionalities of active information gathering systems and proposed a new active information gathering system called Intelligent Tickers. At this moment, the system is still at the initial stage of concept development. We need to continue to develop the system as a useful system in the real world.

References

[1] M. Klein. XML, RDF, and Relatives. *IEEE Intelligent Systems*, 16(2):26–28, 2001.

[2] D. Fensel and M.A. Musen. The Semantic Web: A Brain for Humankind. *IEEE Intelligent Systems*, 16(2):24–25, 2001.

[3] F. Douglis, T. Ball, Y.-F. Chen and E. Koutsofios. The AT&T Internet Difference Engine: Tracking and Viewing Changes on the Web. *World Wide Web*, 1:27–44, 1998.

[4] Y.-F. Chen, F. Douglis, H. Huang and K.-P. Vo. TopBlend: An Efficient Implementation of HtmlDiff in Java. In *WebNet'00*. 2000.

[5] N. Adam, I. Adiwijaya, T. Critchlow and R. Musick. Detecting Data and Schema Changes in Scientific Documents. In *IEEE Advances in Digital Library*. 2000.

[6] T. Critchlow, K. Fidelis, M. Ganesh, R. Musick and T. Slezak. DataFoundry: Information Management for Scientific Data. *IEEE Trans Inf Technol Biomed*, 4(1):52–57, 2000.

[7] B. Nguyen, S. Abiteboul, G. Cobena and M. Preda. Monitoring XML Data on the Web. In *ACM SIGMOD*. 2001.

[8] L. Liu, C. Pu, W. Tang and W. Han. CONQUER: A Continual Query System for Update Monitoring in the WWW. *International Journal of Computer Systems, Science and Engineering*, 14(2):99-112. 1999.

[9] J. Chen, D.J. DeWitt, F. Tian and Y. Wang. Niagara CQ: A Scalable Continuous Query System for Internet Database. In *SIGMOD Conference*, pages 379–390, 2000.

[10] S. Yamada, T. Murata and Y. Kitamura. Intelligent Web Information System (in Japanese). *Journal of Japanese Society for Artificial Intelligence*, 16(4):495-502. 2001.

[11] Y. Kitamura, T. Noda and S. Tatsumi. A Dynamic Access Planning Method for Information Mediator. In *Cooperative Information Agents IV, Lecture Notes in Artificial Intelligence 1860*, pages 39–50, Springer, 2000.

[12] V. Lesser, B. Horling, F. Klassner, A. Raja, T. Wagner and S.X. Zhang. BIG: An agent for resource-bounded information gathering and decision making. *Artificial Intelligence*, 118(1-2):197-244. 2000.

II

USER CENTERED MINING

Active Mining
H. Motoda (Ed.)
IOS Press, 2002

Discovery of Concept Relation Rules Using an Incomplete Key Concept Dictionary

Shigeaki Sakurai, Yumi Ichimura, and Akihiro Suyama
{shigeaki.sakurai,yumi.ichimura,akihiro.suyama}@toshiba.co.jp
Corporate Research & Development Center
Toshiba Corporation

Abstract. We have proposed a method that acquires concept relation rules automatically. The method generates a training example from both concepts extracted from a report and a text class given by a user. Also, the method applies a fuzzy inductive learning algorithm, IDF, to generated training examples. The method regards a report with more than one concept in a concept class as a contradictory report and excludes a training example based on the report. If a key concept dictionary, which extracts key concepts from texts, is ill defined, such report does not always have contradictory contents and the exclusion of the report leads to the lack of an important rule. On the other hand, it is difficult to create a complete key concept dictionary at first. It is necessary to deal with a report with more than one concept in a concept class. Fortunately, IDF can process a value set as an attribute value by defining an appropriate membership function for the set. Thus, the paper proposes a method that defines the membership function and shows the efficiency of the method through numerical experiments using daily business reports in retailing.

1 Introduction

A huge amount of textual information can be stored in a computer. However, the information is not always used efficiently. Text mining methods [1, 2, 3, 4] have been proposed to overcome this problem.

The method [1] finds characteristic relations among words by using their hierarchical structure given by a human expert and classifies texts. The method deals with texts written in English and does not deal with texts written in a language without a space that separates words. It is not possible for the method to analyze texts written in Japanese. Also, another method [4] deals with texts written in Japanese and extracts a meaning. The method uses both lexical analysis and structure that represents connection between words. It considers only a concept given by the connection. It is not possible for the method to analyze meaning that arises due to the combination of more than one concept. On the other hand, the method [3] visualizes the relationship among texts by using two-dimensional map. The distance on the map reflects similarity of texts. The method cannot interpret the meaning of the map. A user has to interpret the meaning.

The method [2] classifies Japanese texts into some classes and displays their statistical information by using two kinds of knowledge dictionary: a key concept dictionary and a concept relation dictionary. Then, the former dictionary hierarchically describes important expressions extracted from texts and the latter one describes the meaning of a combination of some key concepts. It is possible for the method to find important combinations. Also, it is possible to classify a text based on a consideration of the meaning of expressions, even if the expressions have different descriptions. The effect of the method depends on both a key concept dictionary and a concept relation dictionary. We have to generate these dictionaries through trial and error. It is difficult to apply the method to many tasks for the generation. The paper [7] has proposed a method that acquires a concept relation dictionary inductively. The method uses a key concept dictionary and generates a concept relation dictionary from texts with text classes given by a user. The paper also shows the effect of the method by means of numerical experiments based on daily business reports in retailing. It is possible for each user to give a different text class to a text. The class represents the intention of each user. Thus, the method performs an active mining. The method assumes that the concept relation dictionary is well defined and that the dictionary extracts at most one key concept corresponding to a concept class from a text. The method regards a report with multiple key concepts in a concept class as a report with contradiction. Also, the method excludes an example given by the report. On the other hand, it is difficult to generate a well-defined key concept dictionary. We have to use an ill-defined key concept dictionary. The ill-defined dictionary extracts multiple key concepts in a concept class from a text. The exclusion of the report may lead to a lack of necessary concept relations.

Thus, the paper proposes a method that processes a text with multiple key concepts in a concept class. Also, the paper proposes a fuzzy inductive learning method incorporating the method. Moreover, the paper shows the effect of the proposed method by comparing it with an old method through numerical experiments based on daily business reports in retailing.

2 Acquisition of a concept relation dictionary

In this section, both the learning method and the inference method of a concept relation dictionary [7] are briefly reviewed.

2.1 Learning method

A relation in a concept relation dictionary is regarded as a kind of rule. The rule is acquired by using an inductive learning method [5, 6], if training examples are gathered. The paper [7] has proposed a method that gathered the examples. The method extracts concepts from a report by using lexical analysis and a key concept dictionary. The method regards a concept class as an attribute, a key concept as an attribute value, and a text class given by a user as a class of a training example. The method generates a training example from a report. Also, the method sums up training examples, applies them to a fuzzy inductive learning algorithm, IDF(**I**nduction **D**ecision tree with **F**uzziness) [6, 7], and acquires a concept relation dictionary described by a fuzzy decision tree. That is, according to the flow shown in Figure 1, a concept relation dictionary is acquired from reports and text classes.

2.2 Inference method

When a new report is given, an inference method extracts some concepts from the report and generates an example that does not have a text class. Also, the method decides a text class by applying the example to an acquired fuzzy decision tree. The decision

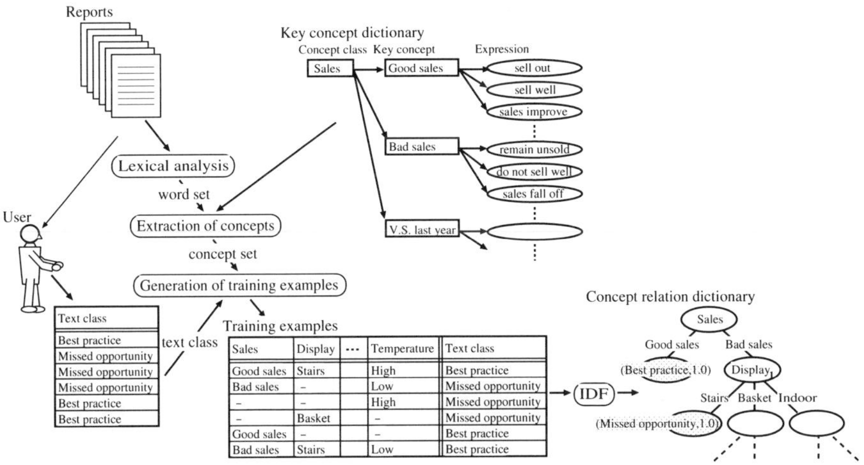

Figure 1: A learning flow

based on the tree takes into consideration imprecision included in the example. That is, in an interior node, the inference method transfers an example with degree of certainty to its lower nodes and performs the inference in the respective nodes even if the interior node does not have a fuzzy class item that is equal to an attribute value of the example. In a terminal node, the method multiplies the degree of certainty corresponding to the transferred example with the one corresponding to each class of the node and sums up the degree for each class. The method obtains the classes with degree of certainty. Normally, the method selects a class that has the maximum degree of certainty and regards the class as a class of the example. So, it is possible for the inference method to decide a text class even if an example to be evaluated has an attribute value that does not occur in the learning phase.

3 Dealing with multiple key concepts

This section shows a problem involving multiple key concepts in a concept class. Also, this section proposes a method that overcomes the problem.

3.1 Disadvantage of the old method

The old method assumes that a well-defined key concept dictionary is given at first. The well-defined dictionary gives at most a key concept for each concept class to a report. If two or more concepts with the same concept class are included in a report, the report is regarded as a report with contradiction. An example generated from the report is excluded in the learning phase. Also, a text class of the report is not inferred in the inference phase. However, a key concept dictionary is not always well defined. It is necessary to spend a lot of time, even if it is possible to generate a well-defined dictionary. Also, if key concepts that are unlikely to occur simultaneously are described in different concept class, it is difficult to understand intuitively a key concept dictionary for a huge number of concept classes. Moreover, a report with multiple key concepts does not always have contradiction in the case of a report with complex structure. So, it is not always appropriate to exclude a report with multiple key concepts in a concept class as a report with contradiction. The exclusion of the report may lead to a problem in that necessary concept relations are not acquired.

Thus, it is necessary to deal with both the report and a report with only a single key concept in a concept class. We propose a method that deals with them in the following section.

3.2 Proposed method

The proposed method uses a fuzzy inductive learning algorithm, IDF, in order to acquire a concept relation dictionary from training examples. The method regards a concept class as an attribute, a key concept as an attribute value, and a text class as a class of an example. It is possible for the IDF to process a label set as an attribute value, if appropriate fuzzy class items are defined for the attribute. Thus, we regard a set of concepts as an attribute value and define fuzzy class items. Also, we deal with a report that has multiple key concepts included in a concept class.

The IDF generates a fuzzy class item for an attribute value existing in the learning phase, when the attribute is discrete. That is, a fuzzy class item is not generated for an attribute value that is not included in training examples assigned to a processing interior node. On the other hand, it is necessary for an acquisition method of a concept relation dictionary to regard an example that has a single key concept in a concept class as a special example, in order to coordinate the example with an example with multiple key concepts. So, it is necessary to generate a fuzzy class item for each key concept existing in the training examples.

Here, we consider a method that defines a membership function representing each fuzzy class item. It is necessary for the function to consider both the importance and the similarity of key concepts for its definition. If we know the importance of key concepts, we evaluate the key concepts by taking into consideration weight based on the importance. However, we do not always know the importance. It is appropriate to give equal weight of the importance to each key concept. On the other hand, an example to be evaluated may have a key concept that is not given in the learning phase. Then, the key concept does not have a corresponding fuzzy class item. If the key concept shares similarities with other key concepts existing in a concept class, we transport examples with weight based on the similarities to their fuzzy class items. However, we do not always know the similarities. It is appropriate to transport examples with equal weight to all fuzzy class items. So, we define a membership function of each fuzzy class item based on these equalities. That is, we give degree of certainty to a fuzzy class item of a given key concept, when the given key concept corresponds with a key concept of the fuzzy class item. Also, we give equal decomposed degree of certainty to all fuzzy class items, when the given key concept does not correspond with any key concepts of the fuzzy class items. That is, the membership function is defined by Formula (1).

$$\begin{aligned}
&\text{If} \quad l_{ikr} \in v_i, \quad \text{then} \quad grade_{ikr} = \tfrac{1}{|v_i|} + \frac{1-\alpha}{|\bigcup_p (l_{ikp})|} \\
&\text{If} \quad l_{ikr} \notin v_i \quad \text{then} \quad grade_{ikr} = \frac{1-\alpha}{|\bigcup_p (l_{ikp})|} \\
&\alpha = \frac{|v_i \bigcap \bigcup_p (l_{ikp})|}{|v_i|}
\end{aligned} \tag{1}$$

Here, v_i is a key concept set included in the i-th attribute of an example to be evaluated, l_{ikr} is a label corresponding to the r-th fuzzy class item assigned to the k-th interior node in interior nodes with the i-th attribute, and $|\cdot|$ is an operation that calculates the number of elements included in a set. The formula shows a case where interior nodes with the same attribute do not always have equal label sets.

Figure 2 shows an outline that deals with an example with multiple key concepts. An interior node N_a is the first interior node in interior nodes with the second concept class "Display". Also, the node has three fuzzy class items, where each item has a key concept "Stairs", "Basket", and "Indoor" respectively and is the first node in interior nodes with the concept class. That is, $\bigcup_p (l_{21p}) = \{$"Stairs", "Basket", "Indoor"$\}$. We consider a case that a report has 1 degree of certainty, and an example to be evaluated has two key concepts "Stairs" and "Outdoor" as key concepts in "Display". That is, $v_i = \{$"Stairs", "Outdoor"$\}$. Each key concept has $\frac{1}{2}$ degree of certainty. One concept "Stairs" is equal to the key concept "Stairs" corresponding to a fuzzy class item. The method gives $\frac{1}{2}$ degree of certainty to a lower node N_b. On the other hand, the other concept "Outdoor" is not equal to any key concepts corresponding to the fuzzy class items. The method gives $\frac{1}{6}(=\frac{\frac{1}{2}}{3})$ degree of certainty to each lower node N_b, N_c, and N_d. So, the example with $\frac{2}{3}$ $(= \frac{1}{2} + \frac{1}{6})$ degree of certainty transfers to the node N_b, and the example with $\frac{1}{6}$ degree of certainty transfers to both the node N_c and the node N_d. The evaluation process is continued in all lower nodes.

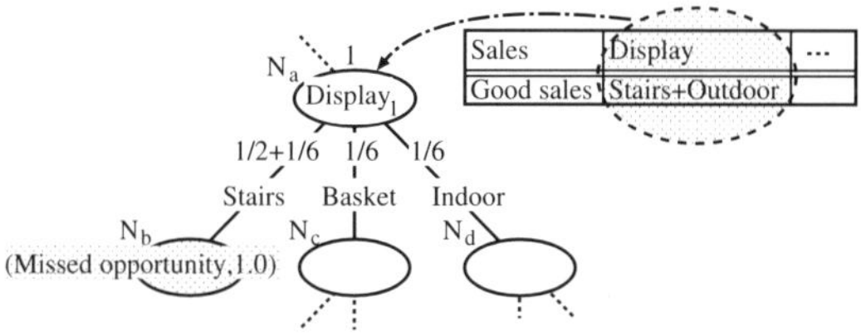

Figure 2: An inference for multiple key concepts

4 Numerical experiments

In this section, we show that the proposed method is efficient through numerical experiments based on daily business reports in retailing.

4.1 Experimental method

A text mining system [2] classifies daily business reports in retailing into three text classes: "Best practice", "Missed opportunity", and "Other". In this system, a key concept dictionary given by a human expert is composed of 13 concept classes and 299 key concepts. Each attribute of a training example has either key concepts included in a concept class or "nothing" as an attribute value, where "nothing" shows a report does not have a key concept in the concept class. Also, a concept relation dictionary decides a class corresponding to an example. Here, the dictionary is given by a human expert and is composed of 349 concept relations with "Best practice" or "Missed opportunity". If a report does not satisfy the relations, a class "Other" is given to the report.

In the experiments, we use 1,029 daily business reports in order to generate examples. 232 reports do not have a key concept extracted by the key concept dictionary and their attribute values are "nothing". We exclude the examples from 1,029 examples and use 797 examples in the experiments, because we consider important information is not described in the excluded examples. In fact, almost all the 232 reports do not describe meaningful contents but only a few reports describe meaningful contents. The key concept dictionary has to be improved in order to extract the information from the reports with the meaningful contents. In the future, we are going to consider a method that improves a key concept dictionary by using the reports that do not have a key concept.

The proposed method is evaluated using 10-fold cross validation in order to avoid the bias of the examples. That is, the examples are decomposed into 10 example subsets. In the learning phase, a concept relation dictionary described by a fuzzy decision tree is acquired from examples included in 9 subsets. In the inference phase, a text class is inferred for each example included in the remaining subset. The inferred text class is compared with the text class pre-assigned to the example. The learning phase and the inference phase are repeated 10 times by replacing the subset used in the inference phase with a different subset. Also, we perform experiments that use C4.5 [5] and the IDF that does not process multiple key concepts [7]. Here, the experiments use examples with only a single key concept in the learning phase. The experiments classify examples with multiple key concepts into "Other" in the inference phase, because "Other" is the most frequent class.

4.2 Experimental results

Table 1 shows average size of 10 generated decision trees for IDF and C4.5. "Interior node" shows average size of interior nodes included in each tree and "Terminal node" shows average size of terminal nodes. Also, "IDF_new" shows results in the case of the IDF that processes multiple key concepts. "IDF_old" shows results in the case of the IDF that does not process them, and "C4.5" shows results in the case of C4.5. In the following, the former IDF is referred to as the old IDF and the latter is referred to as the new IDF.

Table 1: Decision tree size

	IDF_new	IDF_old	C4.5
Interior node	21.0	10.9	10.9
Terminal node	90.5	45.7	362.3
Total	111.5	56.6	373.2

Table 2 shows error ratio, defined by Formula (2). "Single key concept" shows error ratios in the case of evaluating examples with only a single key concept, "Multiple key concepts" shows error ratios in the case of evaluating examples with multiple key concepts, "Average" shows error ratios in all examples. But, in the case of both the old IDF and C4.5, the most frequent class "Other" is inferred for an example with multiple key concepts.

$$\text{Error ratio} = \frac{\text{Number of misclassified examples}}{\text{Number of evaluated examples}} \qquad (2)$$

On the other hand, Figure 3 shows the trend of error ratios accumulated in 10 experimental sessions. Here, the x-axis gives the number of the experimental session, the y-axis gives the accumulated error ratios. Bar graphs corresponding to IDF_new, IDF_old, and C4.5 show the accumulated error ratios of inductive learning algorithms,

Table 2: Error ratio

	IDF_new	IDF_old	C4.5
Single key concept	0.00594	0.00149	0.04012
Multiple key concepts	0.08065	(0.21260)	(0.21260)
Average	0.01757	(0.03513)	(0.06901)

respectively. Also, each figure shows the trend corresponding to each text class. That is, Figure 3(a) shows the whole trend regardless of the classes; Figure 3 (b) shows the trend in the class "Best practice"; Figure 3 (c) shows the class "Missed opportunity"; and Figure 3 (d) shows the class "Other".

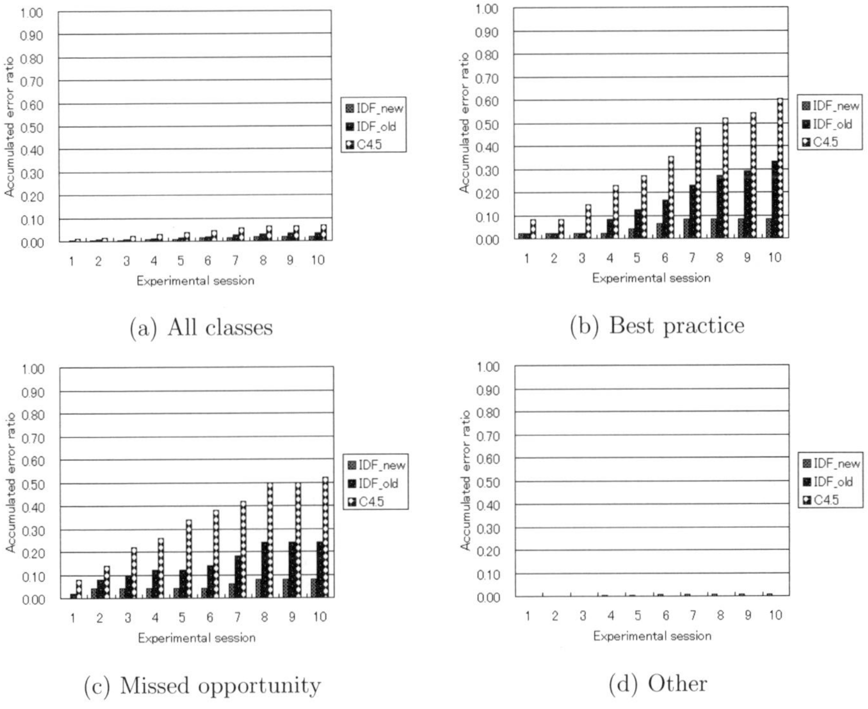

(a) All classes (b) Best practice

(c) Missed opportunity (d) Other

Figure 3: Accumulated error ratio

4.3 Discussions

4.3.1 Error ratio

In the case of "Single key concept", the old IDF gives lower error ratio than the new IDF does. The old IDF is apt to acquire fuzzy decision trees dependent on examples with a single key concept, because the old IDF does not use examples with multiple key concepts. So, the fuzzy decision trees infer the examples with a single key concept with lower error ratio.

In the case of "Multiple key concepts", error ratios in the old IDF and C4.5 are lower, because the number of examples with "Other" is much larger than the number of examples with "Best practice" or "Missed opportunity". The new IDF gives lower error ratio than the ratios. In the case of "Average", the new IDF also gives the lowest error ratio in the ratios. We consider that fuzzy decision trees based on the new IDF infer all examples with low error ratio.

Figure 3 shows that the new IDF more correctly infers examples with "Best practice" and "Missed opportunity". In this task, it is appropriate that the examples are classified more correctly than are examples with "Other", because almost all users are interested in reports with "Best practice" or "Missed opportunity". The new IDF gives a more appropriate result.

So, we consider the new IDF acquires a concept relation dictionary with low error ratio from training examples.

4.3.2 Size

C4.5 generates interior nodes corresponding to all attribute values, even if an attribute value is not given in the learning phase. On the other hand, IDF generates only interior nodes corresponding to attribute values given in the learning phase. So, IDF generates a more compact concept relation dictionary than C4.5 does. In fact, IDF generates more compact dictionaries as Table 1 shows.

The old IDF uses training examples with only a single key concept and does not acquire relations given by training examples with multiple key concepts. The old IDF may generate a concept relation dictionary lacking some relations. So, we consider that the size of the old IDF is smaller than that of the new IDF.

4.3.3 Selected attribute

The new IDF selects the concept class "Sales" as the top attribute in all generated fuzzy decision trees. That is, the text classes have a strong relation with "Sales". This result corresponds to the knowledge of a human expert as "Sales" is the most important concept class in this task. Thus, we consider that the new IDF acquires a concept relation dictionary corresponding to the feelings of the expert.

4.3.4 Useable examples

It is possible for the new IDF to use a report with multiple key concepts in both the learning phase and the inference phase. The method expands the variety of useable examples. Also, the method allows an ill-defined key concept dictionary that extracts an example with multiple key concepts in a concept class. Thus, we consider that the method reduces the burden for generation of a key concept dictionary.

In summary, the new IDF acquires a correct relation dictionary by using training examples with multiple key concepts. Also, the new IDF acquires a comparatively compact one by generating only branches corresponding to key concepts given in the learning phase

5 Summary and future works

The paper proposed a method that processes a report with multiple key concepts included in a concept class. The method regards a set of key concept as an attribute value by defining appropriately its membership functions. The method deals with the attribute values by using a fuzzy inductive learning algorithm, IDF. Also, the paper applied the proposed method to daily business reports in retailing. The paper showed that the method acquired concept relations corresponding to feelings of a human expert and that a key concept dictionary with high precision ratio is acquired.

In the future, we intend to consider a method that acquires a new key concept and a method that acquires a concept relation dictionary without using a key concept dictionary. This is because there are cases in which a report exists from which a key concept is not extracted even if the report is meaningful. Also, we intend to apply the proposed method to a different task and verify its effect. Now, we are planning to apply it to the task of classifying electric mail sent to a customer center.

References

[1] R. Feldman, I. Dagan, and H. Hirsh. Mining text using keyword distributions. *Journal of Intelligent Information Systems*, 10(3):281–300, 1998.

[2] Y. Ichimura, Y. Nakayama, M. Miyoshi, T. Akahane, T. Sekiguchi, and Y. Fujiwara. Text mining system for analysis of a salesperson's daily reports. In Proceedings of the PACLING 2001, pages 127–135, Kitakyuushu, 2001.

[3] K. Lagus, T. Honkela, S. Kaski, and T. Kohonen. Websom for textual data mining. *Artificial Intelligence Review*, 13(5/6):345–364, 1999.

[4] M. Morohashi, T. Nasukawa, and T. Nagano. An application of text mining technology to call-takers' reports. In *Proceedings of the ISM Symposium on Data Mining and Knowledge Discovery in Data Science*, pages 127–136, Tokyo, 1999.

[5] J.R. Quinlan. C4.5:Programs for machine learning. Morgan Kaufmann, 1992.

[6] S. Sakurai and D. Araki. The improvement of a fuzzy inductive learning algorithm. *T. IEE Japan*, 116(9):1057–1063, 1996 (*in Japanese*).

[7] S. Sakurai, Y. Ichimura, A. Suyama, and R. Orihara. Acquisition of a knowledge dictionary for a text mining system using an inductive learning method. In *Proceedings of the IJCAI 2001 workshop on Text Learning: Beyond Supervision*, pages 45–52, Seattle, 2001.

Active Mining
H. Motoda (Ed.)
IOS Press, 2002

Mining Frequent Substructures from Web

Kenji Abe†, Shinji Kawasoe†, Tatsuya Asai†, Hiroki Arimura†,‡,
Hiroshi Sakamoto†, Setsuo Arikawa†
† Department of Informatics, Kyushu University
6-10-1 Hakozaki Higasi-ku, Fukuoka 812-8581, JAPAN
‡ PRESTO, JST, JAPAN
{k-abe, s-kawa, t-asai, arim, hiroshi, arikawa}@i.kyushu-u.ac.jp

Abstract. In this paper, we present an efficient pattern mining algorithm FREQT for discovering all frequent tree patterns from a large collection of semi-structured data, and describe applications of this algorithm to knowledge discovery from Web and XML data on the internet. Our algorithm is based an efficient enumeration technique particularly designed for mining long patterns by extending the itemset enumeration technique by Bayardo (SIGMOD'98). Experiments on real datasets show the utility of our algorithm in Web and XML data mining.

1 Introduction

By rapid progress of network and storage technologies, a huge amount of electronic data such as Web pages and XML data [15] has been available on intra and internet. These electronic data are heterogeneous collection of ill-structured data that have no rigid structures, and often called *semi-structured data* [1]. Hence, there have been increasing demands for automatic methods for extracting useful information, particularly, for discovering rules or patterns from large collections of semi-structured data, namely, *semi-structured data mining* [5, 7, 11, 12, 14, 17].

In this paper, we model such semi-structured data and patterns by *labeled ordered trees*, and study the problem of discovering all frequent tree-like patterns that have at least a *minsup* support in a given collection of semi-structured data. We present an efficient pattern mining algorithm FREQT for discovering all frequent tree patterns from a large collection of labeled ordered trees, and describe applications of this algorithm to knowledge discovery from Web and XML data on the internet.

There are a body of researches on semi-structured data mining [7, 8, 11, 12, 16, 17, 18]. Previous algorithms for finding tree-like patterns basically adopted a straightforward generate-and-test strategy [12, 16]. In contrast, we devise an efficient enumeration technique for ordered trees by generalizing the itemset enumeration tree by Bayardo [9] combined with a incremental computation of their rightmost occurrences similar to vertical layout technique [13]. By these technique, our algorithm scales well on real semi-structured datasets such as HTML documents or XML data.

To evaluate the usefulness of our algorithm, we develop a prototype system of the algorithm and apply it to data mining problems from semi-structured data. The first application is regularity discovery from CGI-generated Web pages. The second application is XML data analysis. These experiments showed the potential capabilities of the proposed algorithm in Web and XML mining.

The rest of paper is organized as follows. In Section 2, we prepare basic notions and definitions. In Section 3, we present our algorithm FREQT for solving the frequent

```
<people>
   <person age="40">
      <name> Alan</name>
      <tel> 7786</tel>
      <tel> 2133</tel>
   </person>
   <person height="155">
      <name>
         <first> Sara</first>
         <last> Green</last>
      </name>
      <tel> 6877</tel>
   </person>
   <person age="33" height="187">
      <name> Fred</name>
   </person>
</people>
```

(a) An XML document

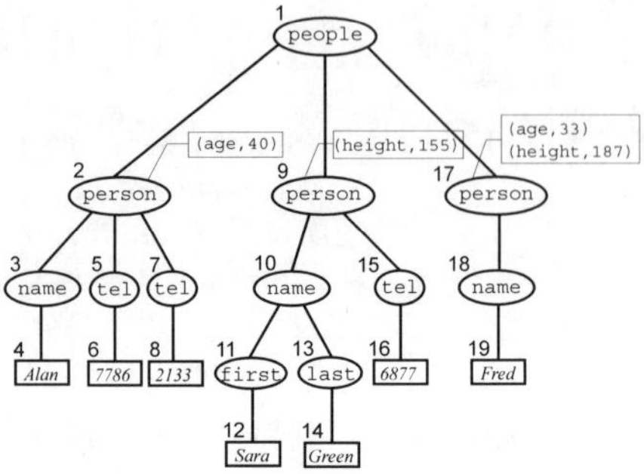

(b) The DOM tree for the left document

Figure 1: XML data expressions

pattern discovery problem for labeled ordered trees. In Section 4, we report applications for Web and XML mining. In Section 5, we conclude.

Very recently, Zaki [18] independently proposed efficient algorithms for the frequent pattern discovery problem for ordered trees, which is essentially same to our rightmost expansion. Also, he reported that a depth-first search algorithm equipped with his enumeration technique performs very well.

2 Preliminaries

2.1 XML and DOM

XML [15] is a textual representation of semi-structured data with hierarchic structure, where standardized effort is done by the World Wide Web Consortium (W3C). An XML document is a tagged text such as the text in Fig. 1(a), where a string surrounded by brackets represents a *tag* and an italic faced text represents a raw text of the document. In an XML documents, any start-tag (e.g., `<person>`) and its corresponding end-tag (e.g.,`</person>`) have to be properly balanced, thus XML documents are considered as trees. A subregion of a text between a pair of a start-tag and its corresponding end-tag, including the tags, is called an *element*, and the subregion without the tags is called the *content* of the element. A content of an element may contain raw texts or *subelements* recursively. Each start-tag may have *attributes* represented by (*name*, *value*) pairs. For example, the third `<person>` tag in the document of Fig. 1(a) has two attributes `(age,33)` and `(height,187)`. Note that any attribute name appears at most once in each tags and attribute-value pairs appearing in any tag are not ordered in the tag. We also note that the attributes are treated as unordered, while the subelements are ordered.

DOM (Document Object Model) is a standard API for manipulating document trees of XML data, and it generates a *DOM tree* by parsing given XML documents. A DOM

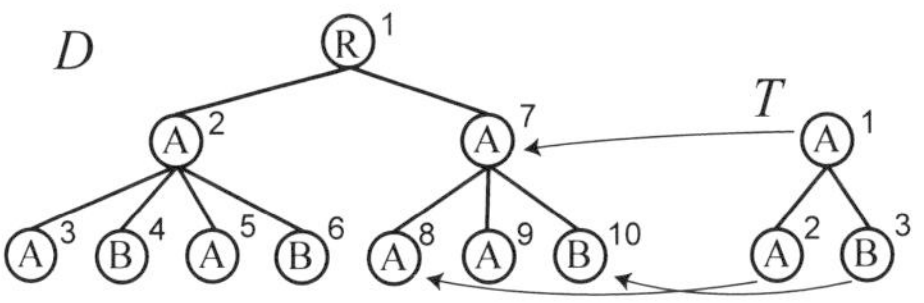

Figure 2: A data tree D and a pattern tree T on the set $\mathcal{L} = \{A, B\}$ of labels

tree mainly consists of two types of *nodes*, namely an *element node* and a *text node*, which respectively represent an element and a raw text of XML documents. Nodes of each types are assigned their unique ID and they have three pointers leading to their parent, leftmost child, and right sibling. An element node has a *label* that represents a tag name for the corresponding element, and the table of its attribute-value pairs. On the other hand, a text node has the corresponding raw text itself as its *text value*, as well as the special label #*text*. In Fig. 1(b), we give the DOM tree for the XML document of Fig. 1(a). In the figure, circles and boxes indicate element nodes and text nodes, where strings in each nodes represent their labels and text values resp., and integers on nodes represent their ID. Pairs in rectangles attached to some element nodes represent their attribute-value pairs.

2.2 Labeled Ordered Trees

As a model of semi-structured databases and patterns such as XML [15] and OEM model [2], we adopt the class of labeled ordered trees defined as follows. For a set A, $\#A$ denotes the cardinality of A. Let $\mathcal{L} = \{\ell, \ell_0, \ell_1, \ldots\}$ be a finite alphabet of labels, which correspond to attributes in semi-structured data or tags in tagged texts.

A *labeled ordered tree on* $\mathcal{L}$ (an ordered tree, for short) is a 6-tuple $T = (V, E, \mathcal{L}, L, v_0, \preceq)$ satisfying the following properties. $G = (V, E, v_0)$ is a tree with the root $v_0 \in V$. If $(u, v) \in E$ then we say that u is a *parent* of v or that v is a *child* of u. The mapping $L : V \to \mathcal{L}$, called a *labeling function*, assigns a label $L(v)$ to each node $v \in V$. The binary relation $\preceq \subseteq V^2$ represents a *sibling relation* for the ordered tree T such that u and v are children of the same parent and $u \preceq v$ iff u is an elder brother of v. We assume that the set V of all the nodes of T is $V = \{1, \ldots, k\}$ where $\#V = k$, and all elements in V are numbered by the preorder traversal [4] of T. Note that the root is $v_0 = 1$ and the rightmost leaf is $v_{k-1} = k$ in T by the assumption. In the above definition, an ordered tree is *not ranked*, that is, a node can have arbitrary many children regardless its label. In what follows, given an ordered tree $T = (V, E, \mathcal{L}, L, v_0, \preceq)$, we refer to V, E, L, and $\preceq$, respectively, as V_T, E_T, L_T, and $\preceq_T$ if it is clear from context.

Let T be a labeled ordered tree. The *size* of T is defined by the number of its nodes $|T| = \#V_T$. The *length* of a path of T is defined by the number of its nodes. For every $p \geq 0$ and a node v, the *p-th parent* of v, denoted by $\pi_T^p(v)$, is the unique ancestor u of v such that the length of the path from u to v has length exactly $p+1$. By definition, $\pi_T^0(v)$ is v itself and $\pi_T^1(v)$ is the parent of v. The *depth* of a node v of T, denoted by $depth(v)$, is defined by the length d of the path $x_0 = v_0, x_1, \ldots, x_{d-1} = v$ from the root v_0 of T to the node v.

In Fig. 2, we show examples of labeled ordered trees, say D and T, on the alphabet $\mathcal{L} = \{A, B\}$, where a circle with the number, say v, at its upper right corner indicates the node v, and the symbol appearing in a circle indicates its label $L(v)$. We also see that the nodes of these trees are numbered consecutively by the preorder.

2.3 Matching of trees

Let T and D be ordered trees on an alphabet $\mathcal{L}$, which are called the *pattern tree* and the *data tree* (or the text tree), respectively. We call such a tree T as a k-pattern tree on $\mathcal{L}$ (or k-pattern, for short), and denote by $\mathcal{T}_k$ the sets of all k-patterns on $\mathcal{L}$. Then, we define the set of patterns on $\mathcal{L}$ by $\mathcal{T} = \bigcup_{k \geq 0} \mathcal{T}_k$.

First, we define the notion of matching functions as follows. A one-to-one function $\varphi : V_T \rightarrow V_D$ from nodes of T to nodes of D is called a *matching function of T into D* if it satisfies the following conditions for any $v, v_1, v_2 \in V_T$:

- φ preserves the parent relation, i.e., $(v_1, v_2) \in E_T$ iff $(\varphi(v_1), \varphi(v_2)) \in E_D$.
- φ preserves the sibling relation, i.e., $v_1 \preceq_T v_2$ iff $\varphi(v_1) \preceq_D \varphi(v_2)$.
- φ preserves the labels, i.e., $L_T(v) = L_D(\varphi(v))$.

A pattern tree T *matches* a data tree D, or T *occurs in* D, if there exists some matching function φ of T into D. Then, the *root occurrence of T in D w.r.t.* φ is the node $Root(\varphi) = \varphi(1) \in V_D$ of D that the root of T maps, where $k = |T|$. For a pattern T, we define $Occ(T) = \{Root(\varphi) \mid \varphi \text{ is a matching function of } T \text{ into } D\}$, that is, the set of the root-occurrences of T in D. Then, the *frequency* (or *support*) of the pattern T in D, denoted by $freq_D(T)$, is defined by the fraction of the number of the distinct root occurrences to the total number of nodes in D, that is, $freq_D(T) = \#Occ(T)/|D|$. For a positive number $0 < \sigma \leq 1$, a pattern T is *σ-frequent* in D if $freq_D(T) \geq \sigma$.

For example, consider the previous example in Fig. 2. In this figure, a matching function, say φ_1, of the pattern T with three nodes into the data tree D with ten nodes is indicated by a set of arrows from the nodes of T. The root-occurrences corresponding to φ_1 is $Root(\varphi) = 7$. Furthermore, there are two root-occurrences of T in D, namely 2 and 7, while there are five matching functions of T into D. Hence, the support of T in D is $freq_D(T) = \#Occ(T)/|D| = 2/10$.

2.4 Problem Statement

Now, we state our data mining problem, called the frequent pattern discovery problem, which is a generalization of the frequent itemset discovery problem in association rule mining [3], as follows.

> **Frequent Pattern Discovery Problem**
>
> Given a set of labels $\mathcal{L}$, a data tree D on $\mathcal{L}$, and a positive number $0 < \sigma \leq 1$, called the *minimum support* (or *minsup*, for short), find all σ-frequent ordered trees $T \in \mathcal{T}$ such that $freq_D(T) \geq \sigma$.

Throughout this paper, we assume the standard *leftmost-child and right-sibling representation* for ordered trees (e.g., [4]), where a node is represented by a pair of pointers to its first child, *child*(), and the next sibling, *next*(), as well as its node label and the parent pointer, *parent*(). This is the same representation of ordered trees with DOM trees.

Algorithm FREQT

Input: A set $\mathcal{L}$ of labels, a data tree D on $\mathcal{L}$, and a *minimum support* $0 < \sigma \leq 1$.

Output: The set $\mathcal{F}$ of all σ-frequent patterns in D.

1. Compute the set $\mathcal{C}_1 := \mathcal{F}_1$ of σ-frequent 1-patterns and the set RMO_1 of their rightmost occurrences by scanning D; Set $k := 2$;
2. While $\mathcal{F}_{k-1} \neq \emptyset$, do:
   2 (a) $\langle \mathcal{C}_k, RMO_k \rangle :=$ Expand-Trees$(\mathcal{F}_{k-1}, RMO_{k-1})$; Set $\mathcal{F}_k := \emptyset$.
   2 (b) For each pattern $T \in \mathcal{C}_k$, do the followings: Compute $freq_D(T)$ from $RMO_k(T)$, and then, if $freq_D(T) \geq \sigma$, then $\mathcal{F}_k := \mathcal{F}_k \cup \{T\}$.
3. Return $\mathcal{F} = \mathcal{F}_1 \cup \cdots \cup \mathcal{F}_{k-1}$.

Figure 3: An algorithm for discovering all frequent ordered tree patterns

2.5 Translating DOM Trees into Labeled Ordered Trees

In this paper, we translate a DOM tree *Dom* into a data tree D as follows. Firstly, all text nodes in *Dom* are substituted by element nodes as follows. For a text node with the label $\#text$ and the text value t_val, the substitutive node is a such element node whose label and attribute-value pair are $\#text$ and $(@text, t\_val)$ respectively, where $@text$ is a special attribute name indicating that an element node with the attribute name $@text$ is originally a text node. Secondly, if an element node v of *Dom* has attribute-value pairs $(name, value)$ then we create a two-node tree consisting of the root and a child labeled with *name* and *value*, respectively. Then, we attach such two-node trees as a subtree of v in the lexicographic order on attribute names. Thus we obtain the labeled ordered tree D from the given DOM tree *Dom*.

3 Mining Algorithms

In this section, we present an efficient algorithm for solving the frequent pattern discovery problem for ordered trees that scales almost linearly in the total size of the maximal frequent patterns.

3.1 Overview of the Algorithm

In Fig. 3, we present our algorithm FREQT for discovering all frequent ordered tree patterns with the frequency at least a given minimum support $0 < \sigma \leq 1$ in a data tree D. As the basic design of the algorithm, we adopted the levelwise search strategy as in [3] and the search space similar to the enumeration tree of [9].

In the first pass, FREQT simply creates the set $\mathcal{F}_1$ of all 1-patterns and stores their occurrences in RMO_1 by traversing the data tree D. In the subsequent pass $k \geq 2$, FREQT incrementally computes a set $\mathcal{C}_k$ of all *candidate* k-patterns and the set RMO_k of the rightmost occurrence lists for the trees in $\mathcal{C}_k$ simultaneously from the sets $\mathcal{F}_{k-1}$ and RMO_{k-1} computed in the last stage by using the rightmost expansion and the rightmost occurrence technique using the sub-procedure Expand-Trees. Repeating this process until no more frequent patterns are generated, the algorithm computes all σ-frequent patterns in D. In the rest of this section, we will describe the details of the

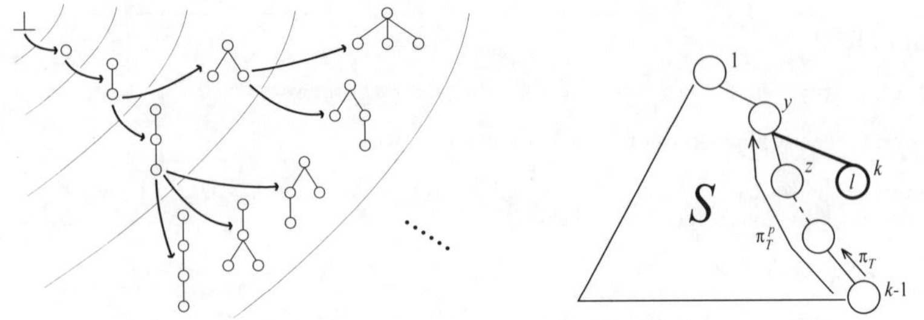

(a) A search graph for ordered trees (b) The (p, ℓ)-expansion of tree S

Figure 4: The rightmost expansion for ordered trees

algorithm.

3.2 *Efficient Enumeration of Ordered Trees*

In this subsection, we present an enumeration technique for generating all ordered trees of normal form without duplicates by incrementally expanding them from smaller to larger. This algorithm is a generalization of the itemset enumeration technique of [9], called the *set-enumeration tree*.

Rightmost expansion. A basic idea of our enumeration algorithm will be illustrated in Fig. 4(a). In the search, starting with the set of trees consisting of single nodes, for every $k \geq 2$ the enumeration algorithm expands a given ordered tree of size $k-1$ by attaching a new node only with a node on the rightmost branch of the tree to yield a larger tree of size k.

Let $k \geq 2$ be an integer and $\mathcal{L}$ be an alphabet. Let S be any $(k-1)$-pattern over $\mathcal{L}$ and $rml(S)$ be the rightmost leaf of S. Then, for a nonnegative integer $0 \leq p \leq d-1$ and a label $\ell \in \mathcal{L}$ where d is the depth of $rml(S)$, the (p,ℓ)*-expansion* of S is the labeled ordered tree T obtained from S by attaching a new node, namely k, to the node $y = {\pi_S}^p(x)$ as the rightmost child of y. The label of k is ℓ (See Fig. 4(b)). A *rightmost expansion* of an ordered tree S is the (p,ℓ)-expansion of S for some integer $p \geq 0$ and some label $\ell \in \mathcal{L}$.

We introduce an *empty tree* $\perp$ of size 0 and assume that all the single node trees in $\mathcal{T}_1$ are the rightmost expansions of $\perp$. Then, all the trees in $\mathcal{T}$ are enumerated without duplicates by repeating a generation of a rightmost expansion, starting from $\perp$ [6].

3.3 *Updating Occurrence Lists*

The key of our algorithm is how to efficiently store the information of a matching φ of each pattern T into the data tree D. Instead of recording the full information $\langle \varphi(1), \ldots, \varphi(k) \rangle$ of φ, our algorithm maintains only the partial information on φ called the rightmost occurrences defined as follows.

Rightmost occurrences. Let $k \geq 0$ be any integer. Let T be any k-pattern and $\varphi : V_T \to V_D$ be any matching function of T into D. Then, the *rightmost occurrence of T in D w.r.t. φ* is the node $Rmo(\varphi) = \varphi(k)$ of D that the rightmost leaf k of T maps. For every T, we define $RMO(T) = \{ Rmo(\varphi) \mid \varphi \text{ is a matching function of } T \text{ into } D \}$,

Algorithm Update-RMO(RMO, p, ℓ)

1. Set RMO_{new} to be the empty list ε and $check := null$.
2. For each element $x \in RMO$, do:
   (a) If $p = 0$, let y be the leftmost child of x.
   (b) Otherwise, $p \geq 1$. Then, do:
   - If $check = \pi_D{}^p(x)$ then skip x and go to the beginning of Step 2 (Duplicate-Detection).
   - Else, let y be the next sibling of $\pi_D{}^{p-1}(x)$ (the $(p-1)$st parent of x in D) and set $check := \pi_D{}^p(x)$.

   (c) While $y \neq null$, do the following:
   - If $L_D(y) = \ell$, then $RMO_{\text{new}} := RMO_{\text{new}} \cdot (y)$; /* Append */
   - $y := next(y)$; /* the next sibling */
3. Return RMO_{new}.

Figure 5: The incremental algorithm for updating the rightmost occurrence list of the (p, ℓ)-expansion of a given pattern T from that of T

the set of the rightmost occurrences of T in D. For example, consider the data tree D in Fig. 2. Then, the pattern tree T has three rightmost occurrences 4, 6 and 10 in D. The root-occurrences 2 and 7 of T can be easily computed by taking the parents of 4, 6 and 10 in D.

Now, we give an inductive characterization of the rightmost occurrences [6].

Lemma 1 (Asai et al., 2002) *Let S be a $(k-1)$-pattern occurring in a data tree D and $\varphi : V_S \to V_D$ be a matching function of S into D. Let T be a (p, ℓ)-expansion of S and $\psi : V_T \to V_D$ be any extension of φ, i.e., $\psi(i) = \varphi(i)$ holds for every $i = 1, \ldots, k-1$. Then, ψ is a matching function of T into D iff ψ satisfies the following (1) and (2):*

(1) *$\psi(k)$ is a child of $\pi^p(\varphi(k-1))$. Moreover, $\psi(k)$ is one of the right (younger) siblings of $\pi^{p-1}(\varphi(k-1))$, if $p \geq 1$ holds.*

(2) *$L_D(\psi(k)) = \ell$.*

Algorithm Update-RMO. Following Lemma 1, the algorithm Update-RMO of Fig. 5 exactly generates the elements in $RMO(T)$ for the (p, ℓ)-expansion T of a pattern S from the rightmost occurrence list $RMO(S)$.

However, the straightforward implementation of Lemma 1 often scans the same nodes more than once if $p \geq 1$ and then the computed list of the elements in $RMO(T)$ may contain some duplicates. By Duplicate-Detection technique [6], we can avoid this kind of duplicates of the rightmost occurrences x by checking the duplicate of their p-th parent $\pi_D^p(x)$ in the scan.

The following lemma [6] insures the correctness of the algorithm Update-RMO. In what follows, $RMO(T)$ means the *list* of the rightmost occurrences of T but not the *set*.

Lemma 2 (Asai et al., 2002) *The algorithm* Update-RMO *(with the Duplicate-Detection technique) enumerates all the nodes of any rightmost occurrence list without repetition,*

Algorithm Expand-Trees($\mathcal{F}, RMO$)

1. $\mathcal{C} := \emptyset$; $RMO_{\text{new}} := \emptyset$;
2. For each tree $S \in \mathcal{F}$, do:
   - For each $(p, \ell) \in \{0, \ldots, d-1\} \times \mathcal{L}$, do the followings, where d is the depth of the rightmost leaf of S:
     - Compute the (p, ℓ)-expansion T of S;
     - $RMO_{\text{new}}(T) :=$ Update-RMO($RMO(S), p, \ell$);
     - $\mathcal{C} = \mathcal{C} \cup \{T\}$;
3. Return $\langle \mathcal{C}, RMO_{\text{new}} \rangle$;

Figure 6: The algorithm for computing the set of rightmost expansions and their rightmost occurrence lists.

if the following two conditions are satisfied: (i) all the elements of $RMO(T)$ are ordered in the preorder of D for any 1-pattern T, and (ii) the algorithm scans all the nodes of $RMO(T)$ in the order of $RMO(T)$.

3.4 Analysis of the Algorithm

We go back to the computation of the candidate set $\mathcal{C}_k$. In Fig. 6, we present the algorithm Expand-Trees that computes the set $\mathcal{C}$ and the corresponding set RMO_k of the rightmost occurrence lists. The set RMO_k is implemented by a hash table such that for each tree $T \in \mathcal{C}$, $RMO_k(T)$ is the list of the rightmost occurrences of T in D.

The correctness of the algorithm FREQT of Fig. 3 have been shown as the following theorem [6].

Theorem 1 (Asai et al., 2002) *Let $\mathcal{L}$ be a label set, D be a data tree on $\mathcal{L}$, and $0 < \sigma \leq 1$ be a minimum support. The algorithm* FREQT *correctly computes all σ-frequent patterns in $\mathcal{T}$ without duplicates.*

The algorithm FREQT generates at most $O(kLM)$ patterns during the computation, where M is the sum of the sizes of the maximal σ-frequent patterns, while a straightforward extension of Apriori [3] to tree patterns may generate exponentially many patterns in k.

3.5 An Example

Consider the data tree D in Fig. 2 of size $|D| = 10$ and assume that the minimum support is $\sigma = 0.2$ and $\mathcal{L} = \{A, B\}$. Fig. 7 shows the patterns generated by FREQT during the computation. In this figure, we see that the tree-enumeration graph based on rightmost expansion is a tree whose root is $\perp$.

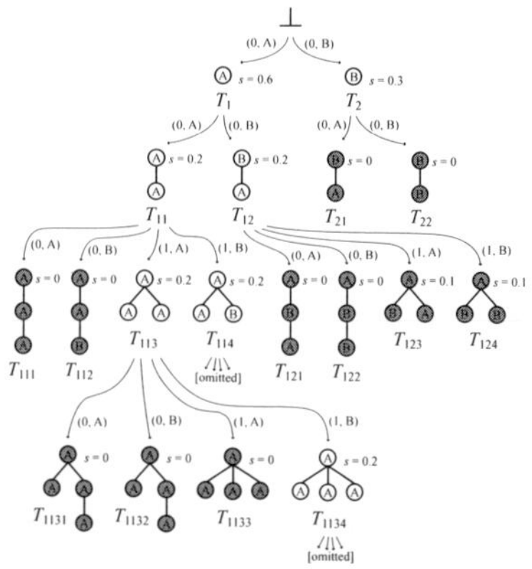

Figure 7: The enumeration tree for patterns on the data tree D in Fig. 2, where the minsup $\sigma = 0.2$. The pair (p, ℓ) attached to an arrow from a pattern S to a pattern T indicates that T is the (p, ℓ)-expansion of S. A white pattern represents a frequent pattern, and a shadowed pattern represents an infrequent pattern. For each pattern, the number s attached it represents its frequency.

4 Applications to Real-life Datasets

In this section, we apply our mining algorithm to Web and XML mining.

The task considered here is substructure discovery from HTML/XML pages, which is to discover a set of frequent substructures as patterns in a large document tree generated from a collection of Web pages gathered from Internet.

4.1 Implementation and Experimental Setup

We developed a prototype system of our mining algorithm FREQT in Section 3 in Java (SUN JDK1.3.1 JIT) using a DOM library (OpenXML). All experiments were run on PC (Pentium III 600MHz) with 512 megabytes of main memory running Linux 2.2.14 or Windows 2000. By experiments, the prototype system discovers around ten maximal frequent patterns with minimum support $\sigma = 3\%$ in 1.57 seconds on a data tree generated from HTML pages of total size 5.6MB [6]. For the details of the performance study on our algorithm, see the paper [6].

4.2 Regularity Discovery in CGI-generated Web Pages

We first applied our prototype system to substructure discovery from a collection of CGI-generated Web pages. We prepared a dataset *Citeseers*, which was a collection of a few hundreds of Web pages from an online bibliographic archive Citeseers[1]. After collecting pages, the dataset was parsed to create the DOM tree, and then we translate it into a labeled ordered tree by the method in Subsection 2.5. After preprocessing, the data tree for *Citeseers* had 89,128 nodes with 2,313 unique tags.

In Fig. 8, we show some interesting patterns, in HTML format, discovered by the system with $\sigma = 1.17\%$. These patterns captures a regularity common in those bibliographic entries. The first pattern with id 68 and size 6 appears in 1162 nodes and represents an HTML link with text in dark gray ("#6F6F6F") that appears at the header

[1] `http://citeseer.nj.nec.com/`

No. 68, Size 6, Hit 1162, Freq 1.30%

```
<a href=_> <font color="#6F6F6F"> #text_1 </font> </a>
```

No. 10104, Size 20, Hit 1039, Freq 1.17%

```
<p> #text_2
<b> #text_3 <!-- CITE --> <font color="green"> #text_4 </font>
#text_5 </b> #text_6 <br /> <br />
<font color="#999999"> #text_7 <i> #text_8 </i> #text_9 </font> </p>
```

Figure 8: Examples of discovered frequent patterns

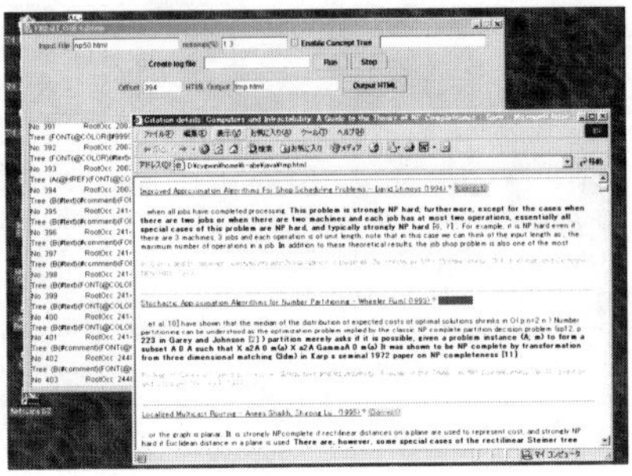

Figure 9: A GUI for the mining system

of a bibliographic entry. The second pattern with id 10104 and of size 20 appears in 1039 nodes and represents a common structure of the body of such an entry that has the root tag `<p>` and consists of a text region in bold face containing a citation with green color (`color="green"`) followed by two newline characters and another text region in light gray (`color="#999999"`) containing a smaller region in italic face.

It is sometimes hard for a human to capture where frequent patterns appear in the data tree. We implemented a GUI for the mining system that displays the occurrences of a chosen frequent pattern by highlighting the corresponding subregion in an HTML page. Fig 9 shows the GUI where the chosen pattern appears in three places.

4.3 Application to XML Data Analysis

The next application of our frequent pattern discovery algorithm is XML data analysis. In XML, the semantic structure of a data set is organized by a specific set of tags. Without knowing the document structure or DTD, it is often hard to see the structure of an XML archive. The goal of this experiment is to quickly capture an overview of the structure and the contents of an XML archive without an apriori knowledge on its tag set and document structure.

We used a small XML data obtained from the Internet Movie Database (IMDb) [2]. The IMDb is an online database of movie information containing more than 300,000 titles. In IMDb all movie entries are organized into HTML documents. HTML files of Movie entries with associated subentries are translated into XML data by extracting their contents using a hand-written perl script. Fig. 10 shows the tree structure of an XML entry, where a tag with + sign indicates that its subtree is not displayed.

First, we used as sample an XML data for the movie titled "God father," of size

[2] `http://www.imdb.com/`

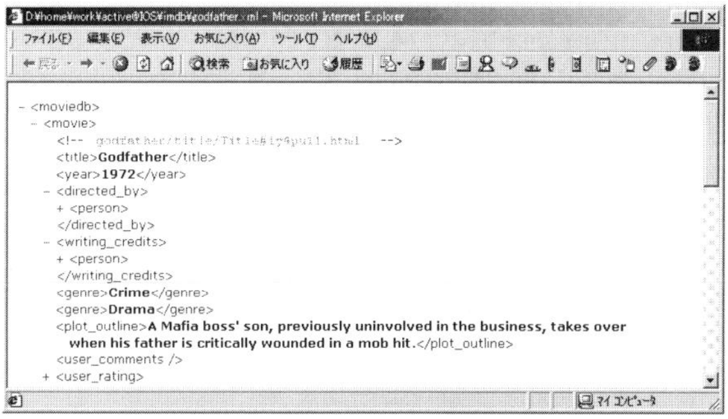

Figure 10: A movie database

No. 3,	Size 2,	Hit 20	`@TEXT("GODFATHER TRILOGY: 1901-1980, THE ...")`
No. 7,	Size 2,	Hit 17	`PERSON(NAME)`
No. 8,	Size 2,	Hit 16	`PERSON(DATE_OF_BIRTH)`
No. 9,	Size 2,	Hit 17	`PERSON(FILMOGRAPHY)`
No. 31,	Size 2,	Hit 15	`ACTOR(TITLE)`
No. 34,	Size 2,	Hit 15	`ACTOR(PERSON)`
No. 17,	Size 3,	Hit 16	`DATE_OF_BIRTH(DAY(#TEXT))`
No. 21,	Size 3,	Hit 16	`DATE_OF_BIRTH(YEAR(#TEXT))`
No. 24,	Size 3,	Hit 16	`DATE_OF_BIRTH(LOCATE(#TEXT))`

Figure 11: Examples of discovered frequent patterns

43KB and the corresponding tree has 7472 nodes with 819 unique labels, where trees are displayed in prefix notation as in first-order terms. Then, we run the prototype system on the XML data with minimum support $\sigma = 0.3\%$.

In Fig. 11, we show the frequent patterns discovered. The first six frequent trees, which are two-nodes trees, told the substructure of some tags that the `PERSON` and `ACTOR` tags are frequent in the XML entry and have {`NAME, DATE_OF_BIRTH`, `FILMOGRAPHY`} and {`TITLE, PERSON`}, respectively, as their children. The text value `"GODFATHER, THE (1972)"`, the title of the movie, also occurs 20 times in the filmographies of its actors. Also, the last three frequent trees told that the `DATE_OF_BIRTH` entry below `PERSON` has subentries `DAY, YEAR, LOCATE`. As summary, we had a quick overview of a given XML data by applying frequent substructure discovery.

5 Conclusion

In this paper, we studied a data mining problem for semi-structured data by modeling semi-structured data as labeled ordered trees. We presented an efficient algorithm for finding all frequent ordered tree patterns from a collection of semi-structured data, which scales almost linearly in the total size of maximal patterns. From the experiment on Web data, our algorithm is useful to extract regular substructures in a large collection of Web pages, and thus, may have applications in information extraction from Web and query modification in semi-structured database languages [10, 14].

In some applications, it is desirable to quickly reach some of maximal frequent patterns in short time by giving up the exhaustiveness of the search. Hence, introduction of stochastic search or beam search into our algorithm would be interesting. Our frequent pattern mining algorithm produces a large amount of frequent patterns with given minimum support thresholds. Thus, it is a future research to develop methods for aggregating these discovered patterns so that it is easy for a human user to understand and evaluate them.

Acknowledgments

The authors would like to thank Shinichi Morishita, Satoru Miyano, Akihiro Yamamoto, Masayuki Takeda, Ayumi Shinohara, and Shinichi Shimozono for the valuable discussions and comments. Hiroki Arimura would like to express his sincere thanks to Heikki Mannila and Esko Ukkonen to direct his attention to this area.

References

[1] S. Abiteboul, P. Buneman, and D. Suciu. *Data on the Web*. Morgan Kaufmann, 2000.

[2] S. Abiteboul, D. Quass, J. McHugh, J. Widom, and J. Wiener. The lorel query language for semistructured data. *Intl. J. on Digital Libraries*, 1(1):68–88, 1997.

[3] R. Agrawal and R. Srikant. Fast algorithms for mining assoiciation rules. In *Proc. the 20th VLDB*, pages 487–499, 1994.

[4] A. V. Aho, J. E. Hopcroft, and J. D. Ullman. *Data Structures and Algorithms*. Addison-Wesley, 1983.

[5] H. Arimura. Efficient learning of semi-structured data from queries. In *Proc. the 12th International Conference on Algorithmic Learning Theory (ALT'01)*, pages 315–331, 2001.

[6] T. Asai, K. Abe, S. Kawasoe, H. Arimura, H. Sakamoto, and S. Arikawa. Efficient substructure discovery from large semi-structured data. In *Proc. the 2nd SIAM Int'l Conf. on Data Mining (SDM2002)*, 2002 (To appear).

[7] L. Dehaspe, H. Toivonen, and R. D. King. Finding frequent substructures in chemical compounds. In *Proc. KDD-98*, pages 30–36, 1998.

[8] A. Inokuchi, T. Washio, and H. Motoda. An apriori-based algorithm for mining frequent substructures from graph data. In *Proc. PKDD 2000*, pages 13–23, 2000.

[9] R. J. Bayardo Jr. Efficiently mining long patterns from databases. In *Proc. SIGMOD98*, pages 85–93, 1998.

[10] N. Kushmerick. Wrapper induction: efficiency and expressiveness. *Artificial Intelligence*, 118:15–68, 2000.

[11] T. Matsuda, T. Horiuchi, H. Motoda, T. Washio, K. Kumazawa, and N. Arai. Graph-based induction for general graph structured data. In *Proc. DS'99*, pages 340–342, 1999.

[12] T. Miyahara, T. Shoudai, T. Uchida, K. Takahashi, and H. Ueda. Discovery of frequent tree structured patterns in semistructured web documents. In *Proc. PAKDD-2001*, pages 47–52, 2001.

[13] J. Sese and S. Morishita. (In Japanese). In *Proc. the second JSSST Workshop on Data Mining*, pages 38–47, 2001.

[14] K. Taniguchi, H. Sakamoto, H. Arimura, S. Shimozono, and S. Arikawa. Mining semi-structured data by path expressions. In *Proc. DS2001*, pages 387–388, 2001.

[15] W3C Recommendation. *Extensibe Markup Language (XML) 1.0*, second edition, 06 October 2000. `http://www.w3.org/TR/REC-xml`.

[16] J. T. L. Wang, B. A. Shapiro, D. Shasha, K. Zhang, and C.-Y. Chang. Automated discovery of active motifs in multiple rna seconary structures. In *Proc. KDD-96*, pages 70–75, 1996.

[17] K. Wang and H. Q. Liu. Discovering structual association of semistructured data. *IEEE Transactions on Knowledge and Data Engineering (TKDE2000)*, 12(3):353–371, May/June 2000.

[18] M. J. Zaki. Efficiently mining frequent trees in a forest. Technical Report PRI-TR01-7-2001, Computer Science Department, Rensselaer Ploytechnic Institute, 2001. `http://www.cs.rpi.edu/~zaki/PS/TR01-7.ps.gz`.

Active Mining
H. Motoda (Ed.)
IOS Press, 2002

Toward the Discovery of Web Communities from Input Keywords to a Search Engine

Tsuyoshi Murata
tmurata@nii.ac.jp
Foundations of Informatics Research Division
National Institute of Informatics
2-1-2, Hitotsubashi, Chiyoda-ku, Tokyo 101-8430, JAPAN

PRESTO, Japan Science and Technology Corporation
1-11-2, Yoyogi, Shibuya-ku, Tokyo 151-0053, JAPAN

Abstract. It is often observed that information needs for new event cause the increase of queries about it on a search engine. Since the phenomena in cyberspace are closely related with those in human world, detecting dynamic changes in Web communities is expected to clarify what is going on in human communities. This paper discusses a method for discovering emerging Web communities by using input keywords to a search engine. In order to detect the topics of emerging Web communities, some of the online resources are available for acquiring input keywords to a search engine. Keyword data acquired from such resources can be collected and used as the input to a discovery system for Web communities that the author has developed previously. By combining the mechanisms of detecting new keywords and discovering Web communities, emerging Web communities are expected to be discovered.

1 Introduction

In order to discover useful knowledge from huge Web network, several attempts have been made for the research of Web mining. There are three main approaches for Web mining: Web content mining, Web structure mining, and Web usage mining [8]. Web structure mining, whose goals are to discover or to rank useful Web pages based on hyperlink graph structure, is very important for engineering such as information retrieval or Web crawling, and also for sociology such as the estimation of Web development or the trends in real world.

According to the previous research of Web structure mining, macroscopic Web structure is like a bow tie [2], and microscopic Web structure is a bipartite core graph [9]. In the research of Web communities, which are composed of related Web pages sharing common interests, modeling their structure and clarifying the mechanisms of their dynamic changes are important for making good use of Web information and for understanding of human communities that correspond to Web communities.

Since the Web is huge and is growing every moment, Web communities are also changing dynamically. In the process of birth, growth and death, Web communities may be merged or divided. We can assume that such dynamic changes of Web communities correspond to the changes of human communities, such as the growing needs for new information and the increase of the people having concern with new event. Therefore, detecting such dynamic changes of Web communities is important as well as modeling static graph structure of Web communities.

As an approach of Web structure mining, the author has developed a system [11] that has abilities of discovering Web communities by using online resources such as search engines. The system discovers Web communities that share common interests with input URLs. In order to evaluate the power of the system, experiments are performed by using top URLs of each topic which are listed in 100hot.com (http://www.100hot.com/) as the inputs to the system. From the input, the system succeeds in discovering 19.8 correct (listed in 100.com ranking) URLs on average.

Since the Web communities that are discovered by the above system are relevant to the contents of input URLs, different inputs will cause different outputs. If a discovery system is really intellectual, it is desirable that the system itself has abilities of acquiring data autonomously for achieving discovery. As an attempt to actualize such discovery, the author has developed a discovery system for plane geometry [13]. The system draws diagrams by itself in order to acquire data that are needed for the discovery of geometrical theorems. In the case of Web community discovery system, autonomous data acquisition is important as well. Observing dynamic changes of Web data and performing experiments for data acquisition enables the system to discover Web communities that meet the needs of real human world.

The most popular online resources that reflect humans' information needs are search engines. After smashing news or terrible accidents, more searches are performed about the news or accidents. For example, the number of search about keywords "World Trade Center" and "Nostradamus" increase rapidly after the terror on September 11 in the United States [1]. And it is also reported that the accesses to washingtonpost.com increases rapidly after the terror, and the accesses to the sites for preventing computer viruses increases rapidly after the prevalence of Nimda viruses.

It is appropriate to consider that dynamic changes in Web communities are closely related to those in human communities. Increase of the number of search about specific keywords corresponds to the increase of people that need the information about the keywords. In such situations, it is expected that more Web pages are created about the keywords and corresponding Web community will grow. By detecting the changes of input keywords to a search engine, dynamic changes of Web communities can be detected. And it also enables better understanding of dynamic changes of human communities. In this paper, a method for discovering method for emerging Web communities are proposed. By providing input URLs to the above Web community discovery system from input keywords to a search engine, discovery of Web communities that meets human's information needs is enabled.

2 Related Work

As mentioned above, Web mining research can be divided into three approaches: Web content mining, Web structure mining, and Web usage mining [8]. The goal of this paper is the combination of Web structure mining and Web usage mining since the discovery is based on both graph structure of hyperlinks and input keywords to a search engine. Related work about Web structure mining is explained first in this section.

Broder's work [2] is one of the famous examples of macroscopic approach for Web structure mining. This work concluded that the structure of the Web is a bow tie from the analysis of the dimensions and connectivities of Web snapshot data as of 1999 (270 million URLs and 2.1 billion hyperlinks). As a microscopic approach, Kumar's Web

Trawling [9] searches bipartite core graph from Web snapshot data. However, attempts for middle approach are not enough. Discovery of Web communities can be regarded as the middle of both macroscopic and microscopic approaches.

Most of these Web structure mining researches regard each Web page as a node and each hyperlink as an edge of graph structure. Chakrabarti, on the other hand, introduces DOM(Document Object Model) structure [3] to the collection of links contained in a hub in order to improve the performance of HITS algorithm [7]. The reasons for taking such approach are as follows:

- Increase of noise in links such as those only for navigational purpose or advertisement: such links are not intended to give authorities to pointed pages.
- Mixture of hubs: in the cases that only a part of link collections in a hub point to the pages related to specific topic, irrelevant Web pages pointed by the links of other parts of the hub may be regarded as authorities of the topic as the result of the HITS algorithm.

It is often pointed out that inappropriate Web pages may be regarded as authorities by HITS algorithm. These are called topic generalization [5] or topic drift [4]. As an approach for Web structure mining, changing the granularity of Web graph is important for avoiding these phenomena.

Li's work [10] is just on the contrary to the above approach. He introduces new concept "information unit" for generating a set of Web pages that cover answers to given question for a question-answering system. This approach is the same as that of Chakrabarti in that both change the granularity of Web graph in order to solve problems that are caused by the gap between physical unit of Web contents (HTML file) and logical unit of Web contents.

Although these attempts are for analyzing Web graph structure statically, there can be another direction of Web mining: by using data about users' Web browsing behaviors, useful information from the Web can be acquired. Web usage mining, which is based on the data of users' Web browsing behaviors, can be divided into the following two approaches: learning user profiles and learning users' browsing patterns [8]. Another classification of Web usage mining is to divide it into the following five: personalization, system improvement, site modification, business intelligence, and usage characterization [14].

In order to perform Web usage mining, there are some available data about Web users' behavior, such as Web audience measurement or input keywords to a search engine. The former is the data about Web users that are randomly sampled just like the analysis of TV audience measurement. Netratings (http://www.nielsen-netratings.com/), ACNielsen eRatings.com (http://eratings.com/), Video Research Netcom (http://www.videor.co.jp/), and Netratings (http://www.netratings.co.jp/) are famous examples of the companies collecting such data. Since most of the data from these companies are summed up per week, it is not easy to use them for detecting dynamic changes of Web usage.

As the latter type of available data (input keywords to a search engine), weekly data in Infoseek (http://www.infoseek.co.jp/) is one of the examples. Although dynamic changes of input keywords for the latest one or two weeks are shown, the data are classified into several predefined genres so they cannot be used for the purpose that we are aiming at in this paper.

As an example of a site that provide real-time input keywords to a search engine, metaspy.com (http://www.metaspy.com/) is a simple and powerful site [6]. In the site, a part of real-time input keywords given to metacrawler.com are shown (figure 1). The page will automatically refresh every 15 seconds. The author observed that the number of keywords regarding "World Trade Center" had increased very rapidly after the terror on September 11, 2001. By using the real-time input keyword data that are open to public at this site, we expect that clues for discovering emerging Web communities can be detected.

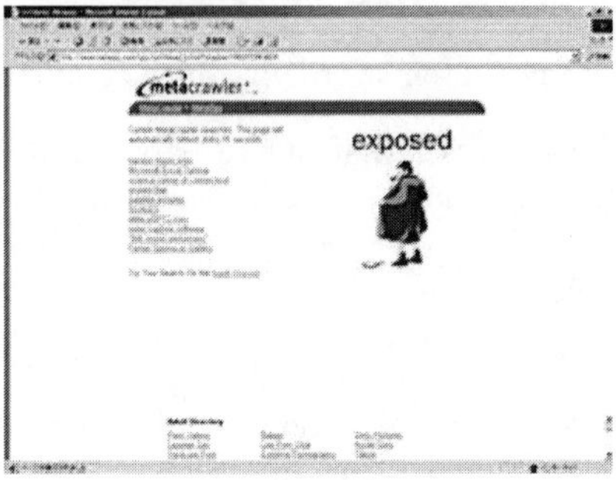

Figure 1: metaspy.com

3 Toward the Discovery of Emerging Web Communities

The steps for discovering emerging Web communities that we are aiming at are described in the following subsections.

- Detecting characteristic keywords from metaspy.com
- Acquisition of URLs about the input keywords
- Discovery of Web communities from the acquired URLs

3.1 Detecting characteristic keywords from metaspy.com

Properties of the keyword data shown in metaspy.com are as follows:

- No information about users is given.
- Not all input keyword data are shown.
- The number of words for a search is mostly from one to three.
- There are many keywords about free music, pictures and services.

We will discuss the strategies for treating such keywords. There are two ways for finding characteristic ones from input keywords to metaspy.com:

- collecting input keywords by accessing metaspy.com with certain time interval and comparing acquired keyword sets in order to detect changes

- collecting input keywords by accessing metaspy.com and compare them with frequent words in the pages of news sites

The former strategy is based on the assumption that comparing keyword sets will clarify dynamic changes of the frequency of keywords rather than analyzing only one keyword set. The latter strategy is based on the assumption that most of the frequent keywords that appear in news sites are about the words of events that occur recently in real world. Therefore, if frequent keywords in metaspy.com also appear frequently in news sites, many users must have interests to the new keywords.

3.2 Acquisition of URLs about the input keywords

After acquiring characteristic keywords from metaspy.com by the above method, URLs about the keywords can be acquired by performing search on a search engine. The URLs are then used as the input URLs of Web community discovery system described below.

3.3 Discovery of Web communities from the acquired URLs

A method for discovering Web communities that the author has proposed previously is explained in this subsection. The goal of this method is to discover a complete bipartite graph containing input URLs that are given from a user as the member of initial Web community. Such complete bipartite graph can be discovered by applying the following two steps iteratively, as shown in Fig. 2:

- searching fans that point all of the centers
- finding a new center which is pointed by most of the fans

The former step is to search Web pages that contain hyperlinks to all the members of centers in order to discover a complete bipartite graph containing all input URLs. With this backlink search, hyperlinks can be followed backward. Acquired URLs that contain hyperlinks to all the members of centers are used as fans.

In the latter step, all the HTML files of the URLs of fans are acquired and all the hyperlinks contained in the HTML files are extracted. The most frequent hyperlink of the extracted hyperlinks point to a URL that is likely to be closely related with the contents of centers. Therefore, the URL is added as a new member of centers, and the above two steps are iteratively repeated in order to discover a Web community that can be represented as a complete bipartite graph.

The above method for discovering Web communities and a method for refining Web communities [12] also proposed by the author are based on the assumption that there are many Web pages sharing common interests with input URLs. However, since such assumption is not always true for Web communities of newly generated topics, it is necessary to discover the central part of Web communities. Keyword acquisition from a search engine enables the discovery of such central part of Web communities, and it also facilitates the characterization of discovered Web communities.

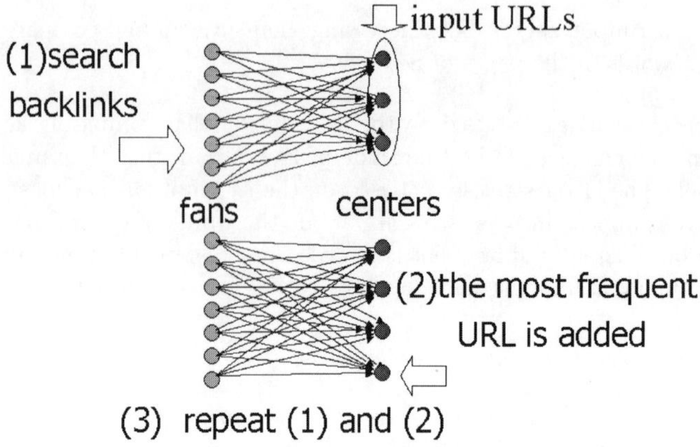

Figure 2: An outline of Web community discovery method

4 Concluding Remark

As an attempt for knowledge discovery from online resources, this paper discusses a method for discovering emerging Web communities based on input keywords to a search engine. We are going to examine each step of the method in more detail in order to develop a discovery system. In addition to search engines and metaspy.com, there are other online resources that enable data acquisition for discovery. By using such resources effectively, systems that have abilities of discovering advanced knowledge are expected to be developed.

Acknowledgement

Prof. Raymond Greenlaw gave me a hint for this research at the conference of Discovery Science 2000. I would like to express my gratitude for him.

References

[1] Asahi.com. News and nostradamus are top-ranking searched keywords http://www.asahi.com/culture/enews/k2001092001803.html, 2001. (In Japanese).

[2] A. Broder, R. Kumar, F. Maghoul, P. Raghavan, S. Rajagopalan, R. Stata, A. Tomkins, and J. Wiener. Graph structure in the web. In *Proc. of the 9th WWW conference*, 2000.

[3] S. Chakrabarti. Integrating the document object model with hyperlinks for enhanced topic distillation and information extraction. In *Proc. of the 10th WWW conference*, pages 211–220, 2001.

[4] S. Chakrabarti, B. E. Dom, R. Kumar, P. Raghavan, S. Rajagopalan, A. Tomkins, D. Gibson, and J. Kleinberg. Mining the web's link structure. *IEEE Computer*, 32(8):60–67, 1999.

[5] D. Gibson, J. Kleinberg, and P. Raghavan. Inferring web communities from link topology. In *Proc. of the 9th Conf. on Hypertext and Hypermedia*, 1998.

[6] R. Greenlaw. personal communication, 2000.

[7] J. Kleinberg, R. Kumar, P. Raghavan, S. Rajagopalan, and A. Tomkins. The web as a graph: Measurements, models, and methods. In *Proc. of the 5th Annual International Conf. on Computing and Combinatorics (COCOON '99)*, LNCS 1627, pages 1–17. Springer, 1999.

[8] R. Kosala and H. Blockeel. Web mining research: A survey. *ACM SIGKDD Explorations*, 2(1):1–15, 2000.

[9] R. Kumar, P. Raghavan, S. Rajagopalan, and A. Tomkins. Trawling the web for emerging cyber-communities. In *Proc. of the 8th WWW conference*, 1999.

[10] W.-S. Li, K. Candan, Q. Vu, and D. Agrawal. Retrieving and organizing web pages by 'information unit'. In *Proc. of the 21st Int'l ACM SIGIR Conf.*, pages 230–244, 2001.

[11] T. Murata. Machine discovery based on the co-occurrence of references in a search engine. In *Proc. of the second Int'l Conf. on Discovery Science (DS99)*, LNCS 1721, pages 220–229. Springer, 1999.

[12] T. Murata. A method for discovering purified web communities. In *Proc. of the 4th Int'l Conf. on Discovery Science (DS2001)*, LNCS 2226, pages 282–289. Springer, 2001.

[13] T. Murata and M. Shimura. Machine discovery based on numerical data generated in computer experiments. In *Proc. of the 13th National Conference on Artificial Intelligence (AAAI-96)*, pages 737–742, 1996.

[14] J. Srivastava, R. Cooley, M. Deshpande, and P.-N. Tan. Web usage mining: Discovery and applications of usage patterns from web data. *ACM SIGKDD Explorations*, 1(2):12–23, 2000.

Active Mining
H. Motoda (Ed.)
IOS Press, 2002

Temporal Spatial Index Techniques for OLAP in Traffic Data Warehouse

Hiroyuki Kawano
kawano@i.kyoto-u.ac.ac.jp
Graduate School of Informatics,
Kyoto University
Yoshida Hommachi, Kyoto 606-8501, JAPAN

Abstract. Technologies of GIS and location service are becoming popular, and huge volume of spatial and geographic data are stored into the clearing houses. We focus on the data from positioning systems, which affect the quality and quantity of traffic management system. For traffic planning, marketing and so on, new advanced spatial queries are required, we try to analyze the actual data in a traffic data warehouse effectively. In this paper, we discuss some basic problems in order to construct actual spatial information systems. Moreover, from the view point of traffic engineering, we present our proposed route estimation method and advanced spatial queries based on the techniques of temporal spatial indices.

1 Introduction

At present, technologies of GIS (Geographic Information System) and spatial databases are growing rapidly[1], and location services by GPS(Global Positioning System) and PHS(Personal Handy-phone System) are becoming popular. Moreover, various kinds of geographic data, numerical map data and the query results provided by many kinds of location services, have been stored into the various kinds of spatial data warehouses.

Furthermore, in the research fields of data mining, a lot of algorithms to discover knowledge in the huge volume of databases are proposed. Many spatial data mining algorithms[8, 6, 2, 9] have been also proposed, these algorithms derive useful and meaningful patterns, trends, rules and knowledge from spatial and geographical data. For example, in order to make clusters effectively, clustering algorithms make full use of the spatial characteristics, such as density, continuity and so on. We also focused on effective clustering algorithms based on the spatial index technologies[11, 12], such as R-Tree, R*-Tree, PR-Quadtree and others[4].

In this paper, in order to analyze the characteristics of traffic flows with some non-spatial attributes, and make city planning and marketing, we discuss fundamental problems in traffic data warehouse and person trip database. By using our previous research results of spatial indexing techniques[4], we can derive trip routes from actual positioning data effectively. This route estimation algorithm[5] plays important roles in our traffic . Furthermore, we also discuss the architecture of location monitoring systems and the temporal spatial queries[7] in order to analyze the database by real-time operations.

In Section 2, we show some basic problems which must be solved before constructing the traffic data warehouse. In Section 3, from the view points of temporal spatial characteristics, we introduce typical spatial and temporal spatial indices in order to store person trip data into traffic data warehouse. In Section 4, we propose the route estimation method shortly, and we discuss typical OLAP queries in our traffic data warehouse. We also evaluate the performance of our proposed method by using actual positioning data provided by PHS location service in Osaka city. Finally, we present some results and future problems in Section 5.

2 Fundamental problems of GIS and a location service

In order to construct traffic data warehouse systems, firstly we have to integrate the technologies of GIS, spatial database and location service effectively. Therefore, in this section, we point out several fundamental problems of geographical information systems and a location service.

- **The purpose of GIS is wide and spread.**

  There are so many geographic information systems which are composed of spatial databases and advanced analytical tools. However, it is not so easy to integrate different information systems in order to develop our proposed traffic data warehouse. Because we need various attributes with detail descriptions of road, transportation and building, which are not described in the present geographic information systems.

- **We don't have adequate clearing warehouses of map data.**

  Clearing warehouses and common spatial data formats, which are sometimes described in XML, are becoming very useful in order to exchange different type of spatial and non-spatial attribute values.

  However, at present, it is very hard to integrate schemata, attributes, accuracy and many other characteristics. For example, even in order to display the location (P) in Fig. 1, we need several conversion programs[1] to transform the positioning data to a point in a different numerical map. Because we have several different spatial coodinates such as WGS-84, ITRF(International Terrestrial Reference Frame) and others.

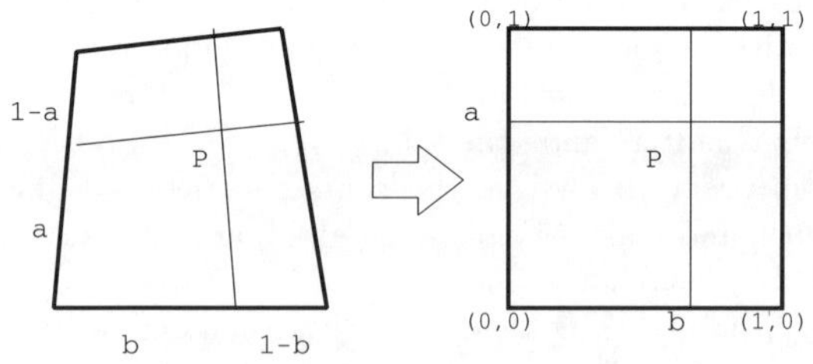

Figure 1: Conversion of a point on different maps

Furthermore, in typical numerical maps, road models are presented by simple attribute values, such as ID, tags, official 4 digits code, latitude, longitude, number

[1]http://vldb.gsi-mc.go.jp/sokuchi/program_s.html

of connecting edges and so on. However, in order to develop our traffic data warehouse, we need much more detail attributes, such as the width of roads, location of signals, regulations regarding types of transportations, the structures and building codes of intersections and sidewalks, and many others.

In addition to these attributes, we need on-site investigation to verify the consistency of some kinds of attribute values. For example, as shown in Fig. 2, we need to store the attribute of road connections. This attribute value frequently changes according to time and a mean of of transportation, such as car, bus, train, bicycle, wheelchair and walking. Therefore, we have to consider the way to present different transportation transparently in our map.

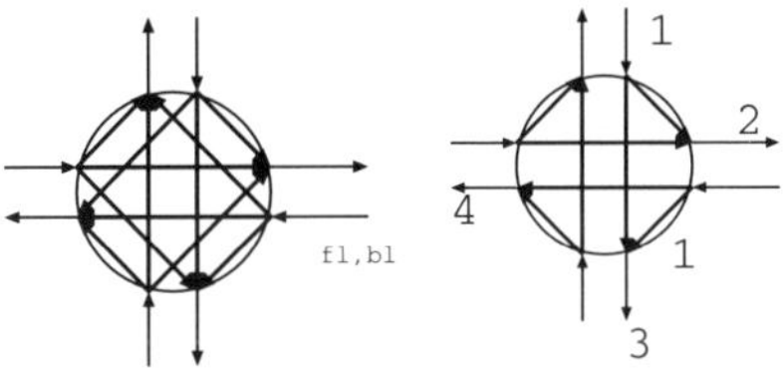

(a) Simple cross-road

(b) No right turn

Figure 2: Expression of road connection at crossroads

- **We need more adequate traffic monitoring systems.**

In Table 1, we cite the results of measurement errors in Osaka city by mobile GPS type terminals[3], and Fig. 3 shows the characteristics of measuring errors by PHS type location service [2] .

Table 1: Characteristics of devices: error of measurement (m)

devices	error	median	standard deviation	samples
PHS	188.7	187.2	41.7	71
GPS	24.4	20.9	10.5	29
DGPS	16.8	17.8	5.7	29

The major error factor of this PHS type location service seems to be caused by the reflection of buildings and constructions. It may be possible to estimate the correct location by using detail building maps and additional attributes provided by PHS location service. Furthermore, for estimating the person trip route correctly, we need portable devices in order to record the location continuously. Therefore, in our first experiment, we adopted the PHS location service. Of course, in recent years, the error is becoming smaller by GPS and pseudolites. Moreover, we have to pay attention to other location services[3]

[2]The original graph in Fig. 3 was drawn by Urban Transport Planning Co., Ltd.
[3]One of important URLs is http://www.fcc.gov/e911/.

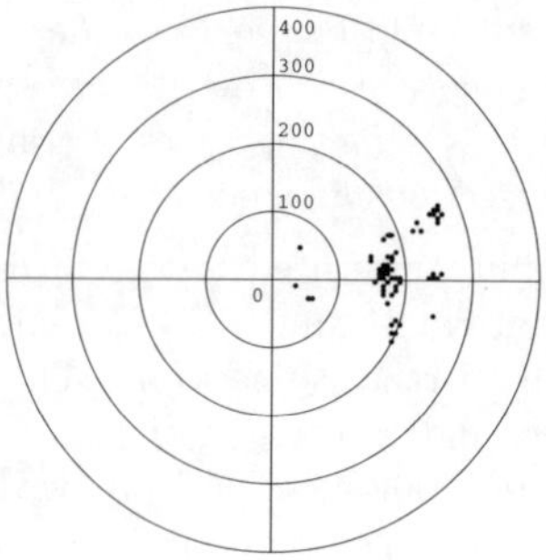

Figure 3: Actual positioning data by PHS in the center of Osaka (m)

3 Spatial and temporal spatial indices for traffic data warehouse

After developing the numerical map, various geographical and spatial attributes, such as nodes of crossroads, road arcs, directions and so on, are stored into our traffic data warehouse. We also accumulate many sequences of location data by using PHS location service.

Therefore, in this section, we reconsider the characteristics of spatial and temporal spatial indices for handling huge volume of moving objects, in order to reduce the execution cost of analytical queries in traffic data warehouse effectively.

3.1 Spatial indices in traffic data warehouse

In our previous researches[4, 5], we focused on the spatial indexing techniques[11, 12] for the effectiveness of execution of complex spatial queries. We discussed the applicability of data mining techniques to a spatial information systems, and we compared the characteristics of spatial indices, such as R-Tree (Fig.4), R*-Tree and Quad-tree (Fig.5).

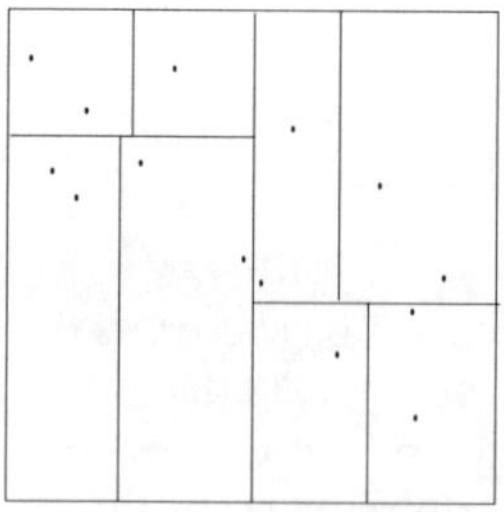

(a) Objects in the region

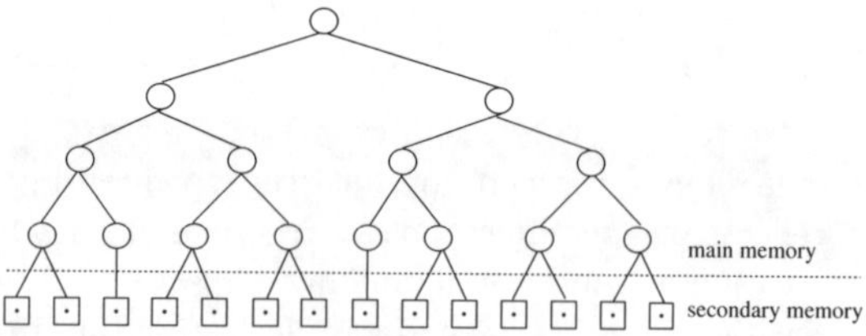

(b) Objects in the R-tree.

Figure 4: R-tree type index structure

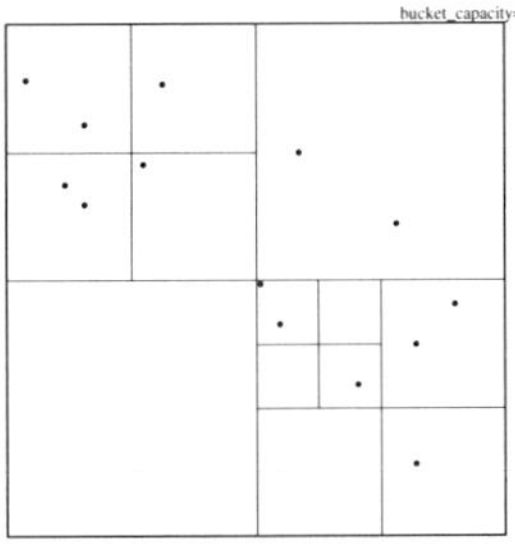

(a) Objects in the region

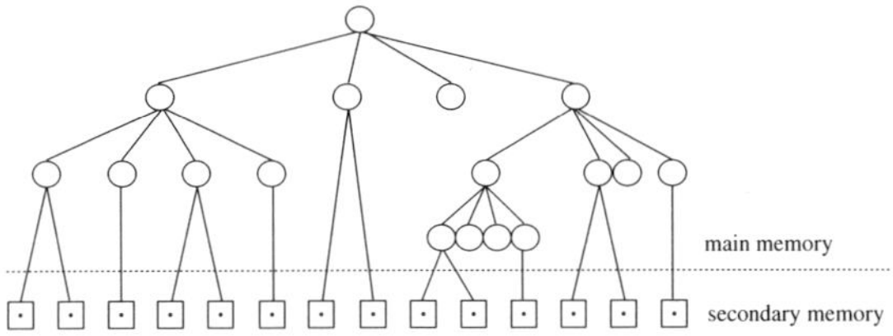

(b) Objects in the Quad-tree.

Figure 5: Quad-tree type index structure

Figures 4 and 5 show the the depth of index and objects that are stored in the spatial database. Generally speaking, the distribution of nodes in the R-tree type index structures is well-balanced and the average depth is less than that of the Quad-tree type index structures, so it is more effective to search a target object by using the R-tree type index structures.

However, there are some exceptional cases where this kind of typical situations does not apply. In the traffic data warehouse, we have to search nearest objects within a circle, oval or rectangular region, furthermore the distribution of objects is sometimes strongly biased. In this case, by using the Quad-tree type index, it is effective to reduce the execution cost in order to discover the rational person trip routes.

3.2 Temporal spatial indices in traffic data warehouse

Furthermore, we have to consider temporal spatial index such as TPR-tree (Time Parameterized R-tree)[10] in order to handle moving objects dynamically[7]. TPR-tree is extension of the structure of R-tree, moving nodes are stored in nodes of TPR-tree index by using the time function. But the characteristic of grouping objects is also different.

For example, when the location of one object is $\mathbf{x}_{ref}$ at time $t = t_{ref}$, by using the speed ($\mathbf{v}$), the location $\mathbf{x}(t)$ is presented by the following equation 1 at time $t (t > t_{ref})$.

$$\mathbf{x}(t) = \mathbf{x}_{ref} + \mathbf{v}(t - t_{ref}) \tag{1}$$

Then, neighboring objects are stored into same node in the space of $(\mathbf{x}, \mathbf{v})$. Thus, we can reduce the temporal spatial computing cost of moving objects, which are stored in the data structure. In Fig.6(a), seven objects (A, B, $\cdots$, G) are moving in the different directions, we can make (A,B,C), (D,E) and (F,G) clusters of neighboring objects in Fig.6(b). When we use R-tree type index, the distance of objects is becoming large fast in Fig.6(c). For moving objects, we have to use temporal spatial index shown in Fig.6(d).

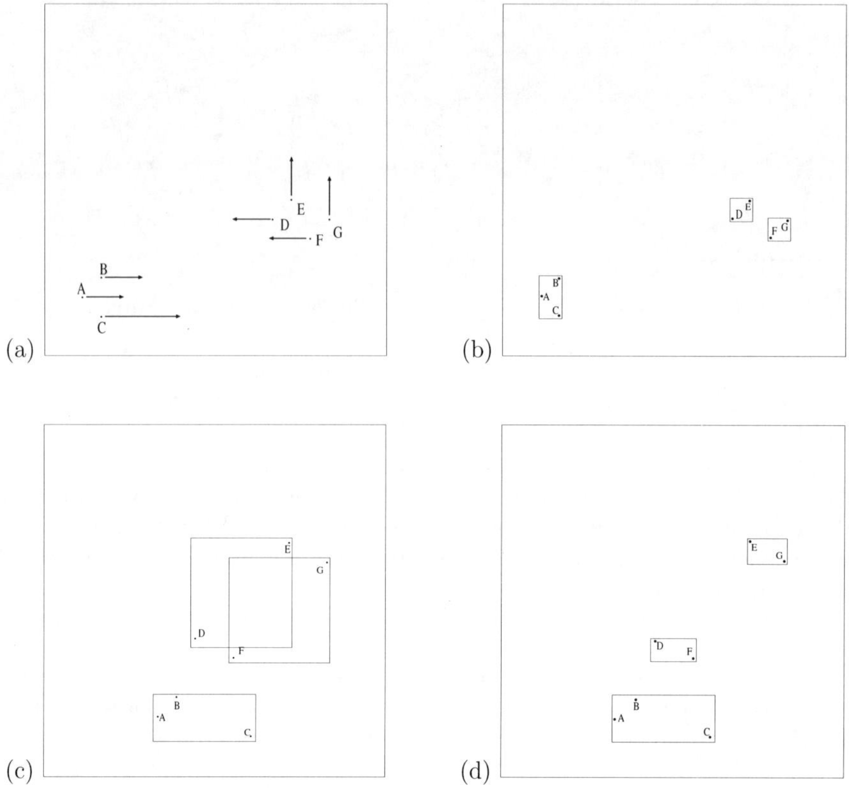

Figure 6: Moving objects in (c) R-tree and (d) TPR-tree

4 Analytical operations in traffic data warehouse

In this section, we consider the problems of OLAP (On-Line Analytical Processing) in a traffic data warehouse. Firstly, we execute the route discovery query after constructing numerical map and sequences of location data indexed by Quad-tree. In order to execute queries effectively, we have to reduce the search region and the number of objects. Secondly, we try to derive characteristics of moving objects in the data warehouse repeatedly. We discuss important analytical queries from the view point of traffic engineering. Thirdly, we try to have simple evaluation of our proposed indices by using actual location data.

4.1 From positioning data to map data

The original location data in Table 2 have several spatial and non-spatial attributes, such as ID, day, time, the number of antennas, flags and others, and the size of one record is 80 byte. Considering the problems discussed in Section 2, we can convert from spatial attributes to a point in our data warehouse in Table 3.

Table 2: Sample data of PHS location service

Latitude	Longitude	ID	Date and Time
34.680616	135.506702	1	1999/09/13 14:32:16
34.680661	135.506702	2	1999/09/13 14:32:33
34.681065	135.506425	3	1999/09/13 14:32:52
34.681939	135.506843	4	1999/09/13 14:33:10
34.681310	135.506096	5	1999/09/13 14:33:33

Table 3: Conversion of raw positioning data to map data

X	Y	ID	Date and Time
7593	8031	1	1999/09/13 14:32:16
7643	8031	2	1999/09/13 14:32:33
8092	7780	3	1999/09/13 14:32:52
9060	8168	4	1999/09/13 14:33:10
8366	7480	5	1999/09/13 14:33:33

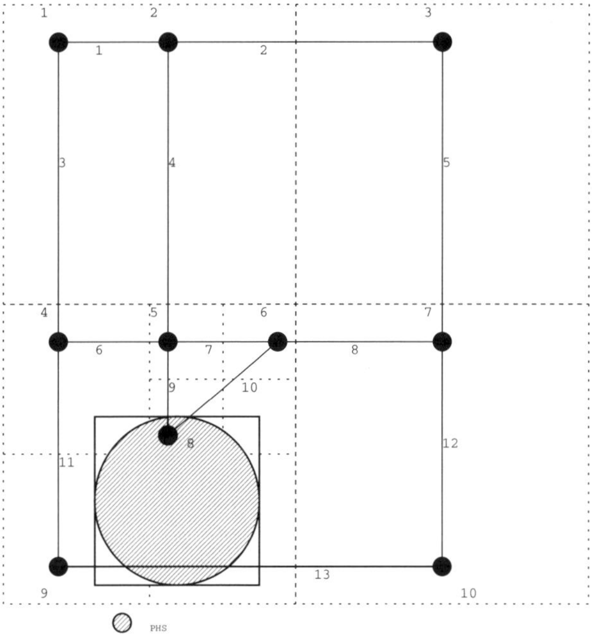

Figure 7: Objects indexed by Quad-tree

4.2 *Estimation of person trip route*

In this subsection, we consider the architecture of traffic data warehouse for traffic management, traffic forecast, person trip analysis and advanced search processings. Firstly, in order to handle location data and numerical map data in spatial database systems effectively, the data of moving objects is stored by using a spatial index.

For example, when we execute the route estimation query in traffic data warehouse, we have to search for neighboring arcs or nodes. In Fig. 7, the region of positioning data provided by PHS location service is presented by a stripe circle. By drawing the MBR of this circle and using Quad-tree index, we can reduce search space to three middle and two small squares. The objects, 4, 8, 9 nodes and 6, 9, 10, 11, 13 paths are within these squares. If we use R-tree index, the number of search regions increases.

Succeedingly, we calculate the distance between the center of the stripe circle and objects in MBR, and we can estimate the rational locations, node 8 and path 13. Depending on the conditions, the paths of 9 and 10 may be candidates. Step-by-step, we decide the sequence of candidate locations and estimate the rational person trip routes, which satisfy the traffic regulations. Finally, our proposed algorithm derive a few candidate of person trip routes according to the minimum length and the characteristics of person trip routes.

4.3 *OLAP queries in the traffic data warehouse*

After we store a huge number of trips into traffic data warehouse, we need typical temporal spatial queries[10] in order to analyze characteristics of traffic flows from the view point of traffic management and analysis. Here, we use definitions of time series, t_1, $t_2 (t_1 < t_2)$, and regions, R_1, R_2, and we discuss three following typical temporal spatial queries.

1. **timeslice query** $Q_{ts} = (R_1, t_1)$: At time point t, objects are searched for in a region R_1.

   (ex.) Based on the results of a query, we can calculate typical traffic flow parameters, such as traffic density, average traffic velocity and others.

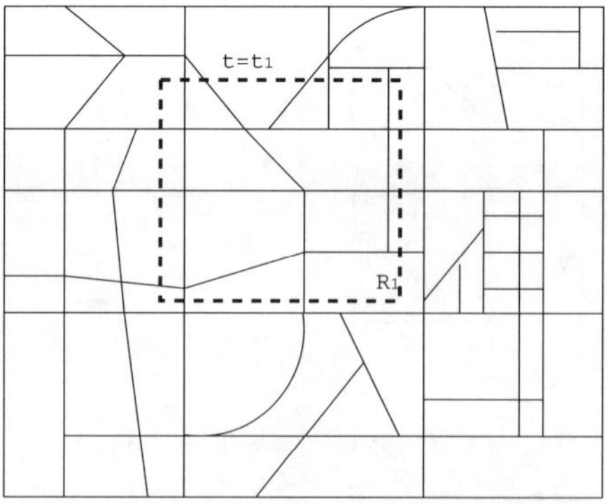

Figure 8: timeslice query

2. **window query** $Q_{win} = (R_1, t_1, t_2)$: Fig. 9 shows that moving objects are searched for in the regionR_1 from t_1 to t_2.

   (ex.) By using the results of window queries, we can calculate time average velocity which has rather stable property in traffic analysis.

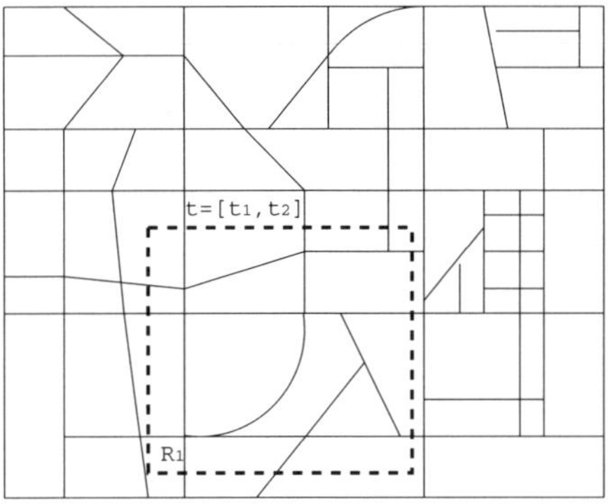

Figure 9: window query

3. **moving query** $Q_{mov} = (R_1, R_2, t_1, t_2)$: In Fig. 10, we search the objects in a moving region, which is covered by a connecting trapezoid of (R_1, t_1) and (R_2, t_2).

   (ex.) Traffic density of moving objects on an expressways is calculated, then the traffic congestion is forecasted by using the density and other specific values.

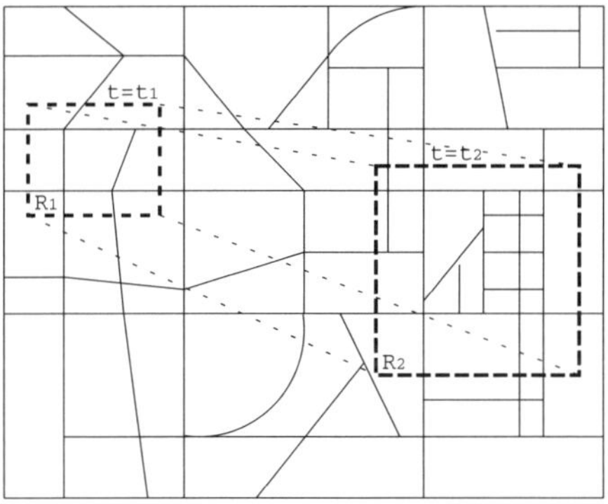

Figure 10: moving query

4.4 *Performance evaluation*

We examine the accuracy and validity of our proposed algorithm based on simulations and examinations using actual PHS location service. In this section, we show one experiment of route estimation based on actual location data provided by PHS location service.

In this experiment, we stored the sequences of positioning data collected during about 30 minutes, and the interval time of positioning detection was 15 seconds. By using conversion programs and Quad-tree index described in Section 2 and 3, we constructed our numerical map of a restricted area in city center. The primary elements of numerical map were extracted from the numerical map on a scale 1 to 2500, and we added several extra attributes in our numerical map.

Furthermore, we assume that the measuring error of PHS location service is within 100m. Fig. 11 (a) and (b) show the comparison with the actual sequences of PHS positioning data and the rational trip route that is derived by our algorithm using Quad-tree index.

In Fig. 11 (a), the numbers show the sequences of location data provided by PHS, pale lines show the roads that should be searched, and the black line with points shows the correct person trip route. In Fig. 11 (b), a pale line with points shows the primary rational person trip route that is estimated, and pale lines mean alternative rational route. In this case, the primary route is entirely consistent with the correct route. But the both of estimated routes are also rational, which satisfy traffic legal restrictions. The alternative route is also the rational person trip route within the range of PHS location service.

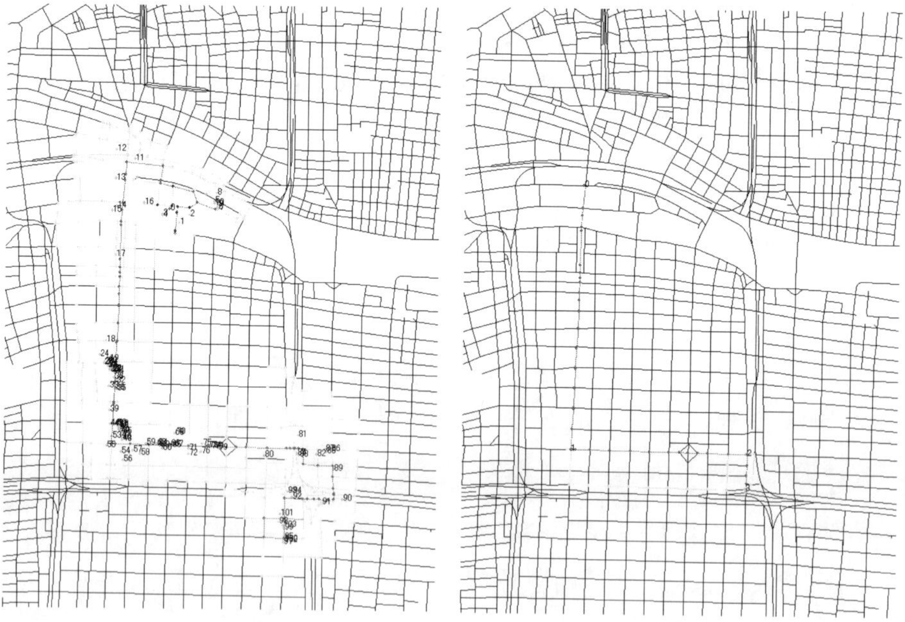

(a) Positioning sequences and correct route

(b) Primary estimated route and alternative route

Figure 11: Comparison between estimated route and correct route

In several experiments, almost of all routes could be specified fast and correctly. However, near the interchange of expressways, the estimated route was mistaken since a measuring error was sometimes so large. We may need to develop the error reduction method for the positioning compensation by obstacles. Of course, from the view point of total cost of location service systems, we consider the integration of different positioning devices, such as GPS cellular phone. This kind of devices will become widespread within a few years.

Furthermore, by integrating several positioning systems, we have other kind of advantages to protect privacy of location by ourself. Then we choose suitable methods, such as an active method (phone) or a passive method (GPS), depending on the public or private situations, and huge volume of our trip data will be stored into traffic data warehouse in order to analyze the traffic congestions and patterns in a statistical level.

5 Results and future problems

In this paper, we proposed the typical queries in traffic data warehouse by using Quad-tree and TPR-tree indices. Then we discuss how to derive characteristics of person trip route correctly and effectively. In near future, we try to apply this analytical and spatial mining techniques in actual traffic data warehouse in order to discover complex patterns.

Acknowledgment

I wish to thank Institute of Urban Transport Planning Co., Ltd., who provides the person trip data investigated in Osaka city. A part of this work is supported by the grant of Mazda Foundation.

References

[1] Böhm, C., Klump, G., and Kriegel, H.-P., "XZ-Ordering: A Space-Filling Curve for Objects with Spatial Extension." Proc. of 6th International Symposium, SSD'99, pp.75–90, Hong Kong, China, July 1999.

[2] Ester, M., Frommelt, A., Kriegel, H.-P., and Sander, J., "Algorithms for Characterization and Trend Detection in Spatial Databases", Proc. of the fourth ACM SIGKDD International Conference (KDD-98), pp.44-50, 1998.

[3] Hanshin Expressway Public Corporation, "Technical Trends of Mobile Location Systems", Technical Report of Mobile Location Service, February, 2000. (In Japanese)

[4] Ito, Y., Kawano, H., Hasegawa, T., "Index based Clustering Discovery query for Spatial Data Mining," Proc. of the 10th Annual Conference of JSAI, pp.231–234, 1996. (In Japanese).

[5] Kawano, H., "Architecture of Trip Database Systems: Spatial Index and Route Estimation Algorithm," XIV International Conf. of Systems Science, Vol. III, pp.110-117, Poland, 2001.

[6] Koperski, K. and Han, J., "Discovery of Spatial Association Rules in Geographic Information Databases," Proc. 4th International Symposium SSD '95, pp. 275–289, 1995.

[7] Minami, T., Tanabe, J. and Kawano, H., "Management of Moving Objects in Spatial Database: Architecture and Performance," Technical Report of IPSJ, Vol.2001, No.70, DBS-125, pp.225–232 (2001). (in Japanese)

[8] Ng, R.T. and Han, J., "Efficient and Effective Clustering Methods for Spatial Data Mining," Proc. 20th VLDB, pp. 144–155, 1994.

[9] Rogers, S., Langley, P., and Wilson, C., "Mining GPS Data to Augment Road Models," Proc. of the fifth ACM SIGKDD International Conference (KDD-99), pp.104–113, 1999.

[10] Saltenis, S., Jensen, C. S., Leutenegger, S. T., and Mario A. Lopez, "Indexing the Positions of Continuously Moving Objects," Proc. of the 2000 ACM SIGMOD International Conference on Management of Data, pp.331-342, USA, 2000.

[11] Samet, H., "Spatial Data structures," Modern Database Systems, (W. Kim, ed.), ACM Press, New York, pp. 361–385, 1995.

[12] Samet, H., "The Design and Analysis of Spatial Data structures," Addison-Wesley, Reading, Mass. New York, 1995.

Active Mining
H. Motoda (Ed.)
IOS Press, 2002

Knowledge Discovery from Structured Data by Beam-wise Graph-Based Induction

Takashi Matsuda, Hiroshi Motoda, Tetsuya Yoshida and Takashi Washio
{matsuda,motoda,yoshida,washio}@ar.sanken.osaka-u.ac.jp
Institute of Scientific and Industrial Research,
Osaka University
Mihogaoka, Ibaraki, Osaka 567-0047, JAPAN

Abstract. A machine learning technique called Graph-Based Induction (GBI) extracts typical patterns from graph data by stepwise pair expansion (pairwise chunking). Because of its greedy search strategy, it is very efficient but suffers from incompleteness of search. We improved its search capability without imposing much computational complexity by incorporating the idea of beam search. Additional improvement is made to extract patterns that are more discriminative than those simply occurring frequently, and to enumerate identical patterns accurately based on the notion of canonical labeling. This new algorithm was implemented (now called Beam-wise GBI, B-GBI for short) and tested against a DNA data set from UCI repository. Since DNA data is a sequence of symbols, representing each sequence by attribute-value pairs by simply assigning these symbols to the values of ordered attributes does not make sense. By transforming the sequence into a graph structure and running B-GBI it is possible to extract discriminative substructures. These can be new attributes for a classification problem. Effect of beam width on the number of discovered attributes and predictive accuracy was evaluated, together with extracted characteristic subsequences, and the results indicate the effectiveness of B-GBI.

1 Introduction

There have been quite a number of research work on data mining in seeking for better performance over the last few years. Better performance includes mining from structured data, which is a new challenge, and there have only been a few work on this subject. Since structure is represented by proper relations and a graph can easily represent relations, knowledge discovery from graph structured data poses a general problem for mining from structured data. Some examples amenable to a graph mining are finding typical web browsing patterns, identifying typical substructure of chemical compounds, finding typical subsequences of DNA and discovering diagnostic rules from patient history records.

Majority of the data mining methods widely used are for data that do not have structure and are represented by attribute-value pairs. Decision tree[10, 11], and induction rules[8, 3] relate attribute values to target classes. Association rules often used in data mining also uses this attribute-value pair representation. However, the attribute-value pair representation is not suitable to represent a more general data structure, and there are problems that need a more powerful representation. Most powerful representation that can handle relation and thus, structure, would be inductive logic programming (ILP) [9] which uses the first-order predicate logic. It can represent general relationship embedded in data, and has a merit that domain knowledge and acquired knowledge can be utilized as background knowledge. However, its state of the art is not so matured that anyone can use the technique easily (e.g. it requires heuristic to make the search efficient.).

AGM (Apriori-based Graph Mining)[6] is one of the representative recent work that can mine the association rules among the frequently appearing substructures in a given graph data set. A graph transaction is represented by an adjacency matrix, and the frequent patterns appearing in the matrices are mined by an extended Apriori algorithm of the basket analysis. This algorithm can extract all connected/disconnected induced subgraphs by complete non-exhaustive search. However, its computation time increases exponentially with input graph size and support threshold. AGM can use only frequency for the evaluation function. SUBDUE[4] is also well known. It extracts a subgraph which can best compress an input graph based on MDL principle. The found substructure can be considered a concept. This algorithm is based on a computationally-constrained beam search. It begins with a substructure comprising only a single vertex in the input graph, and grows it incrementally expanding a node in it. At each expansion it evaluates the total description length (DL) of the input graph, and stops when the substructure that minimizes the total description length is found. After the optimal substructure is found and the input graph is rewritten, next iteration starts using the rewritten graph as a new input. This way, SUBDUE finds a more abstract concept at each round of iteration. As is clear, the algorithm can find only one substructure at each iteration.

Graph-Based Induction (GBI) [13, 7] is a technique which was devised for the purpose of discovering typical patterns in a general graph data by recursively chunking two adjoining nodes. It can handle a graph data having loops (including self-loops) with colored/uncolored nodes and links. There can be more than one link between any two nodes. GBI's expressiveness lies in between the attribute-value pair representation and the first-order logic. The computation time for GBI is very short because of its greedy search, yet it does not lose any information of graph structure after chunking. GBI can use various evaluation functions based on frequency. It is not, however, suitable for pattern extraction from a graph structured data where many nodes share the same label because of its greedy recursive chunking without backtracking. However, it is still effective in extracting patterns from such graph structured data where each node has a distinct label (*e.g.,* World Wide Web browsing data) or where some typical structures exist even if some nodes share the same labels (*e.g.,* chemical structure data containing benzene rings etc).

Efficiency of GBI comes from its greedy search in exchange of search incompleteness. There is no guarantee that it can find all the important typical patterns although our past application to various domains produced acceptable results [7]. In this paper we report how we attacked this problem. We improved its search capability without imposing much computational complexity by incorporating the idea of beam search. Furthermore, two other improvements are made: one for criterion to define typical patterns in a more natural way and the other for accurate enumeration of typical patterns based on the notion of canonical labeling. This new algorithm was implemented (now called Beam-wise GBI, B-GBI for short) and tested against a DNA data set from UCI repository and its results were evaluated in terms of predictive accuracy and discovered typical subsequences, showing the effectiveness of the improvement.

The paper is organized as follows. In section 2, we briefly describe the framework of GBI and its improvement made to extracting discriminative patterns. In section 3, we describe B-GBI, which is the main contribution over the existing GBI, followed by canonical labeling treatment. In section 4, we show the experimental results of B-GBI. In section 6 we conclude the paper by summarizing the results and the future work.

2 Graph-Based Induction

2.1 GBI revisited

GBI employs the idea of extracting typical patterns by stepwise pair expansion as shown in Fig. 1. In the original GBI an assumption is made that typical patterns represent some concepts/substructure and "typicality" is characterized by the pattern's frequency or the value of some evaluation function of its frequency. We can use statistical indices as an evaluation function, such as frequency itself, Information Gain [10], Gain Ratio [11] and Gini Index [2], all of which are based on frequency.

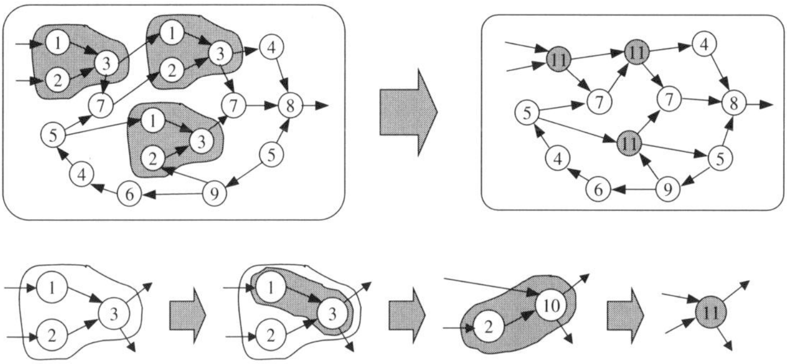

Figure 1: The basic idea of the GBI method

The stepwise pair expansion (pairwise chunking) repeats the following three steps until no more typical patterns are found.

Step 1 Extract all the pairs consisting of connected two nodes in the graph.

Step 2 Select the most typical pair based on the criterion from among the pairs extracted in Step 1 and register it as the pattern to chunk. If either or both nodes of the selected pair have already been rewritten (chunked), they are restored to the original patterns before registration. Stop when there is no more pattern to chunk.

Step 3 Replace the selected pair in Step 2 with one node and assign a new label to it. Rewrite the graph by replacing all the occurrence of the selected pair with a node with the newly assigned label. Go back to Step 1.

It is possible to extract typical patterns of various sizes by repeating the above three steps. Note that the search is greedy. No backtracking is made. This means that in enumerating pairs no pattern which has been chunked into one node is restored to the original pattern. Because of this, all the "typical patterns" that exist in the input graph are not necessarily extracted. The problem of extracting all the isomorphic subgraphs is known to be NP-complete. Thus, GBI aims at extracting only meaningful typical patterns of a certain size. Its objective is not finding all the typical patterns nor finding all frequent patterns.

2.2 Extracting Discriminative Patterns

GBI can use any criterion that is based on the frequency of paired nodes. However, in order to find a pattern that is of interest, any of its subpatterns must be of interest because of the nature of repeated chunking. Frequency measure satisfies this monotonicity. However, if the criterion chosen does not satisfy this monotonicity, repeated chunking may not find good patterns even though the best pair based on the criterion is selected at each iteration. This motivated us to improve GBI allowing to use two criteria, one for frequency measure for chunking and the other for finding discriminative patterns after chunking. The latter criterion does not necessarily hold monotonicity property. Any function that is discriminative can be used, such as Information Gain [10], Gain Ratio [11] and Gini Index [2], and some others (*e.g.* the one used in 4).

The improved stepwise pair expansion repeats the following four steps until chunking threshold is reached (normally minimum support value is used as the stopping criterion).

Step 1 Extract all the pairs consisting of connected two nodes in the graph.

Step 2a Select all the typical pairs based on the criterion from among the pairs extracted in Step 1, rank them according to the criterion and register them as typical patterns. If either or both nodes of the selected pairs have already been rewritten (chunked), they are restored to the original patterns before registration.

Step 2b Select the most frequent pair from among the pairs extracted in Step 1 and register it as the pattern to chunk. If either or both nodes of the selected pair have already been rewritten (chunked), they are restored to the original patterns before registration. Stop when there is no more pattern to chunk.

Step 3 Replace the selected pair in Step 2b with one node and assign a new label to it. Rewrite the graph by replacing all the occurrence of the selected pair with a node with the newly assigned label. Go back to Step 1.

The output of the improved GBI is a set of ranked typical patterns extracted at Step 2a. These patterns are typical in the sense that they are more discriminative than non-selected patterns in terms of the criterion used.

3 Beam-wise Graph-Based Induction

3.1 Algorithm of B-GBI

The improved GBI still has disadvantages. When each node has a distinct label in the input graph, no ambiguity arises in selecting a pair to be chunked and GBI performs well. However, since the search in GBI is greedy, when the same label is shared by plural nodes in the input graph, there arises ambiguity when there are ties in the frequency or there is a chain of nodes of the same label. For example, in the case of the structure like $a \rightarrow a \rightarrow a$, we don't know which $a \rightarrow a$ is best to chunk. Further, as chunked nodes are never restored to its original patterns for succeeding pair evaluation process, search is incomplete anyway even if there is no such ambiguity.

To relax this ambiguity and seach incompletness problem, a beam search is incorporated to GBI within the framework of greedy search. A certain fixed number of pairs ranked from the top are allowed to be chunked in parallel. To prevent each branch from growing exponentially, the total number of pairs to chunk is fixed at each level of branch. Thus, at any iteration step, there is always a fixed number of chunking that is performed in parallel.

The new stepwise pair expansion repeats the following four steps.

Step 1 Extract all the pairs consisting of connected two nodes in all the graphs.

Step 2a Select all the typical pairs based on the criterion from among the pairs extracted in Step 1, rank them according to the criterion and register them as typical patterns. If either or both nodes of the selected pairs have already been rewritten (chunked), they are restored to the original patterns before registration.

Step 2b Select, from among the pairs extracted in Step 1, a fixed number of frequent pairs from the top and register them as the patterns to chunk. If either or both nodes of the selected pairs have already been rewritten (chunked), they are restored to the original patterns before registration. Stop when there is no more pattern to chunk.

Step 3 Replace each of the selected pairs in Step 2b with one node and assign a new label to it. Delete a graph for which no pair is selected and branch (copy) a graph for which more than one pair are selected. Rewrite each remaining graph by replacing all the occurrence of the selected pair in the graph with a node with the newly assigned label. Go back to Step 1.

An example of state transition of B-GBI is shown in Fig.2 in case that the beam width is 5. The initial condition is the single state cs. All the pairs in cs are enumerated and ranked according to both the frequency measure and the typicality measure. Top 5 pairs according to the frequency measure are selected, and each of them is used as a pattern to chunk, branching into 5 children c_{11}, c_{12}, $\ldots$, c_{15}, each rewritten by the chunked pair. All the pairs within these 5 states are enumerated and ranked according to the two measures, and again the top 5 ranked pairs according to the frequency measure are selected. The state c_{11} is split into two states c_{21} and c_{22} because two pairs are selected, but the state c_{12} is deleted because no pair is selected. This is repeated until the stopping condition is satisfied. This increase in the search space improves the pattern extraction capability of GBI.

3.2 Canonical Labeling

Another improvement made in conjunction with B-GBI is canonical labeling. GBI assigns a new label for each newly chunked pair. Because it recursively chunks pairs, it happens that the new pairs that have different labels happen to be the same pattern (subgraph). A simple example is shown in Fig. 3.

To identify whether the two pairs represent the same pattern or not, each pair is represented by canonical label[12, 5] and only when the label is the same, they are regarded as identical. The basic procedure of canonical labelling is as follows. Nodes in the graph are grouped according to their labels (node colors) and the degrees of node (number of links attached to the node) and ordered lexicographically. Then an adjacency matrix is created using this node ordering. When the graph is symmetric, the upper triangular elements are concatenated scanning either horizontally or vertically to codify the graph. When the graph is asymmetric, all the elements in both triangles are used to codify the graph in a similar way. If there are more than one node that have identical node label and the degrees of node, the ordering which results in the maximum

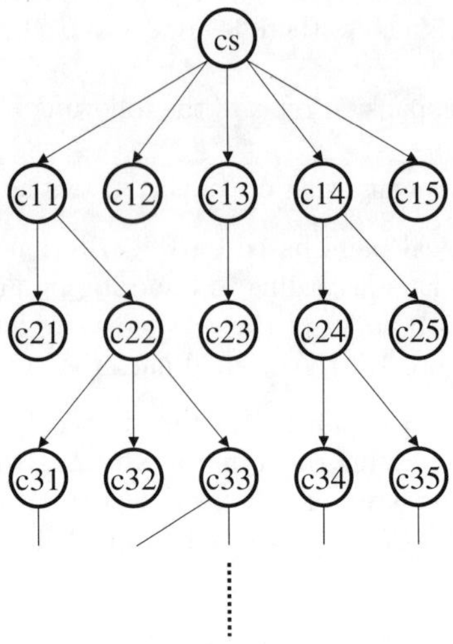

Figure 2: An Example of State Transition of B-GBI when the beam width = 5

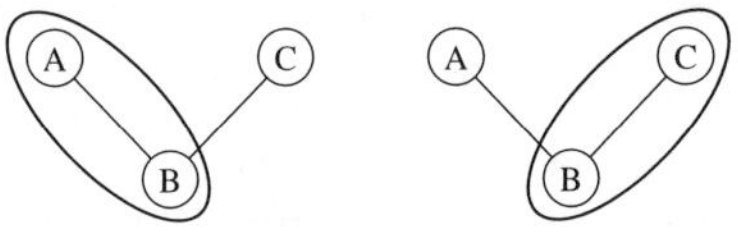

Figure 3: Two Different Pairs Representing Identical Pattern

(or minimum) value of the code is searched. The corresponding code is the canonical label. Let M be the number of nodes in a graph, N be the number of groups of the nodes, and $p_i (i = 1, 2, \dots, N)$ be the number of the nodes within group i. The search space can be reduced to $\prod_{i=1}^{N}(p_i!)$ from $M!$ by using canonical labeling. The code of an adjacency matrix for the case in which elements in the upper triangle are vertically concatenated is defined as

$$
\begin{aligned}
A &= \begin{pmatrix} a_{11} & a_{12} & \cdots & a_{1n} \\ & a_{22} & \cdots & a_{2n} \\ & & \ddots & \vdots \\ & & & a_{nn} \end{pmatrix} \\
code(A) &= a_{11}a_{12}a_{22}a_{13}a_{23}\dots a_{nn} \qquad (1) \\
&= \sum_{j=1}^{n}\sum_{i=1}^{j}(2^{\sum_{k=j+1}^{m} k+j-i} a_i j). \qquad (2)
\end{aligned}
$$

It is possible to further prune the search space. We choose the option of vertical concatenation. Elements of the adjacency matrix of higher ranked nodes form higher bits of the code. Thus, once the locations of higher ranked nodes in the adjacency matrix are fixed, corresponding higher bits of the code are also fixed and are not affected by the order of elements of lower ranks. For example, in Eq. 1 elements that the first two ranked nodes can decide are the first 3 bits in the $code(A)$ and no bits corresponding to the nodes of the lower ranks are included. This reduces the search space of $\prod_{i=1}^{N}(p_i!)$ to $\sum_{i=1}^{N}(p_i!)$.

However, there is still a problem of combinatorial explosion for a case where there are many nodes of the same labels and the same degrees of node such as the case of chemical compounds because the value of p_i becomes large. What we can do is to make the best of already determined nodes of higher ranks. Assume that the nodes $v_i \in V(G)(i = 1, 2, \ldots, N)$ are already determined in a graph G. Consider finding the order of the nodes $u_i \in V(G)(i = 1, 2, \ldots, k)$ of the same group that gives the maximum code value. The node that comes to v_{N+1} is the one in $u_i(i = 1, \ldots, k)$ that has a link to the node v_1 because the highest bit that v_{N+1} can make is a_{1N+1} and the node that makes this bit 1, that is, the node that is linked to v_1 gives the maximum code. If there are more than one node or no node at all that has a link to v_{N+1}, the one that has a link to v_2 comes to v_{N+1}. Repeating this process determines which node comes to v_{N+1}. If no node can't be determined after the last comparison at v_N, permutation within the group is needed.

This is explained using an example in Fig. 4. Assume that nodes 1, 2 and 3 have already been determined. Nodes 4, 5 and 6 are in the same group. The fourth node is the node that has a link to the highest ranked node 1, which is the node 4. Likewise, the fifth and the sixth nodes are the nodes 5 and 6 respectively. In this case, node ordering is uniquely determined. If l nodes can be determined by this procedure, the search space can be reduced from $k!$ to $(k-l)!$.

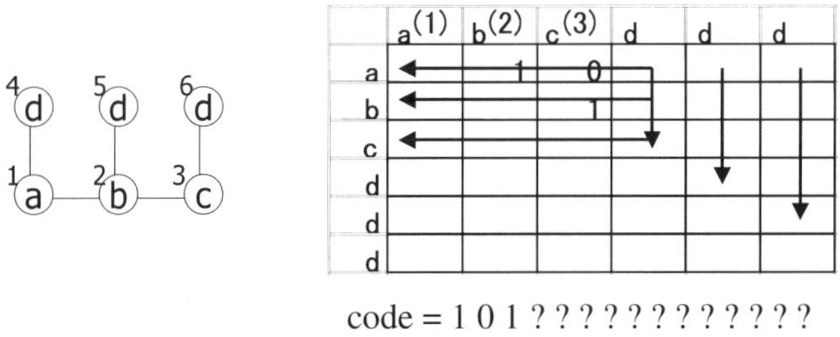

Figure 4: Determination of Node Ordering within a Group

4 Experimental Evaluation of B-GBI

The proposed method is implemented as B-GBI and tested against the promoter dataset in UCI Machine Learning Repository [1]. A promoter is a genetic region which initiates the first step in the expression of an adjacent gene (*transcription*). The promoter dataset consists of strings that represent nucleotides (one of A, G, T or C). The input features are 57 sequential DNA nucleotides and the total number of instances is 106 including 53 positive instances (sample promoter sequences) and 53 negative instances (non-promoter sequence). Direct encoding of this data to the standard attribute-value format assigning the n-th attribute to the n-th nucleotide in the sequence does not make sense. Encoding the data this way and running C4.5[11] gives a predictive error of 16.0% by leaving one out cross validation. Randomly shifting the sequence by 3 elements gives 21.7% and by 5 elements 44.3%. Graph representation resolves this problem. Since it is not known in advance how strong the interaction among the elements are for determining the class label, it is assumed that an element interacts up to 10 elements on both sides (See Fig. 5.). Each sequence results in a graph with 57 nodes and 515 links.

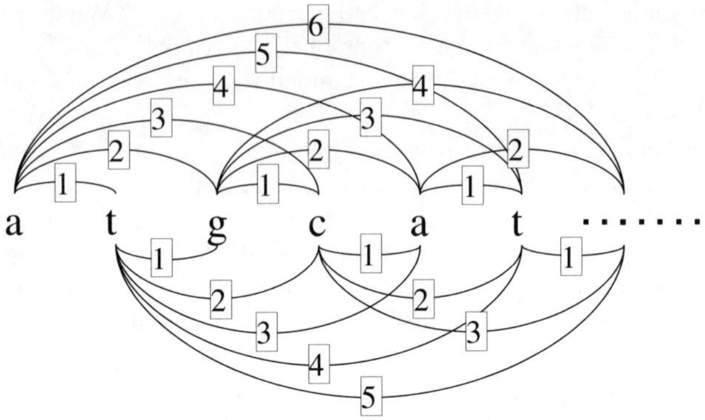

Figure 5: Conversion of DNA Sequence Data to a graph

The minimum support for chunking is set at 20%. The normalized class probability of Eq. 3 is used as a criterion to select typical patterns. Here, p and n indicate respectively the number of positive and negative instances that have a typical pattern and P and N the total number of positive and negative instances ($i.e.$, $P = N = 53$).

$$Max. \left\{ \frac{\frac{p}{P}}{\frac{p}{P} + \frac{n}{N}}, \frac{\frac{n}{N}}{\frac{p}{P} + \frac{n}{N}} \right\} \tag{3}$$

Three threshold values are used: 0.6, 0.7 and 0.8. The beam width is set at 1, 5, 10, ..., 50. The number of typical patterns and their sizes (nodes) are listed in Table 1.

Table 1: Number of Typical Patterns and Average Sizes

Beam width		1	5	10	15	20	25	30	35	40
Thres.	Pattern	355	1442	2125	3174	3688	4733	4697	5604	6293
0.6	Size	3.8	3.9	3.8	3.8	3.9	3.9	3.8	3.9	3.9
Thres.	Pattern	81	282	399	590	645	896	815	992	1136
0.7	Size	3.9	4.1	4.0	4.0	4.1	4.0	4.0	4.1	4.0
Thres.	Pattern	16	41	38	72	73	117	115	145	163
0.8	Size	4.1	4.3	4.0	4.1	4.1	4.2	4.2	4.3	4.2

Beam width		45	50
Thres.	Pattern	6780	7342
0.6	size	3.9	3.9
Thres.	Pattern	1162	1249
0.7	size	4.1	4.1
Thres.	Pattern	161	178
0.8	size	4.2	4.3

To evaluate these patterns, they are used as binary attributes of each sequence to build a classifier by C4.5[11]. Predictive error rate is evaluated by leaving one out. Results are shown in Fig. 6.

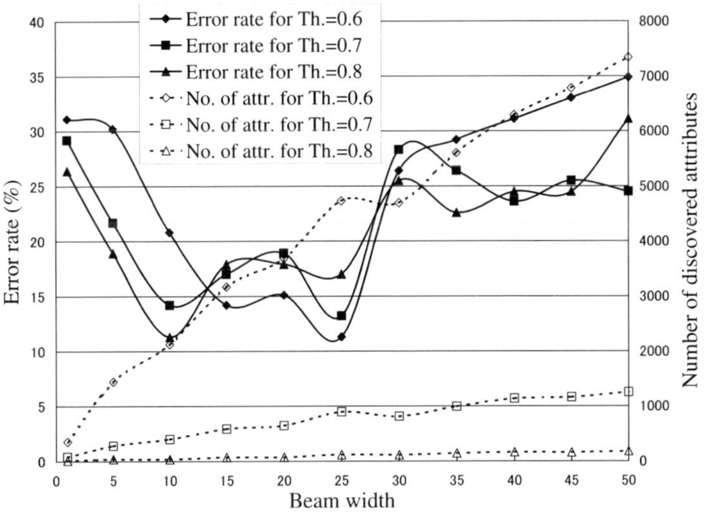

Figure 6: Effect of Beam Width for the Final Decision Tree

It is noted that the error reduces as the beam width is increased from 1 to 10, levels off to 25 and then increases. Number of extracted patterns increases monotonically with the beam width. Too many beams affect adversely due to overfitting/oversearching. Sharp reduction of the error up to beam width of 10 indicates that important typical patterns that contribute to discriminative power are indeed extracted.

The induced decision tree for which the error is minimum (11.3%) is shown in Fig. 7. The threshold of Eq. 3 is 0.8 and the beam width is = 10. There are 38 typical patterns in this case and C4.5 chooses 6 patterns from among them. These are shown in the nodes of the tree. The tree shown in Fig. 8 is for the case where the threshold of Eq. 3 is 0.6 and the beam width is = 10. The error of this tree is 20.8%. There are 2125 typical patterns but C4.5 chooses only 7 patterns, out of which 4 patterns also appear in the first tree. These 4 are the most discriminative patterns. Graph representation is not affected by random shifting of sequences.

5 Conclusion

Finding typical patterns in a graph structured data set lend itself to an important class of data mining problems. Graph based induction GBI is meant to searve as a practical tool for this kind of probem. In this paper some drawbacks of GBI is discussed and a solution is reproted. GBI is improved in three aspects by incorporating: 1) two criteria, one for chunking and the other for task specific criterion to extract more discriminative patterns, 2) beam search to enhance search capability and 3) canonical labeling to accurately count identical patterns. The improved B-GBI is applied to a classification problem of DNA promoter sequence and the results indicate that it is possible to extract discriminative patterns which otherwise are hard to extract.

Immediate future work includes to use feature selection method to filter out less useful patterns.

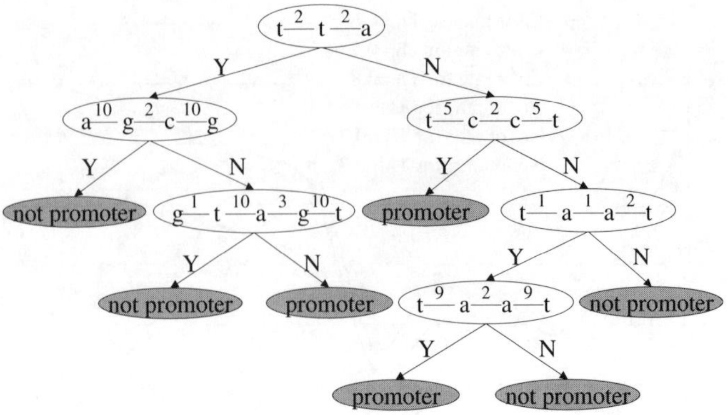

Figure 7: Decision Tree (Threshold value of Eq. 3 = 0.8 and Beam width = 10)

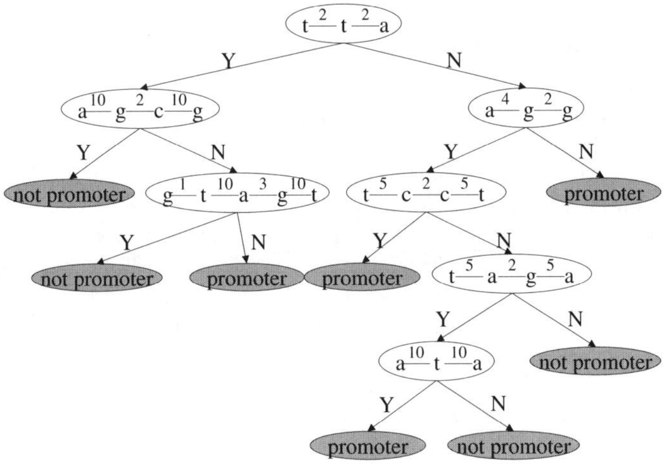

Figure 8: Decision Tree (Threshold value of Eq. 3 = 0.6 and Beam width = 10)

Acknowledgement

This work was partially supported by the grant-in-aid for scientific research on priority area "Active Mining" funded by the Japanese Ministry of Education, Culture, Sport, Science and Technology.

References

[1] C. L. Blake, E. Keogh, and C.J. Merz. Uci repository of machine leaning database, 1998. http://www.ics.uci.edu/~mlearn/MLRepository.html.

[2] L. Breiman, J. H. Friedman, R. A. Olshen, and C. J. Stone. *Classification and Regression Trees*. Wadsworth & Brooks/Cole Advanced Books & Software, 1984.

[3] L. Breiman, J. H. Friedman, R. A. Olshen, and C. J. Stone. The cn2 induction algorithm. *Machine Learning*, 3:261–283, 1989.

[4] D. J. Cook and L. B. Holder. Graph-based data mining. *IEEE Intelligent Systems*, 15(2):32–41, 2000.

[5] S. Fortin. The graph isomorphism problem, 1996.

[6] A. Inokuchi, T. Washio, and H. Motoda. An apriori-based algorithm for mining frequent substructures from graph data. In *Proc. of the 4th European Conference on Principles of Data Mining and Knowledge Discovery*, pages 13–23, 2000.

[7] T. Matsuda, T. Horiuchi, H. Motoda, and T. Washio. Extension of graph-based induction for general graph structured data. In *Knowledge Discovery and Data Mining: Current Issues and New Applications, Springer Verlag, LNAI 1805*, pages 420–431, 2000.

[8] R. S. Michalski. Learning flexible concepts: Fundamental ideas and a method based on two-tiered representaion. *In Machine Learning, An Artificial Intelligence Approiach*, 3:63–102, 1990.

[9] S. Muggleton and L. de Raedt. Inductive logic programming: Theory and methods. *Journal of Logic Programming*, 19(20):629–679, 1994.

[10] J. R. Quinlan. Induction of decision trees. *Machine Learning*, 1:81–106, 1986.

[11] J. R. Quinlan. *C4.5:Programs For Machine Learning.* Morgan Kaufmann Publishers, 1993.

[12] R. C. Read and D. G. Corneil. The graph isomorphism disease. *Journal of Graph Theory*, 1:339–363, 1977.

[13] K. Yoshida and H. Motoda. Clip : Concept learning from inference pattern. *Journal of Artificial Intelligence*, 75(1):63–92, 1995.

Active Mining
H. Motoda (Ed.)
IOS Press, 2002

PAGA Discovery: A Worst-Case Analysis of Rule Discovery for Active Mining

Einoshin Suzuki
suzuki@ynu.ac.jp
Electrical and Computer Engineering,
Yokohama National University
79-5 Tokiwadai, Hodogaya, Yokohama 240-8501, Japan

Abstract. In this paper, we perform a worst-case analysis of rule discovery. A rule is defined as a probabilistic constraint of true assignment to the class attribute of corresponding examples. In data mining, a rule can be considered as representing an important class of discovered patterns. We accomplish the aforementioned objective by extending a fundamental version of PAC (Probably Approximately Correct) learning, which represents a worst-case analysis for classification. Our analysis consists of two cases: the case in which we try to avoid finding a bad rule, and the case in which we try to avoid overlooking a good rule. Discussions on related works are also provided for PAC learning, multiple comparison, analysis of association rule discovery, and simultaneous reliability evaluation of a discovered rule. We also present how our results can be applied to rule discovery in the context of active mining.

1 Introduction

Data mining [4] can be defined as extraction of useful knowledge from massive data, and is gaining increasing attention due to advancement of various information technologies. Data mining can be regarded as advanced data analysis, and a typical process of analysis consists of several steps [4]. Pattern extraction represents an important step in such a process. A rule is defined as a probabilistic constraint inherent in a data set, and is widely recognized as representing one of the most important patterns in data mining.

Although rule discovery has been extensively studied in data mining, its theoretical analyses are surprisingly rare. Several exceptions include Agrawal et al.'s analysis of association rule discovery [1] and our analysis of a discovered rule based on simultaneous reliability evaluation [13]. However, these studies ignore the total number of rules that can be discovered from a data set. This fact represents that these studies fail to relate the size of a discovery problem to the number of examples needed for successful discovery, and suggests that a more solid foundation of data mining should be established. Moreover, clarifying the sample complexity of rule discovery would be useful for active mining, which roughly represents an active process in data mining.

As a first step toward this objective, we extend a fundamental version of PAC (Probably Approximately Correct) learning [10], which represents a worst-case analysis of classification [15]. Our analysis consists of two cases: the case in which we try to avoid finding a bad rule, and the case in which we try to avoid overlooking a good rule. We also discuss about related works including PAC learning [7, 10], Jensen and Cohen's multiple comparison [6], Agrawal et al.'s analysis of association rule discovery [1], and our previous analysis of a discovered rule based on simultaneous reliability evaluation [13]. Application of our results in active mining will be also provided.

2 Rule Discovery Problem

2.1 Rule

Let a data set contain m examples each of which is expressed by b discrete attributes and a class attribute. Typically rule discovery assumes no specific class attribute unlike classification. However, for the sake of formalization, we consider a rule which predicts a specific class attribute to be true.

Let a value V assignment $A = V$ to an attribute A be an atom. In this paper, we regard a given data set as a result of sampling with replacement from a true data set. We call the probability of occurrence of examples each of which satisfies a propositional logical formula f the true probability $\Pr(f)$ of f. Similarly, an estimated probability which is obtained from a given data set for $\Pr(f)$ is represented by $\widehat{\Pr}(f)$. Note that $\widehat{\Pr}(f)$ can be calculated by the Laplace estimate or simply by the ratio of examples which satisfy f in the data set. We employ the latter method in this paper.

A rule r is represented as follows with a premise y which is represented by a propositional formula of atoms, and a conclusion x which is represented by a true assignment to the class attribute.

$$r : y \rightarrow x$$

An intuitive interpretation of r is that many examples satisfy y and those examples are likely to satisfy x with high probability. We define $\Pr(y)$ and $\Pr(x|y)$ as the generality and the accuracy of r respectively. Similarly, we call $\widehat{\Pr}(y)$ and $\widehat{\Pr}(x|y)$ the estimated generality and the estimated accuracy of r respectively.

2.2 Related Classes of Rules

This section presents several classes of rules which are related to ours. A probabilistic if-then rule [12] is defined as follows, where y_i represents a single atom.

$$y_1 \wedge y_2 \wedge \cdots \wedge y_K \rightarrow x$$

In [13], a probabilistic if-then rule is called a conjunction rule, and this paper follows this paraphrasing. A conjunction rule can be regarded as a special case of our rule: the premise is restricted to either a single atom or a conjunction of atoms.

Since a premise of a conjunction rule is represented by a combination of atoms, the number $|R|$ of possible conjunction rules is typically huge. The following gives $|R|$, where a data set contains b attributes and each of these attributes can have one of a values.

$$|R| = (a+1)^b - 1 \tag{1}$$

This formula can be explained by the fact that each of b attributes can either have one of a values or be excluded from the premise. A typical value for $|R|$ is huge: for example, $|R| =$ 3,486,784,400 for a data set of 20 binary attributes. A realistic measure would be to restrict the number of atoms allowed in the premise to at most K. The possible number $|R_K|$, in this case, is given as follows.

$$|R_K| = \sum_{i=1}^{K} a^i \binom{b}{i} \tag{2}$$

Note that (1) can be also derived by settling $K = b$ in (2) and considering the binary coefficients.

In association rule discovery [1], a data set is restricted to a transactional data set which consists of binary attributes. A true assignment to a binary attribute is called an item. Let an itemset be either a single item or a conjunction of items. An association rule, in its original form, consists of a premise and a conclusion each of which is represented by an itemset. In our framework of subsection 2.1, an association rule can be regarded as a special case of our conjunction rule: only the value "true" is allowed. The cases of $|R|$ and $|R_K|$ for association rule discovery are obtained by settling $a = 1$ in (1) and (2).

2.3 Discovery Problem

In this paper, the objective of a user is to obtain, with high probability $1 - \delta$, a rule of which generality and accuracy are no smaller than $1 - \zeta$ and $1 - \epsilon$ respectively. Typically multiple rules are obtained in rule discovery, but we restrict ourselves to single-rule discovery for the sake of analysis.

$$\begin{aligned} \text{Objective}: \quad & \text{Find } y \rightarrow x \text{ which satisfies} \\ & \Pr\left[\Pr(y) \geq 1 - \zeta,\ \Pr(x|y) \geq 1 - \epsilon\right] \geq 1 - \delta \\ & \text{where } \zeta,\ \epsilon,\ \delta\ >\ 0 \end{aligned} \tag{3}$$

A discovery algorithm to be analyzed obtains a rule of which generality and accuracy are no smaller than user-given thresholds θ_{S} and θ_{F} respectively. As stated in subsection 2.1, since a given data set is a result of sampling with replacement from a true data set, the user employs thresholds $\theta_{\mathrm{S}} \neq 1 - \zeta,\ \theta_{\mathrm{F}} \neq 1 - \epsilon$ in applying the algorithm.

$$\begin{aligned} \text{Algorithm}: \quad & \text{Find } y \rightarrow x \text{ which satisfies} \\ & \widehat{\Pr}(y) \geq \theta_{\mathrm{S}},\ \widehat{\Pr}(x|y) \geq \theta_{\mathrm{F}} \end{aligned} \tag{4}$$

An interesting problem here is to obtain the required number of examples (i.e. the sample complexity [3]) to accomplish (3) under (4). This problem can be named as PAGA (Probably Approximately General and Accurate) discovery after the well-known PAC learning [7, 10], and can be regarded as a foundation of active mining, discovery science, and data mining.

3 Case 1: Exclusion of a Bad Rule

In this section, we obtain a sample complexity for the problem defined in the previous section. An assumed condition is to avoid finding a bad rule. This condition can be considered as important in several domains where reliability represents a crucial concern.

3.1 Preliminaries

First we introduce preliminaries which are needed in subsequent analyses. If the domain of a probabilistic variable X is $\{0, 1, \cdots, \nu\}$ and the probability distribution of the variable is represented as follows, X is said to follow a binary distribution [5]

$$\Pr(X = k) \quad = \quad \mathrm{B}(k; \nu, p)$$

$$= \binom{\nu}{k} p^k (1-p)^{\nu-k} \tag{5}$$

where p represents a constant $0 < p < 1$ and $k = 0, 1, \cdots, \nu$. The Chernoff bound states that the following holds for an arbitrary constant $q > p$ [1].

$$\Pr(X > \nu q) < \exp[-2\nu(q-p)^2] \tag{6}$$

3.2 Theoretical Analysis

From (3), a bad rule $r_{\rm b} : y \to x$ satisfies

$$\Pr(y) < 1 - \zeta \text{ or } \Pr(x|y) < 1 - \epsilon. \tag{7}$$

Since we assume, in this section, that we avoid finding a bad rule, the employed thresholds for generality and accuracy are relatively large. This assumption together with (3) and (4) necessitate the following.

$$\theta_{\rm S} > 1 - \zeta \text{ and } \theta_{\rm F} > 1 - \epsilon \tag{8}$$

From (7) and (8),

$$\theta_{\rm S} > \Pr(y) \text{ or } \theta_{\rm F} > \Pr(x|y). \tag{9}$$

Since $r_{\rm b} : y \to x$ is discovered,

$$\widehat{\Pr}(y) \geq \theta_{\rm S} \text{ and } \widehat{\Pr}(x|y) \geq \theta_{\rm F}. \tag{10}$$

Let the number of examples in the given data set be m. If and only if y and xy are satisfied by at least $\lceil m\theta_{\rm S} \rceil$ and $\lceil m\widehat{\Pr}(y)\theta_{\rm F} \rceil$ examples respectively in the data set, $r_{\rm b}$ happens to be discovered. Since each of the numbers of examples which satisfy y and xy follows a binary distribution,

$$\begin{aligned}
\Pr\ & (r_{\rm b} \text{ discovered}) \\
& \leq \mathrm{MAX}\left[\sum_{k=\lceil m\theta_{\rm S}\rceil}^{m} \mathrm{B}(k; m, \Pr(y)), \sum_{k=\lceil m\widehat{\Pr}(y)\theta_{\rm F}\rceil}^{m\widehat{\Pr}(y)} \mathrm{B}(k; m\widehat{\Pr}(y), \Pr(x|y))\right] && (11) \\
& < \mathrm{MAX}\left\{\exp\left[-2m\left(\frac{\lceil m\theta_{\rm S}\rceil}{m} - \Pr(y)\right)^2\right],\right. \\
& \qquad \left.\exp\left[-2m\widehat{\Pr}(y)\left(\frac{\lceil m\widehat{\Pr}(y)\theta_{\rm F}\rceil}{m\widehat{\Pr}(y)} - \Pr(x|y)\right)^2\right]\right\} && (12) \\
& < \mathrm{MAX}\left\{\exp\left[-2m(\theta_{\rm S} - 1 + \zeta)^2\right], \exp\left[-2m\theta_{\rm S}(\theta_{\rm F} - 1 + \epsilon)^2\right]\right\}. && (13)
\end{aligned}$$

Note that, in (11), we consider separately the case in which a bad rule $r_{\rm b1}$ in terms of generality is discovered and the case in which a bad rule $r_{\rm b2}$ in terms of accuracy is discovered. The first and second terms correspond to the left inequality and the right inequality of (7) respectively. Since $\Pr(r_{\rm b1})$ and $\Pr(r_{\rm b2})$ are unknown, we upper-bound $\Pr(r_{\rm b}$ discovered) by $\mathrm{MAX}[\Pr(r_{\rm b1}$ discovered), $\Pr(r_{\rm b2}$ discovered)]. In (12), the Chernoff bound (6) is employed from (9). Finally in (13), we employ (7) and the left inequality of (10).

Let the set of all rules and the set of all bad rules be R and $R_{\rm b}$ respectively, and let the cardinality of a set S be $|S|$. The probability of discovering a bad rule satisfies the following inequalities.

$$\begin{aligned} \Pr\ & (R_{\rm b} \text{ contains a discovered rule}) \\ & < |R_{\rm b}|\text{MAX}\left\{\exp\left[-2m(\theta_{\rm S}-1+\zeta)^2\right], \exp\left[-2m\theta_{\rm S}(\theta_{\rm F}-1+\epsilon)^2\right]\right\} \quad (14) \\ & \leq |R|\text{MAX}\left\{\exp\left[-2m(\theta_{\rm S}-1+\zeta)^2\right], \exp\left[-2m\theta_{\rm S}(\theta_{\rm F}-1+\epsilon)^2\right]\right\} \quad (15) \end{aligned}$$

Note that we allow to count multiple times the cases in which several bad rules satisfy the discovery condition in (14), and (15) uses $|R| \geq |R_{\rm b}|$. In order to avoid finding a bad rule, we require the following with respect to a sufficiently small δ.

$$|R|\text{MAX}\left\{\exp\left[-2m(\theta_{\rm S}-1+\zeta)^2\right], \exp\left[-2m\theta_{\rm S}(\theta_{\rm F}-1+\epsilon)^2\right]\right\} \leq \delta \quad (16)$$

We obtain a sample complexity for rule discovery in which finding a bad rule is avoided with a high probability.

$$m \geq \frac{\ln\left(\frac{|R|}{\delta}\right)}{2\text{MIN}\left[(\theta_{\rm S}-1+\zeta)^2, \theta_{\rm S}\left(\theta_{\rm F}-1+\epsilon\right)^2\right]} \quad (17)$$

The above inequality describes influence of each parameter to the sample complexity quantitatively. As we have seen in subsection 2.2, $|R|$ is typically large and is thus important even if its influence is tolerated by a logarithmic function. The second most important factors are $\theta_{\rm S}-1+\zeta$ and $\theta_{\rm F}-1+\epsilon$. Since they influence the required number of examples by the inverse of their squares, they can be problematic when they are small. Since each of these terms represents the difference of a threshold and the user-expected value, $\theta_{\rm S}-1+\zeta$ and $\theta_{\rm F}-1+\epsilon$ can be named as the margin of generality and the margin of accuracy respectively. In a typical setting of rule discovery, we can assume $\theta_{\rm S} = 0.1$, and $(\theta_{\rm S}-1+\zeta) = 10^{-1}$ or 10^{-2}. We can also assume that $(\theta_{\rm F}-1+\epsilon) = 10^{-1}$ or 10^{-2}. Under these assumptions, the denominator is either $2*10^{-3}$ or $2*10^{-5}$. Finally, δ can be considered as a moderately important factor in a typical situation $\delta = 0.01$ - 0.05 since it appears only as a denominator of $|R|$.

3.3 Application to Conjunction Rule Discovery

From (1) and (17), the sample complexity is given as follows if we restrict the discovered rule to a conjunction rule.

$$m \geq \frac{\ln\left[(a+1)^b - 1\right] + \ln\left(\frac{1}{\delta}\right)}{2\text{MIN}\left[(\theta_{\rm S}-1+\zeta)^2, \theta_{\rm S}\left(\theta_{\rm F}-1+\epsilon\right)^2\right]} \quad (18)$$

Note that settling $a = 1$ gives the case of association rule discovery.

Firstly, $\ln(1/\delta)$ can be typically ignored when $\delta = 0.01$ - 0.05 from $\ln[(a+1)^b - 1] \gg \ln(1/\delta)$, thus the sample complexity is approximately proportional to b. Secondly, since the number a of possible values for an attribute only affects the right-hand side through a logarithmic function, a is typically not so important as b and margins of generality and accuracy. We show, in figure 1, a plot of the required number of examples against

$\text{MIN}[(\theta_S - 1 + \zeta)^2, \theta_S(\theta_F - 1 + \epsilon)^2]$ for $b = 10^2, 10^3, 10^4$, where we settled $a = 2$ and $\delta = 0.05$. Note that each of the x axis and the y axis is represented by a logarithmic scale.

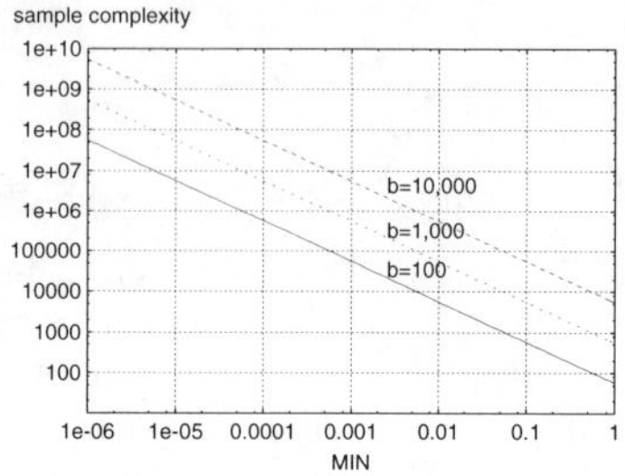

Figure 1: Required number of examples needed for conjunction rule discovery without finding a bad rule. In the figure, MIN represents $\text{MIN}[(\theta_S - 1 + \zeta)^2, \theta_S(\theta_F - 1 + \epsilon)^2]$.

We discuss about the required number of examples for a typical setting with figure 1. The examples described in subsection 3.2 state $\text{MIN}[(\theta_S - 1 + \zeta)^2, \theta_S(\theta_F - 1 + \epsilon)^2] = 10^{-3}$ or 10^{-5}. For these cases, the sample complexity is approximately $5.6 * 10^4$ - $5.6 * 10^6$ or $5.6 * 10^6$ - $5.6 * 10^8$ for $b = 10^2$ - 10^4. These results indicate that the required number of examples for successful discovery can be prohibitively large for small margins. Note that large margins represent large thresholds, and no rules are usually discovered for large thresholds. A realistic and effective measure to this problem would be to adjust thresholds according to a discovery process such as [14]. It should be anyway noted that our analysis in this paper corresponds to the worst case, and the required number of examples in a real discovery problem can be much smaller than those mentioned above.

From (2) and (17), the sample complexity is given as follows if we restrict the discovered rule to a conjunction rule with at most K atoms in its premise.

$$m \geq \frac{\ln\left[\sum_{i=1}^{K} a^i \binom{b}{i}\right] + \ln\left(\frac{1}{\delta}\right)}{2\text{MIN}\left[\left(\theta_S - 1 + \zeta\right)^2, \theta_S\left(\theta_F - 1 + \epsilon\right)^2\right]} \tag{19}$$

Note that settling $a = 1$ gives the case of association rule discovery.

Similarly as we did in figure 1, we show, in figure 2, two plots of the sample complexity for $a = 2$ and $\delta = 0.05$. The left plot represents a case in which we varied $b = 10^2$, 10^3, 10^4 under $K = 2$, and in the right plot we varied $K = 1, 2, 3, 4, 100 (= b)$ under $b = 10^2$.

From the left plot of figure 2, we see that the influence of b is relatively small for $K = 2$. On the other hand, the right plot shows that, for $K \leq 4$, the required number of examples is smaller by approximately an order of magnitude than the case of considering all conjunction rules ($K = b = 100$). It is widely accepted that a rule with a short premise exhibits high readability, and the above results suggest that they are also attractive in terms of the required number of examples.

4 Case 2: Inclusion of a Good Rule

In this section, we derive another sample complexity for the problem defined in subsection 2.3. An assumed condition is to avoid overlooking a good rule. This condition can

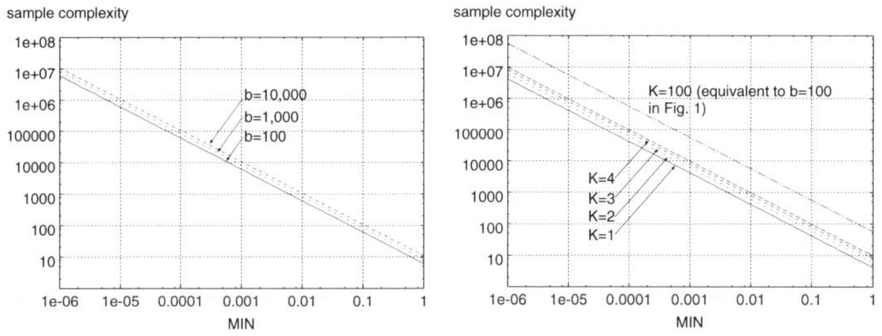

Figure 2: Required number of examples needed for conjunction rule discovery without finding a bad rule, where at most K atoms are allowed in the premise. The left and right plots assume $K = 2$ and $b = 100$ respectively.

be considered as important in several domains where possibility is considered as highly important.

From (3), a good rule $r_{\rm g} : y \rightarrow x$ satisfies

$$\Pr(y) \geq 1 - \zeta \text{ and } \Pr(x|y) \geq 1 - \epsilon. \tag{20}$$

Since we assume, in this section, that we avoid overlooking a good rule, the employed thresholds for generality and accuracy are relatively small. This assumption together with (3) and (4) necessitate the following.

$$\theta_{\rm S} < 1 - \zeta \text{ and } \theta_{\rm F} < 1 - \epsilon \tag{21}$$

From (20) and (21),

$$\theta_{\rm S} < \Pr(y) \text{ and } \theta_{\rm F} < \Pr(x|y). \tag{22}$$

Let the number of examples in the given data set be m. If and only if y is satisfied by at most $\lceil m\theta_{\rm S} \rceil - 1$ examples or xy is satisfied by at most $\lceil m\widehat{\Pr}(y)\theta_{\rm F} \rceil - 1$ examples in the data set, $r_{\rm g}$ happens to be undiscovered. Since each of the numbers of examples which satisfy y and xy follows a binary distribution,

$$\begin{aligned}
\Pr \quad & (r_{\rm g} \text{ undiscovered}) \\
& \leq \mathrm{MAX}\left[\sum_{k=0}^{\lceil m\theta_{\rm S}\rceil - 1} \mathrm{B}(k; m, \Pr(y)), \sum_{k=0}^{\lceil m\widehat{\Pr}(y)\theta_{\rm F}\rceil - 1} \mathrm{B}(k; m\widehat{\Pr}(y), \Pr(x|y))\right] && (23) \\
& = \mathrm{MAX}\Bigg[\sum_{k=\phi(m,\theta_{\rm S})}^{m} \mathrm{B}(k; m, 1 - \Pr(y)), \\
& \quad \sum_{k=\phi(m\widehat{\Pr}(y),\theta_{\rm F})}^{m\widehat{\Pr}(y)} \mathrm{B}(k; m\widehat{\Pr}(y), 1 - \Pr(x|y))\Bigg] && (24) \\
& < \mathrm{MAX}\Bigg\{\exp\left[-2m\left(\frac{\phi(m,\theta_{\rm S})}{m} - 1 + \Pr(y)\right)^2\right], \\
& \quad \exp\left[-2m\widehat{\Pr}(y)\left(\frac{\phi(m\widehat{\Pr}(y),\theta_{\rm F})}{m\widehat{\Pr}(y)} - 1 + \Pr(x|y)\right)^2\right]\Bigg\} && (25) \\
& < \mathrm{MAX}\left\{\exp\left[-2m(-\theta_{\rm S} + 1 - \zeta)^2\right], \exp\left[-2m\theta_{\rm S}(-\theta_{\rm F} + 1 - \epsilon)^2\right]\right\}. && (26)
\end{aligned}$$

where $\phi(m, \theta) \equiv m - \lceil m\theta \rceil + 1$.

Note that we consider separately the cases in which the generality and the accuracy of a good rule are below the respective thresholds in (23). We represent such rules as r_{g1} and r_{g2} respectively. Since $\Pr(r_{g1})$ and $\Pr(r_{g2})$ are unknown, we use the same technique as in subsection 3.2. Note that (24) corresponds to replacement of p by $1-p$ in (5). In (25), the Chernoff bound (6) is employed from (22). Finally in (26), we employ (20) and $\widehat{\Pr}(y) \geq \theta_S$. Note that the last inequality holds in the second term since r_g is undiscovered due to apparently low accuracy.

Similarly to subsection 3.2, the following can be obtained as a sample complexity for rule discovery in which overlooking a good rule is avoided with a high probability.

$$m \geq \frac{\ln\left(\frac{|R|}{\delta}\right)}{2\mathrm{MIN}\left[(-\theta_S + 1 - \zeta)^2,\ \theta_S(-\theta_F + 1 - \epsilon)^2\right]} \tag{27}$$

Note that (27) is equivalent to (17), and similar discussions as subsections 3.2 and 3.3 hold. Note that large margins ($1-\zeta-\theta_S$ and $1-\epsilon-\theta_F$ in this case) represent small thresholds in this case, and small thresholds typically result in a large number of candidates of the discovered rule to be inspected. The automatic adjustment of thresholds [14] can be also a realistic measure for this problem.

5 Discussions on Related Topics

5.1 PAC Learning

PAC learning represents a worst-case analysis for classification, and has numerous excellent results. Our results in section 3 can be considered as an extension to a fundamental version of PAC learning [10]. First, a classifier ignores generality since it predicts the class attribute for all examples. This is the reason $\Pr(y)$, $\widehat{\Pr(y)}$, θ_S are not considered in [10]. Actually, the objective of learning in [10] is represented as a simplification of (3) as follows.

$$\begin{aligned}\text{Objective}:\ & \text{Find a classifier } h \text{ which satisfies}\\ & \Pr\left[\Pr(h \text{ correctly predicts } x) \geq 1-\epsilon\right] \geq 1-\delta\\ & \text{where } \epsilon,\ \delta > 0\end{aligned} \tag{28}$$

Next, [10] assumes that a classification algorithm returns a classifier which is consistent to all training examples. This corresponds to assuming $\theta_F = 1$.

To sum up, compared to our study, [10] ignores the case of learning a classifier with low generality and the case of learning a classifier which is inconsistent to the training examples. In this case, application of the Chernoff bound can be skipped, and for a bad classifier h_b, we obtain $\Pr(h_b \text{ learned}) = (1-\epsilon)^m$. In [10], the sample complexity is given by the following, where H represents a set of all classifiers.

$$m \geq \frac{\ln\left(\frac{|H|}{\delta}\right)}{\epsilon} \tag{29}$$

Note that (29) resembles to (17): it only ignores generality ($\theta_S = 1$ and no ζ), assumes $\theta_F = 1$, and omits the squares in ϵ^2 and 2 in the denominator. The last two omissions are due to skipping the application of the Chernoff bound.

Although [3] analyzes the case with low accuracy, it ignores the case with low generality unlike our study. Studies for the PAC learnability of functional dependencies [2, 8] are related to ours, but they deal with a different discovery problem.

5.2 Jensen and Cohen's Multiple Comparison

Jensen and Cohen's multiple comparison [6] proposes a prudent view of classification. Its essential point can be stated as a probabilistic explanation that the more candidates of classifiers are inspected in a learning algorithm, the smaller accuracy is exhibited by the obtained classifier. The multiple comparison provides a comprehensive unified view of several previous studies including overfitting [11] and oversearching [9], and [6] also proposes several realistic measures.

Since this study deals with classification as PAC learning, it ignores generality. This corresponds to considering only the second term in (11). Since [6] considers the case of $\theta_{\rm F} < 1$, it provides a more realistic framework to learning than [10]. The multiple comparison differs from our study in that it directly calculates, based on a binary distribution without using the Chernoff bound, the probability for a bad classifier to satisfy at least $\lceil m\theta_{\rm F} \rceil$ examples. Moreover, they calculate exactly the probability that no bad classifier is learned while we, in (14), allow counting multiples times the cases in which more than one bad rules satisfy the discovery condition. Let the set of all bad classifiers be $H_{\rm b}$, then the probability in [6] is given by the following.

$$\Pr(H_{\rm b} \text{ contains a learned classifier}) = 1 - [1 - \Pr(h_{\rm b} \text{ learned})]^{|H|} \tag{30}$$

Pursuing strictness in calculation can be compared to a double-edged sword. Jensen and Cohen give no analytical solutions to the required number of examples for successful learning. We attribute this reason to the fact that resolving (30) for m is relatively difficult. We have employed several approximations in our theoretical analyses since we believe that they are necessary to bound m analytically. Another difference between [6] and our analyses is rather philosophical: while they are pessimistic about classification, we are realistic about rule discovery. The study in [6] emphasizes that $|H|$ is huge, and demonstrates various examples in which it is difficult to avoid learning a bad classifier. We also recognize that $|R|$ is huge, but obtain the required number of examples analytically with respect to $|R|$.

5.3 Theoretical Analysis of Association Rule Discovery

Analyses of association rule discovery [1] are threefold: a lower bound of the number of queries under the use of a database system, the expected number of itemsets each of which is satisfied by at least a required number of examples in a random data set, and the number of examples satisfied by an itemset in a sampled data set. The third analysis is highly related to our study in that both of the two deal with the case of sampling ν examples with replacement from a true data set in rule discovery.

The analysis provides a specification of the Chernoff bound (6), where X is regarded as $\nu\widehat{\Pr}(f)$ for an itemset f. It first regards the right-hand side $\exp[-2\nu(q-p)^2]$ as the upper bound of the probability for $\widehat{\Pr}(f)$ to deviate at least $q-p$ from its value $p\ (= \Pr(f))$ in the true data set. Next, it gives several examples of values for $q-p$ and δ in $\exp[-2\nu(q-p)^2] = \delta$, and shows the corresponding values of ν in a table.

The discovery algorithm employed in [1] first obtains, by an algorithm called Apriori, a set of all imtemsets f each of which satisfies $\widehat{\Pr}(f) \geq \theta_{\rm S}$. Then, it generates a set of association rules from this set. One of the motivations of the above analysis is to reduce the run-time of Apriori by the use of a sampled data set. Due to this motivation, [1] ignores accuracy unlike our study. Moreover, since it ignores total number of association rules, the study fails to relate the size of a discovery problem (e.g. the number of attributes in the data set, the representation of a discovered rule) to the number of examples needed for successful discovery.

5.4 Simultaneous Reliability Evaluation of a Discovered Rule

Simultaneous reliability evaluation of a discovered rule [13] also deals with the case of sampling m examples with replacement from a true data set in rule discovery as in subsection 5.3 and our study. Unlike the analysis in subsection 5.3, this study considers both generality and accuracy.

The objective considered in [13] is identical to ours, and is represented by (3). However, the analysis fixes m and employs neither θ_{S} nor θ_{F}. Let $\overline{x}$ represent the negation of x. It assumes that $(m\Pr(x,y), m\Pr(\overline{x},y))$ follows a two-dimensional normal distribution, and obtains the necessary and sufficient condition for accomplishing the objective analytically. This is a different framework from ours: we use a discovery algorithm with fixed thresholds θ_{S}, θ_{F} in (4) and bound the number m of sampled examples. The problem dealt in [13] can be reduced to the problem of deriving and analyzing two tangent lines of an ellipse, and applying Lagrange's multiplier method gives the following analytical solutions.

$$\left(1-\beta(\delta)\sqrt{\frac{1-\widehat{\Pr}(y)}{m\widehat{\Pr}(y)}}\right)\widehat{\Pr}(y) \geq 1-\zeta \tag{31}$$

$$\left(1-\beta(\delta)\sqrt{\frac{\widehat{\Pr}(\overline{x},y)}{\widehat{\Pr}(x,y)\{(m+\beta(\delta)^2)\widehat{\Pr}(y)-\beta(\delta)^2\}}}\right)\widehat{\Pr}(x|y) \geq 1-\epsilon \tag{32}$$

Here $\beta(\delta)$ represents a positive constant which defines the size of a $1-\delta$ confidence region i.e. the ellipse for $(m\Pr(x,y), m\Pr(\overline{x},y))$, and can be obtained by a simple numerical integration [13]. Note that (31) and (32) represent conditions for generality and accuracy respectively. Each of them states that the corresponding estimated probability multiplied by a coefficient which is related to the size of the confidence region is no smaller than the corresponding user-expected value ($1-\zeta$ or $1-\epsilon$).

Since the study [13] assumes a specific distribution to the simultaneous occurrence of random variables, it does not fall in the category of worst-case analysis. Similarly to the analysis in subsection 5.3, the study fails to relate the size of a discovery problem to the number of examples needed for successful discovery since it ignore total number of rules.

6 Application to Active Mining

Our results (17), (18), (19), and (27) are useful in active mining, which roughly represents an active process in data mining. They provide guidelines for the required number of examples; selection of attributes; representation of discovered rules; required level ζ, ϵ, δ of discovery; and application $\theta_{\mathrm{S}}, \theta_{\mathrm{F}}$ of the algorithm.

For instance, in a situation described in the right plot of figure 2, suppose we have $MIN = 0.1$, $K = 1$, and $m = 100$. The plot certificates successful discovery in this case. Here, assume we obtain 1000 new examples, and wish to settle our requirement to $MIN = 0.001$ and $K = 2$. From the plot, we see that successful discovery is not guaranteed in this case, but we can settle $K = 4$ if we accept $MIN = 0.01$. Such judgments would be necessary especially when the conditions of discovery are related to costs. Judging from these discussions, we can safely conclude that our results in this paper suggest various useful policies in active mining.

7 Conclusions

The main contribution of this paper is threefold. 1) We formalized a worst-case analysis of rule discovery. The proposed framework employs thresholds θ_S, θ_F for generality and accuracy which are different from user-expected values $1-\zeta$, $1-\epsilon$ respectively. We considered the case in which we try to avoid finding a bad rule, and the case in which we try to avoid overlooking a good rule. 2) We derived sample complexities for the problems by using probabilistic formalization and appropriate approximations. Quantitative analysis of a sample complexity revealed that the total number $|R|$ of rules, the margin $\theta_S - 1 + \zeta$ for generality, and the margin $\theta_F - 1 + \epsilon$ for accuracy are important. 3) We analyzed the sample complexity for a set of specific problems of conjunction rule discovery. Various useful insights are obtained by inspecting sample complexities for a set of typical settings.

The contribution of 1) represents that this paper has provided, in rule discovery, a framework which corresponds to PAC learning. This framework has been named as PAGA (Probably Approximately General and Accurate) discovery. PAGA discovery can be regarded as promising as a theoretical foundation of active mining, which can request new examples in a discovery process. As we described in section 6, the contributions of 2) and 3) suggest various useful policies in applying various rule algorithms in practice. Such policies also include sampling/extending a data set and modification of the class of discovered rules. We can safely conclude that our comprehension to rule discovery has deepened with these contributions and discussions in section 5. Ongoing work focuses on analyses of more realistic algorithms, especially an algorithm which discovers multiple rules with various conclusions.

Acknowledgement

We are grateful to Setsuo Arikawa for enabling us to initiate this study by suggesting us to pursue the relationship of one of our previous studies and PAC learning. This work was partially supported by the grant-in-aid for scientific research on priority area "Active Mining" from the Japanese Ministry of Education, Culture, Sports, Science and Technology.

References

[1] R. Agrawal et al. Fast discovery of association rules. In U. M. Fayyad et al., editors, *Advances in Knowledge Discovery and Data Mining*, pages 307–328. AAAI Press, Menlo Park, Calif., 1996.

[2] T. Akutsu and A. Takasu. On PAC learnability of functional depencies. *New Generation Computing*, 12(4):359 – 374, 1994.

[3] L. Devroye, L. Györfi, and G. Lugosi. Vapnik-Chervonenskis theory. In *A Probabilistic Theory of Pattern Recognition*, pages 187–213. Springer-Verlag, New York, 1996.

[4] U. M. Fayyad, G. Piatetsky-Shapiro, and P. Smyth. From data mining to knowledge discovery: An overview. In U. M. Fayyad et al., editors, *Advances in Knowledge Discovery and Data Mining*, pages 1–34. AAAI/MIT Press, Menlo Park, Calif., 1996.

[5] W. Feller. *An Introduction to Probability Theory and Its Applications.* John Wiley & Sons, New York, 1957.

[6] D. D. Jensen and P. R. Cohen. Multiple comparisons in induction algorithms. *Machine Learning*, 38(3):309–338, 2000.

[7] M. J. Kearns and U. V. Vazirani. *An Introduction to Computational Learning Theory.* MIT Press, Cambridge, Mass., 1994.

[8] J. Kivinen and H. Mannila. Approximate inference of functional dependencies from relations. *Theoretical Computer Science*, 149(1):129 – 149, 1995.

[9] J. R. Quinlan and R. Cameron-Jones. Oversearching and layered search in empirical learning. In *Proc. Fourteenth Int'l Joint Conf. on Artificial Intelligence (IJCAI)*, pages 1019 – 1024, 1995.

[10] S. Russel and P. Norvig. *Artificial Intelligence, a Modern Approach.* Prentice Hall, Upper Saddle River, N. J., 1995.

[11] C. Schaffer. Overfitting avoidance as bias. *Machine Learning*, 10(2):153 – 178, 1993.

[12] P. Smyth and R. M. Goodman. An information theoretic approach to rule induction from databases. *IEEE Trans. Knowledge and Data Engineering*, 4(4):301 – 316, 1992.

[13] E. Suzuki. Simultaneous reliability evaluation of generality and accuracy for rule discovery in databases. In *Proc. Fourth Int'l Conf. on Knowledge Discovery and Data Mining (KDD)*, pages 339 – 343, 1998.

[14] E. Suzuki. Scheduled discovery of exception rules. In *Discovery Scicence, LNAI 1721 (DS)*, pages 184–195. Springer, 1999.

[15] E. Suzuki. Worst-case analysis of rule discovery. In *Discovery Science, LNAI 2226 (DS)*, pages 365 – 377. Springer-Verlag, 2001. (Erratum: http:// www.slab.dnj.ynu.ac.jp/erratumds2001.pdf).

Active Mining
H. Motoda (Ed.)
IOS Press, 2002

Evaluating the Automatic Composition of Inductive Applications Using StatLog Repository of Data Set

Hidenao Abe and Takahira Yamaguchi
hidenao@ks.cs.inf.shizuoka.ac.jp,yamaguti@cs.inf.shizuoka.ac.jp
School of Informatics, Shizuoka University
3-5-1 Johoku, Hamamatsu, Shizuoka 432-8011, JAPAN

Abstract. Here is presented CAMLET that is a platform for automatic composition of inductive applications with method repositories that organize many inductive learning methods. After having implemented CAMLET on UNIX platforms with Perl and C languages,we have done the case studies of constructing inductive applications for eight different data sets from the StatLog repository and have compared the accuracies of the inductive applications composed by CAMLET with all the accuracies from popular inductive learning algorithms. The results have shown us that the inductive applications composed by CAMLET take the best accuracy on the average. Furthermore, we have analyzed how the specification for inductive applications changed for better performance and proposed several heuristics to refine them.

1 Introduction

In recent years, end-users of inductive applications are faced with a major problem: model selection, i.e., selecting the best model to a given data set. Conventionally, this problem is resolved by trial-and-error or heuristics such as selection-table for ML algorithms. This solution sometimes takes much time. So automatic and systematic guidance for constructing inductive applications is really required.

From the above background, it is the time to decompose inductive learning algorithms and organize inductive learning methods (ILMs) for reconstructing inductive learning systems. Given such ILMs, we may construct a new inductive application that works well to a given data set by re-interconnecting ILMs. The issue is to meta-learn an inductive application that works well on a given data set. This paper focuses on specifying ILMs into a method repository and how to compose inductive applications with ILMs. Thus we design a computer aided machine (inductive) learning environment called CAMLET and evaluate the competence of CAMLET using some data sets from the Statlog project[2], held by LIACC. Furthermore, we examine about efficient search for better meta-learning.

2 Repository for Inductive Learning

Considerable time and efforts have been devoted to analyzing the following popular inductive learning algorithms: Version Space [5], AQ15, ID3 [6], C4.5 [7], Classifier Systems [1], Back Propagation Neural Networks, Bagged C4.5 and Boosted C4.5 [8]. The analysis results first came up with just unstructured documents to articulate what inductive learning methods are in the above popular inductive learning algorithms. We did it under the condition of that the inputs and outputs of inductive learning methods are data sets or rule sets. When just a datum or rule is input or output of processes, they were too fine to be methods. Here in this paper, a method repository is an explicit specification of a conceptualization about ILMs and a data type hierarchy is the organization of objects manipulated by ILMs. In structuring many inductive learning methods into a method repository, we have identified the following generic methods: "generating training and validation data sets", "generating a classifier set", "evaluating data and classifier sets","modifying a training data set","modifying a classifier set" and "selecting and evaluating / evaluating classifier sets" that compose the top-level control structure of inductive applications, as shown in Fig. 1.

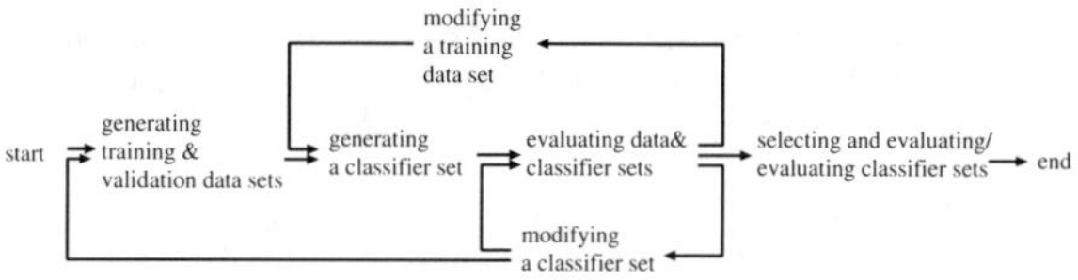

Figure 1: Top-level Control Structure of Inductive Applications

Thus these six generic methods have been placed on the upper part in the hierarchy structure of the method repository, as shown in Fig. 2.

2.1 Method Repository

In order to specify the hierarchy structure of a method repository, it is important how to branch down methods. Because the upper part is related with the generic methods included in the top-level control structure of inductive applications and the lower part with specific methods included in inductive applications, it is necessary to set up different ways to make up the hierarchy structure, depending on the hierarchy level.

In specifying the lower part down from the upper part of the hierarchy, the generic methods have been divided down by the characteristics with each method. Thus we can construct the hierarchy structure of the method repository, as shown in Fig. 2. In Fig. 2, leaf nodes get into the library of executable program codes that have been written in C language.

On the other hand, in order to specify the method scheme, we have identified the method scheme including the following roles: "input", "output" and "reference" from the point of objects manipulated by the methods, and then "pre-method" just before the defined method and "post-method" just after the defined method.

2.2 Data Type Hierarchy

In order to specify the hierarchy structure of data type, we use the way to branch down the data structures manipulated by the methods. The leaf nodes with the data type hierarchy get into the following method scheme roles: input, output and reference.

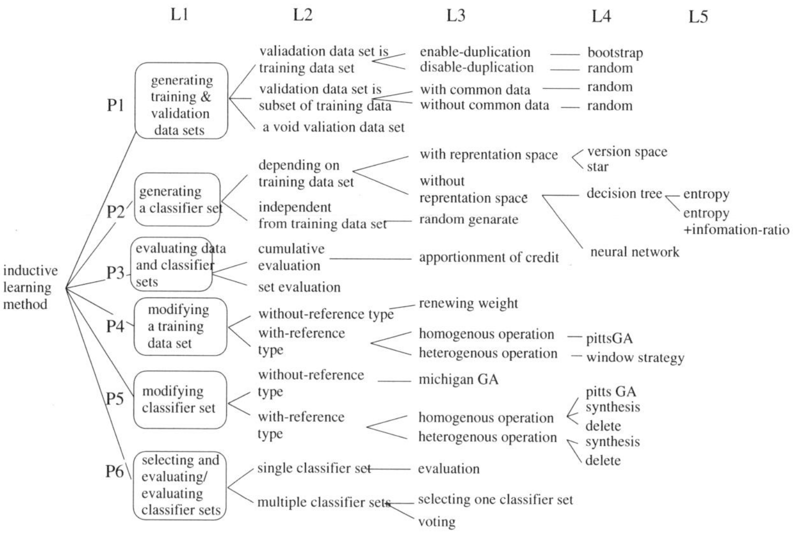

Figure 2: Hierarchy of Method Repository

3 Basic Design of CAMLET

Fig. 3 shows us the basic activities CAMLET, bases on constructing knowledge systems with problem solving methods (PSMs) [9]. In this section, we apply the activities to constructing inductive applications with inductive learning methods.

At the construction activity, CAMLET takes a top-level control structure randomly by selecting any path from "start" to "end" in Fig. 1 and constructs an initial specification to a given data set. At the instantiation activity, CAMLET gets down from the generic methods included in the initial specification into the leaf-level methods, and instantiates the initial specification, getting the data types from a given data set into input and output roles of the leaf-level methods. The values of other roles, such as reference, pre-method and post-method, have not been instantiated but come directly from the method schemes. Thus CAMLET can make up the instantiated specification. At the compilation activity, CAMLET transforms the instantiated specification into executable codes with the library for ILMs. When the method is connected to another method at the level of implementation details, the specification for I/O data types must be unified. To do so, the compilation activity has such a data conversion facility that converts a decision tree into a rule set. The test activity tests if the executable codes goes beyond the goal accuracy from a user. If not so, the refinement activity comes up, changing the initial specification into another one.

Furthermore, in order to deduce the time to execute many inductive applications, we have implemented a distributed CAMLET that takes one processing element to just compose specifications of inductive applications and other processing elements to execute them.

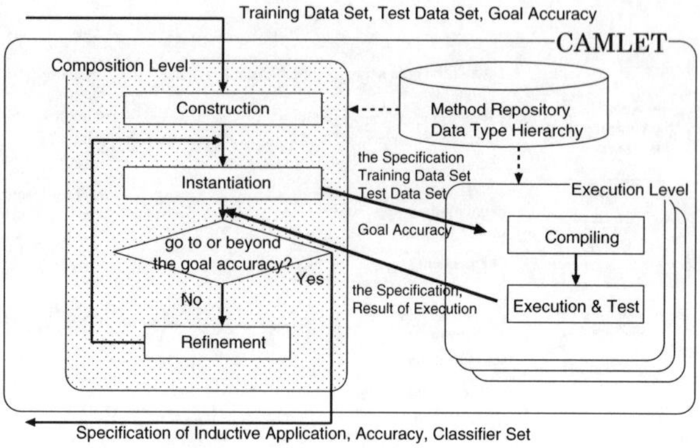

Figure 3: Basic Activities for Constructing Inductive Applications

4 Case Studies Using Statlog Data Sets

We have implemented CAMLET with Perl language, including twenty components in the method repository implemented with C language. We have done two different case studies in oder to evaluate CAMLET. One case study is to compare all the accuracies of inductive applications composed by CAMLET with those of popular inductive and statistical algorithms, using eight different data sets from the Statlog project[1] . The other is for looking for how to search the specification space efficiently, taking meta rules to change the specifications better (The meta rules are called 'spec refinement rules' later).

4.1 Evaluating the Automatic Compositions of Inductive Applications

The goal of meta-learning by CAMLET is to compose inductive applications that go beyond the accuracy of the best learning algorithms from the Statlog project. However, taking time into consideration, in the case of that the CAMLET does not get to the goal after having composed inductive applications one hundred times, CAMLET chooses the inductive application with the highest accuracy among all the composed inductive applications. Furthermore, in the case of getting two or more inductive applications with the highest accuracy, CAMLET chooses one with the least learning cost. As we can take just one personal computer with dual processors and fifty four engineering workstations as hardware resources in our laboratory, we have set up the following environment to do case studies: assigning the personal computer to compose the specifications of inductive applications and engineering workstations to execute all the composed inductive applications. Table 1 shows how to use data sets from the StatLog project while evaluating inductive and statistical algorithms. We take the same way as Table 1 while evaluating CAMLET.

Table 2 shows us all the results about accuracy comparison between the inductive applications composed by CAMLET and the best learning algorithms from the StatLog

[1]The StatLog project has taken twenty four popular learning algorithms and statistical systems with ten common data sets. However, we have not taken two data sets because of having cost matrix evaluation. Refer to the URL of http://borba.ncc.up.pt/niaad/statlog/

Table 1: Data Set Description in the StatLog Project

Data Set Name	Training Data Set	Test Data Set
Credit Research for Credit Cards in Australia	10-fold cross validation	
Diabetes of Pima-Indians	12-fold cross validation	
Splice-junction Recognition of DNA Sequence	assigned	assigned
Letter Recognition	assigned	assigned
LANDSAT Satellite Image Recognition	assigned	assigned
Image Segmentation	10-fold cross validation	
Shuttle Control	assigned	assigned
Vehicle Recognition Using Silhouettes	9-fold cross validation	

project. The inductive applications composed by CAMLET take the first best accuracy on the average, taking the first to tenth best accuracy to each data set. CAMLET goes beyond given goal accuracy to the following two data sets: Australian and Shuttle. To the other six data sets, CAMLET composes inductive application one hundred times and chooses the best one from them.

Table 2: Accuracy Comparison between Inductive Applications Composed by CAMLET and the best Learning Algorithm from the StatLog Project, and Search Times (Acc.:accucacy,Gen.:number of generation,Time:search time)

Data Set	CAMLET					StatLog		C4.5
	the best			search				
	Acc.(%)	Gen.	Time(sec)	Gen.	Time(sec)	Acc.(%)	Algorithm	Acc.(%)
Australian	87.3	77	17,700	77	17,700	86.9	Cal5	84.5
Diabetes	76.4	21	2,031	100	20,551	77.7	LogDisc	73.0
DNA	95.0	20	19,569	100	381,689	95.9	Radial	92.4
Letter	82.0	53	151,444	100	174,106	93.6	Alloc80	86.8
Satimage	88.3	13	214,208	100	423,704	90.6	KNN	85.0
Segment	96.1	41	71,250	100	108,608	97.0	Alloc80	96.0
Shuttle	99.8	2	3,893	2	3,893	99.0	NewId	90.0
Vehicle	79.2	23	4,034	100	41,451	85.0	QuaDisc	73.4
Average	88.0							85.1

Fig. 4 to 11 show us all the best specifications of the inductive applications composed by CAMLET. Looking at the control structure of them, they take the control structure to handle multiple classifier sets with one or more feedback loops, except the data set of Shuttle control. Furthermore, they take the method of "selecting the best classifier set"[2] as P6 generic method (in Fig. 2), except the data set of Image segmentation. It is an issue to make sure why "selecting the best classifier set" works better than "voting"[3] . However, in this case study, generating different types of training data sets with "bootstrap", it makes learned classifier sets smaller models and so avoids over-fitting problems. Thus they seems to work better to given test data set.

[2]selecting just one classifier set from generated classifier sets using their accuracies to each validation data set generated from given data set.

[3]the method from Boosted C4.5

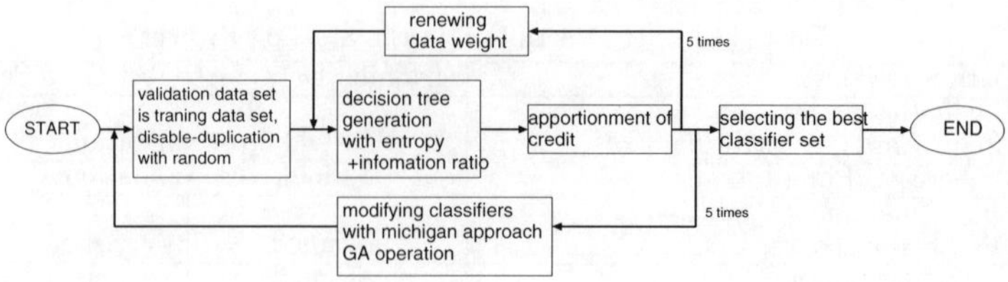

Figure 4: The inductive application composed by CAMLET with Credit Research for Credit Cards in Australia

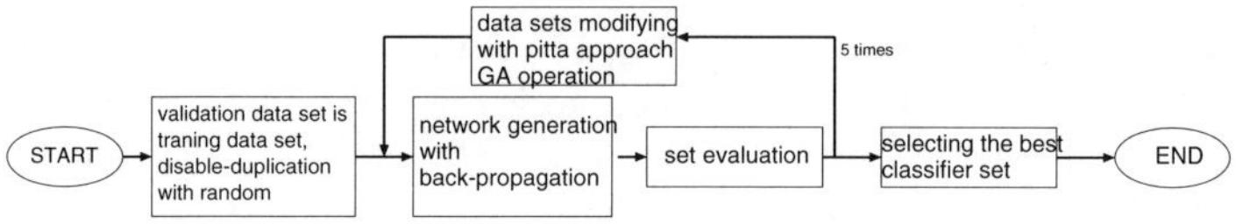

Figure 5: The inductive application composed by CAMLET with Diabetes of Pima-Indians

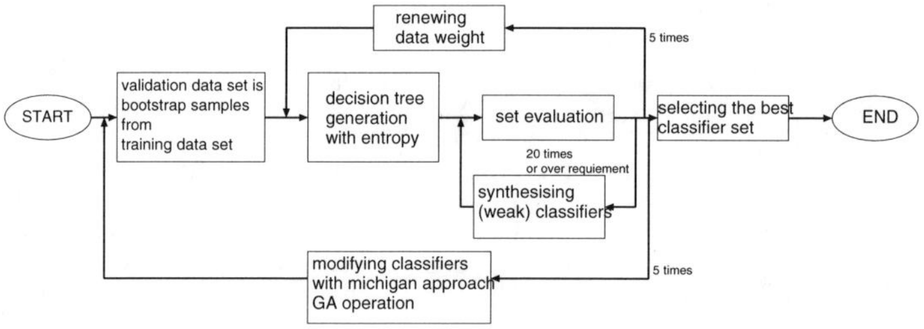

Figure 6: The inductive application composed by CAMLET with Splice-junction Recognition of DNA Sequence

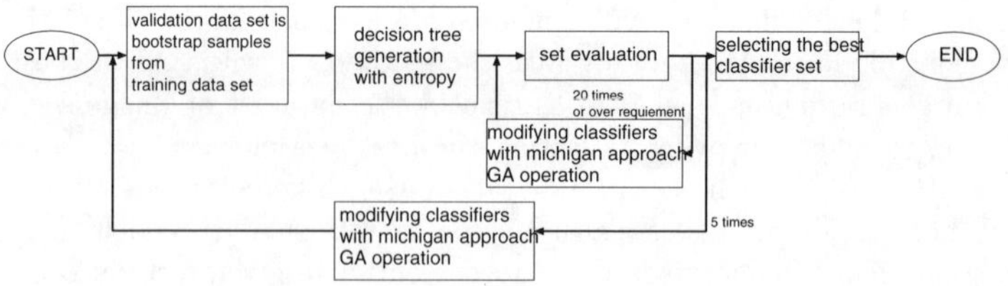

Figure 7: The inductive application composed by CAMLET with Letter Recognition

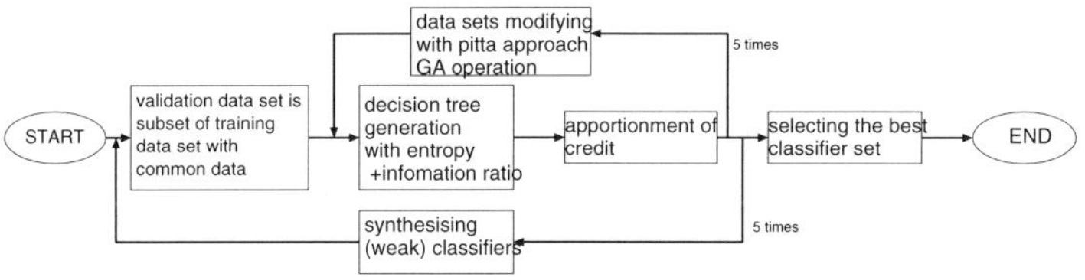

Figure 8: The inductive application composed by CAMLET with LANDSAT Satellite Image Segmentation

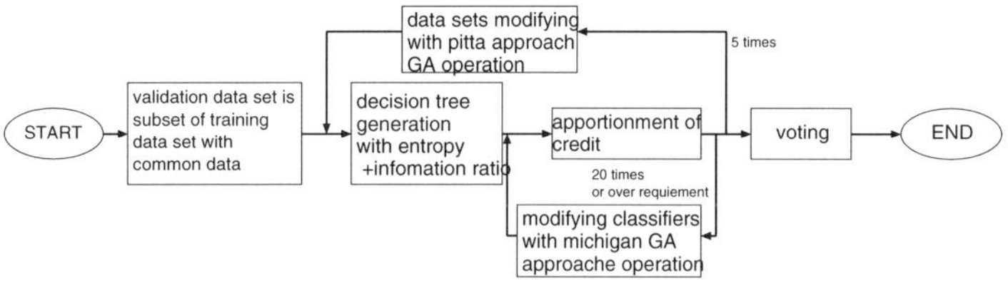

Figure 9: The inductive application composed by CAMLET with Image Segmentation

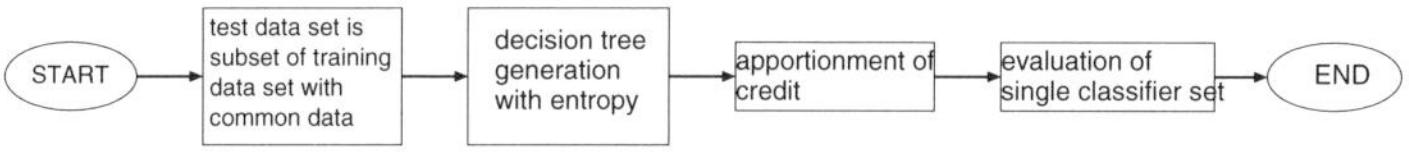

Figure 10: The inductive application composed by CAMLET with Shuttle Control

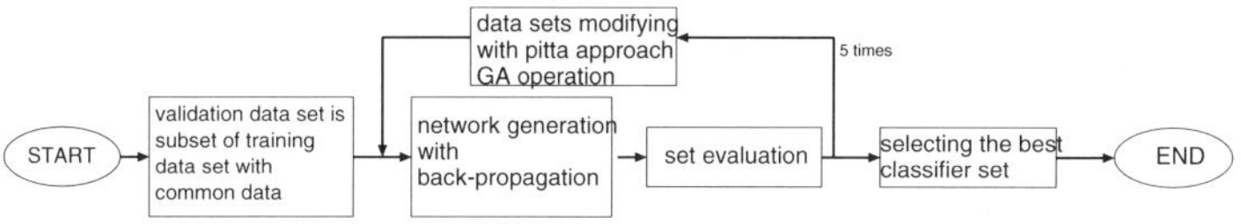

Figure 11: The inductive application composed by CAMLET with Vehicle Recognition Using Silhouettes

4.2 Specification Refinement Rules Learning for Meta-Learning

While the specifications of inductive applications have been changed at random in the above case study, some method needed to efficiently search for the best specification to given data set. This has led us to assemble data relating to generated the specification and the accuracy, and construct a data set pairing spec data prior to changing and spec data after changing. We have attempted to learn spec refinement rules from the data set.

Specifically, CAMLET has executed one thousand different generated inductive applications to the data set of "Credit Research for Credit Cards in Australia". We have set up training data sets pairing the pre-refinement and post-refinement information regarding the specifications and the accuracy, and made Apriori algorithm learn association rules. We have applied the association rules to the other data sets, and evaluated how they work. In the rest of this section we will examine this idea in details.

4.2.1 Generating Training Data Sets for Specification Refinement Rules Learning

To generate training data sets for spec refinement rules learning, we have derived two specs at random from among one thousand generated specs, and treat them as pre-refinement spec *Spec*1 and post-refinement spec *Spec*2. As shown in Fig. 12, each spec is represented as a list. The first element is the control structure type number, while the second through eighth elements are allocated to method numbers included in the control structure (If an element doesn't have a method, then zero is allocated). Next, we have generated a new list by combining corresponding elements from these two lists, and this becomes our training data for learning spec refinement rules (in terms of actual implementation, each element have been regarded as an attribute, the data record has been set up as attribute-value).

Spec1: 1,15,24,31,0,0,0,52 → 80.3%

Spec2: 4,13,26,31,0,0,49,54 → 84.6%

Instance : 1->4,15->13,24->26,31->31,0->0,0->0,0->49,52->54

Goal accuracy : 86.9%
→ classed "small rise" , into Data Set(Step I)

Figure 12: Detail of Training Data Set for Learning of Spec Refinement Rules

However, since our objective is to increase the accuracy, we have generated training data sets that limited to spec pairs in which *Spec*2 has a higher accuracy than *Spec*1. Thus although one thousand specifications with accuracies have been generated, the total number of instances for training data sets have been less than ${}_{1000}C_2$. We also assume that a stepwise procedure is required since it is generally rare to go from a low accuracy to high goal accuracy in a single step. As shown in Fig. 13, we have set the number of steps for applying the spec refinement rules based on the accuracy in *Spec*1, then have allocated the training data sets to the steps based on the accuracy of the pre-refinement specs. In addition, class values have been assigned based on the three levels of increase in percentage correct shown below. The class distribution and scale of training data sets generated by the above procedures are summarized in Table 3.

- Rise of less than half a step (small rise).

- Rise of more than half a step (moderate rise).
- Rise of more than the other two classes (large rise).

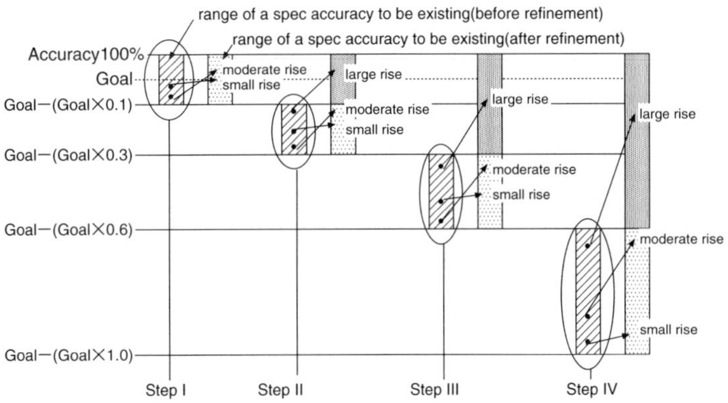

Figure 13: Step Allocation with Accuracy of *Spec*1, and Classes

Table 3: Size of Training Data Sets

	large rise	moderate rise	small rise	Total instances
Step I	—	23,537	70,737	94,274
Step II	1,530	94,212	145,109	240,852
Step III	80,963	21,294	11,035	113,293
Step IV	40,919	2,014	2,722	45,656

4.2.2 Specification Refinement Rules Learning

We have taken Apriori algorithm to learn the spec refinement rules. The minimum supports have been set as shown in Table 4 as a percent of the total number of instances for each step. They have been varied in this way to minimize the effects of the class distribution, and have been set to 1.0% for number of instances with the small rise per step class, and have been set to ratios of the number of instances for the other classes. Note that the small rise minimum support value has been an adjusted figure determined after running the trial a number of times. The minimum confidence has been set at 75% for all the data sets.

The number of rules learned under these conditions is shown in Table 5.

Fig. 14 shows us how a spec refinement rule works. The square area marked off by the dotted line is the part where a spec refinement rule is not included in the control structure that is referred to, while the asterisk (*) signifies parts that can be applied to any element and/or control structure.

Table 4: Minimum support Ratio of Data Sets of Each Step to Learn Each Class(Consequence)

	large rise(%)	moderate rise(%)	small rise(%)
Step I	—	0.33	1.00
Step II	0.65	0.65	1.00
Step III	7.34	1.93	1.00
Step IV	15.03	0.73	1.00

Table 5: Number of Learned Rules

	large rise	moderate rise	small rise
Step I	—	9	416
Step II	0	3	49
Step III	12	0	2
Step IV	8	0	0

4.2.3 Evaluation of the Specification Refinement Rules

We have attempted to apply the spec refinement rules learned by Apriori to the other data sets from the StatLog project, and have evaluated the results. We solve any conflicts among rules by adopting the rule with the highest accuracy. As with the case study described earlier, here the goal accuracies have been the highest accuracies stipulated for each data set provided by the StatLog project. Typical results are shown in Fig. 15 and Fig. 16 (the left figure shows us the change in accuracy for random changing, while the right figure shows us the change in accuracy using spec refinement rules).

We can see in Fig. 15 that using the spec refinement rules has done a better job of reducing the separation between the maximum accuracy and the minimum accuracy of generated specs than the random changing. Yet it is also clear that the spec refinement rules have not always increased the accuracy, and indeed we have not been able to find any specs that exceeded the goal accuracy even after updating the spec 50 times. However turning to Fig. 16, here we have achieved the goal accuracy (97.0%) after refining the spec thirty two times(97.1%). Yet we must note again that the spec refinement rules have done not always raise the accuracy; indeed, about half the spec changes have failed to raise the accuracy.

From the above results we can conclude that, at this point, a spec refinement rule-based mechanism for refining specifications have a not very large effect. To make this approach more effective, we are going to extend and improve the representation for the spec refinement rules. More specifically, we are pursuing an extended approach that would apply not just the current spec but also more background of changes leading up to the current specification when a specification refinement is executed.

5 Conclusions

Taking eight kinds of data sets from the StatLog project, we have succeeded in generating the inductive applications with higher average accuracy by CAMLET, compared with twenty four representative inductive and statistical algorithms surveyed by the StatLog project. We have also implemented a spec refinement rules based on associa-

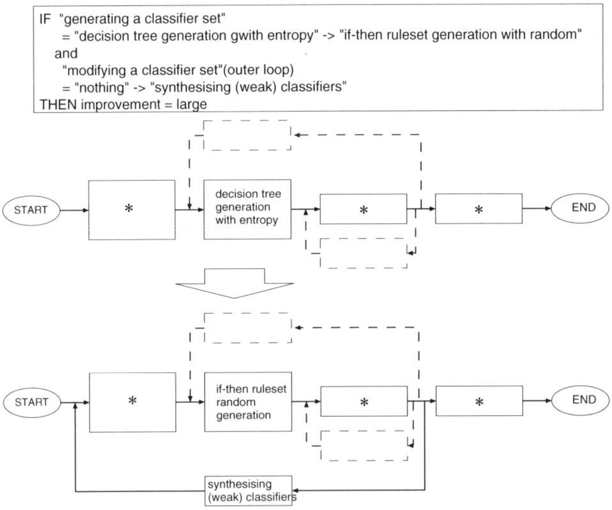

Figure 14: Applying a Specification Refinement Rule

tion rules as a way to achieve efficient specification refinements. While we have achieved more stable spec changes than could be achieved by random searching with this approach, we need to extend the representation for spec refinement rules in order to get to better refinement. Although the parameters with control structure and ILMs have a major effect on performance, CAMLET doesn't search for these parameters. So we will put how to search for the parameters in the framework of a distributed CAMLET.

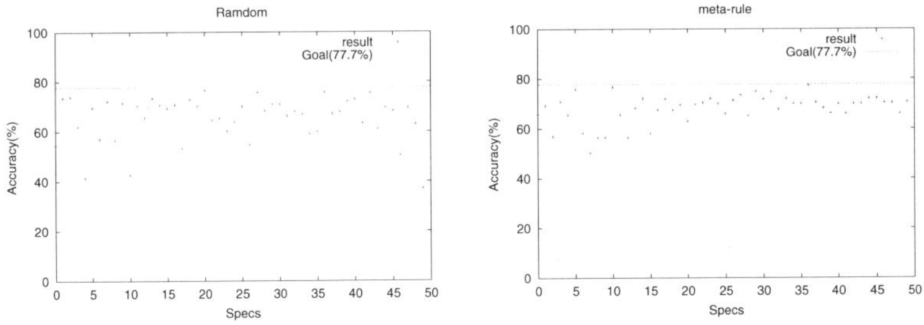

Figure 15: Comparison between Random Search and Using Spec Refinement Rules Search to Diabetes of Pima-Indians

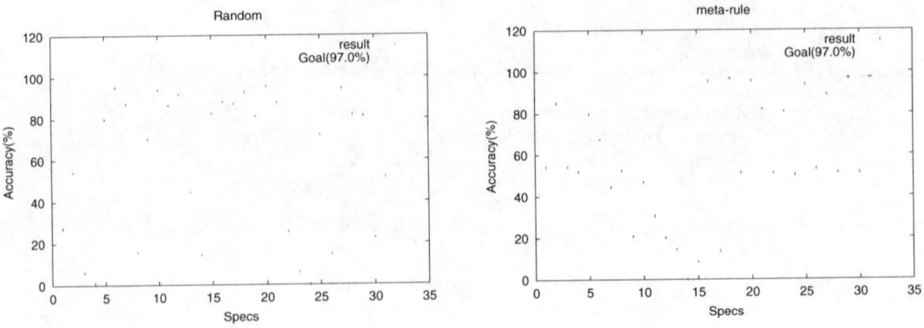

Figure 16: Comparison between Random Search and Using Spec Refinement Rules Search to Image Segmentation

References

[1] Booker, L. B., Holland, J. H. and Goldberg, D. E., "Classifier Systems and Genetic Algorithms", *Artificial Intelligence*, 40, pp.235–282 (1989).

[2] Brazdil,P. and Henery,R.: "Chapter 10, Analysis of Results", in Machine Learning, Neural and Statistical Classification , D. Michie, D.J.Spiegelhalter and C.C.Taylor (eds.), Ellis Horwood, pp.175–212, (1994).

[3] Hinton, G. E., "Learning distributed representations of concepts", *Proceedings of 8th Annual Conference of the Cognitive Science Society*, Amherest, MA. REprinted in R.G.M.Morris (ed.) (1986).

[4] Michalski, R., Mozetic, I., Hong, J. and Lavrac, N., "The AQ15 Inductive Learning System: An Over View and Experiments", *Reports of Machine Learning and Inference Laboratory*, No.MLI-86-6, George Mason University (1986).

[5] "Generalization as Search", *Artificial Intelligence*, 18(2), pp.203-226 (1982).

[6] Quinlan, J. R., "Induction for Decision Tree", *Machine Learning*, Vol.1, Morgan Kaufmann, pp.81–106 (1986).

[7] Quinlan, J. R., *Programs for Machine Learning*, Morgan Kaufmann (1992).

[8] Quinlan, J. R., "Bagging, Boosting and C4.5", *Proceedings of American Association for Artificial Intelligence* (1996).

[9] van Heijst,G.: The Role of Ontologies in Knowledge Engineering. phD thesis. University of Amsterdam (1995)

Active Mining
H. Motoda (Ed.)
IOS Press, 2002

Fast Boosting Based on Iterative Data Squashing

Yuta Choki and Einoshin Suzuki
{choki, suzuki}@slab.dnj.ynu.ac.jp
Electrical and Computer Engineering
Yokohama National University
79-5 Tokiwadai, Hodogaya, Yokohama 240-8501, Japan

Abstract. This paper proposes a fast boosting method which employs iterative data squashing. Boosting represents a learning method which constructs a highly accurate classification model by combining multiple classification models. Boosting requires long computational time since it constructs multiple classification models. Data squashing, which speeds-up a learning method by abstracting the training data set to a smaller data set, typically lowers accuracy. Our SB (Squashing-Boosting) loop, based on a series of data squashing and boosting, iteratively refines an SF (Squashed-Feature) tree, which provides an appropriately squashed data set. Experimental evaluation with artificial data sets and the KDD Cup 1999 data set clearly shows superiority of our method compared with conventional methods. We have also empirically evaluated our distance measure, and found it superior to alternatives.

1 Introduction

Boosting represents a learning method which constructs a highly accurate classification model by combining multiple classification models, each of which is called a weak learner [5]. Since boosting constructs multiple classification models, it requires long computational time. It is possible to reduce computational time by using data squashing [4] to decrease the number of examples in the data set. Since data squashing typically lowers accuracy, we propose to apply data squashing iteratively based on example weights of boosting. Moreover, we consider distribution of examples in the process by using our projected SVD distance as the distance measure for data squashing. Effects of the iterative data squashing and the distance measure are empirically evaluated through experiments with artificial and real-world data sets.

This paper is structured as follows. In section 2, we review boosting especially AdaBoost.M2 [5], which we employ throughout this paper. Section 3 explains previous research for fast learning based on data squashing. In section 4, we propose our SB (Squashing-Boosting) loop, and evaluate it through experiments in section 5. Section 6 describes concluding remarks.

2 Boosting

Boosting constructs a "strong learner" which demonstrates high accuracy by combining a sequence of "weak learners" each of which has accuracy slightly higher than random learning. AdaBoost.M2 deals with a classification problem with no less than 3 classes, and constructs each weak learner by transforming the original classification problem to a binary classification problem in terms of an original class. AdaBoost.M2 assumes an example weight for each example and a model weight for each weak learner. An example weight represents the degree of importance for the example in constructing a weak learner, and is initialized uniformly before learning the first weak learner. An example weight is increased when the obtained weak learner misclassifies the example, and vice versa. A model weight represents the degree of correctness of the corresponding weak learner, and a weak learner with a high model weight is regarded as important in the final classification model. AdaBoost.M2 iterates construction of a weak learner and update of example weights T rounds, and thus constructs T weak learners. We describe its brief outline below.

A training data set $(\mathbf{x_1}, y_1), (\mathbf{x_2}, y_2), \cdots, (\mathbf{x}_m, y_m)$ consists of m examples, where the domain of a class y_i is described as $\{1, 2, \cdots, c\}$, and $\mathbf{x}_i$ is a vector in an n-dimensional space. An example weight of $(\mathbf{x_i}, y_i)$ is represented as $D_t(i, y)$, where t is the number of rounds and $t = 1, 2, \cdots, T$. An initial value for an example weight $D_1(i, y)$ is given by

$$D_1(i, y) = \frac{1}{mc}. \tag{1}$$

An example weight is updated based on a weak learner $h_t(\mathbf{x}, y)$ which is obtained by a weak learning algorithm. A weak learner $h_t(\mathbf{x}, y)$ outputs 1 or -1 as a predicted class. In this paper, we employ a decision stump which represents a decision tree of depth one as a weak learner. In boosting, a pseudo-loss ϵ_t of a weak learner h_t is obtained for all examples $i = 1, 2, \cdots, m$ and all classes $y = 1, 2, \cdots, c$.

$$\epsilon_t = \frac{1}{2} \sum_{i=1}^{m} \sum_{y=1}^{c} (1 - h_t(\mathbf{x}_i, y_i) + h_t(\mathbf{x}_i, y)) \tag{2}$$

From this, β_t is obtained as follows.

$$\beta_t = \frac{\epsilon_t}{1 - \epsilon_t} \tag{3}$$

The example weight is updated to $D_{t+1}(i, y)$ based on β_t, where Z_t represents the add-sum of all example weights and is employed to normalize example weights.

$$D_{t+1}(i, y) = \frac{D_t(i, y)}{Z_t} \beta_t^{\frac{1}{2}(1 + h_t(\mathbf{x}_i, y_i) - h_t(\mathbf{x}_i, y))} \tag{4}$$

$$\text{where } Z_t = \sum_{i=1}^{m} D_t(i, y) \beta_t^{\frac{1}{2}(1 + h_t(\mathbf{x}_i, y_i) - h_t(\mathbf{x}_i, y))} \tag{5}$$

AdaBoost.M2 iterates this procedure T times to construct T weak learners. The final classification model, which is given by (6), predicts the class of each example by a weighted vote of T weak learners, where a weight of a weak learner h_t is given by $\log(1/\beta_t)$.

$$h_{\text{fin}}(\mathbf{x}) = \arg\max_y \sum_{t=1}^{T} \left(\log \frac{1}{\beta_t} \right) h_t(\mathbf{x}, y) \tag{6}$$

Experimental results show that AdaBoost.M2 exhibits high accuracy. However, it is relatively time-consuming since its time complexity is given by $O(2^c Tmn)$ even if it employs a decision stump as a weak learner. Moreover, when the data set contains many outliers, boosting is known to produce a classification model which is highly dependent on the outliers. As the result, the accuracy of the classification model is typically low due to overfitting [3].

3 Fast Learning Based on Data Squashing

3.1 BIRCH

The main stream of conventional data mining research has concerned how to scale up a learning/discovery algorithm to cope with a huge amount of data. Contrary to this approach, data squashing [4] concerns how to scale down such data so that they can be dealt by a conventional algorithm. Here we present a fast clustering [7] algorithm BIRCH [13], which is based on data squashing.

Data reduction methods can be classified into feature selection [8] and instance selection [9]. In machine learning, feature selection has gained greater attention since it is more effective in improving time-efficiency. We, however, have adopted instance selection since it can deal with massive data which do not fit in memory, and crucial information for classification is more likely to be lost with feature selection than instance selection.

BIRCH takes a training data set $\mathbf{x_1}, \mathbf{x_2}, \cdots, \mathbf{x}_m$ as input, and outputs its partition $\gamma_1, \gamma_2, \cdots, \gamma_{n+1}$, where each of $\gamma_1, \gamma_2, \cdots, \gamma_n$ represents a cluster, and γ_{n+1} is a set of noise. A training data set is assumed to be so huge that it is stored on a hard disk, and cannot be dealt by a global clustering algorithm since it does not fit in memory. Data squashing, which transforms a given data set to a much smaller data set by abstraction, can be considered to speed up learning in this situation. BIRCH squashes the training data set stored on a hard disk to obtain a CF (clustering feature) tree, and applies a global clustering algorithm to squashed examples each of which is represented by a leaf of the tree.

A CF tree represents a height-balanced tree which is similar to a B+ tree [2]. A node of a CF tree represents a CF vector, which corresponds to an abstracted expression of a set of examples. For a set of examples $\mathbf{x_1}, \mathbf{x_2}, \cdots, \mathbf{x}_{N_\phi}$ to be squashed, a CF vector $\mathbf{CF}_\phi$ consists of the number N_ϕ of examples, the add-sum vector $\mathbf{LS}_\phi$ of examples, and the squared-sum SS_ϕ of attribute values of examples.

$$\mathbf{CF}_\phi = (N_\phi, \mathbf{LS}_\phi, SS_\phi) \tag{7}$$

$$\mathbf{LS}_\phi = \sum_{i=1}^{N_\phi} \mathbf{x}_i \tag{8}$$

$$SS_\phi = \sum_{i=1}^{N_\phi} \|\mathbf{x}_i\|^2 \tag{9}$$

Since the CF vector satisfies additivity and can be thus updated incrementally, BIRCH requires only one scan of the training data set. Moreover, various inter-cluster distance measures can be calculated with the corresponding two CF vectors only. This signifies that the original data set need not be stored, and clustering can be performed with their CF vectors only.

A CF tree is constructed with a similar procedure for a B+ tree. When a new example is read, it follows a path from the root node to a leaf, then nodes along this path are updated. Selection of an appropriate node in this procedure is based on a distance measure which is specified by a user. The example is assigned to its closest leaf if the distance between the new example and the examples of the leaf is below a given threshold L. Otherwise the new example becomes a novel leaf. Note that a large CF tree is obtained with a small L, and vice versa. For more details, please refer to [13].

3.2 Application of Data Squashing to Classification and Regression

Data squashing has been applied to various learning problems. We briefly review the existing approaches in this section.

DuMouchel proposed to add moments of higher orders to the CF vector, and applied his data squashing method to regression [4]. Pavlov applied data squashing to support vector machine: a classifier which maximizes margins of training examples under a similar philosophy to boosting [11]. Nakayasu substituted a product-sum matrix for the CF vector, and applied their method to Bayesian classification [10]. They proposed a tree structure similar to the CF tree, and defined the squared add-sum of eigenvalues of the covariance matrix for each squashed example as information loss.

4 Proposed Method

4.1 SB loop

Data squashing, which we explained in the last section, typically represents a single squashing of a training data set based on a distance measure. Several pieces of work including that of Nakayasu consider how examples are distributed, and can be considered to squash a data set more appropriately than an approach based on a simple distance measure. However, we believe that a single squashing can consider distribution of examples only insufficiently.

In order to cope with this problem, we propose to squash the training data set iteratively. Since a boosting procedure outputs a set of example weights each of which represents difficulty of prediction of the corresponding example, we considered to use them in data squashing. By using these example weights, we can expect that examples which are difficult to be predicted would be squashed moderately, and examples which are easy to be predicted would be squashed excessively. Alternatively, our approach can be viewed as a speed-up of AdaBoost.M2 presented in section 2 with small degradation of accuracy.

Note that a simple application of a CF tree, which was originally proposed for clustering, would squash examples belonging to different classes to an identical squashed example. We believe that such examples should be processed separately, and thus propose an SF (Squashed-Feature) tree which is similar to a CF tree but separates examples belonging to different classes in its root node. Figure 1 shows an example of an SF tree for a 3-class classification problem.

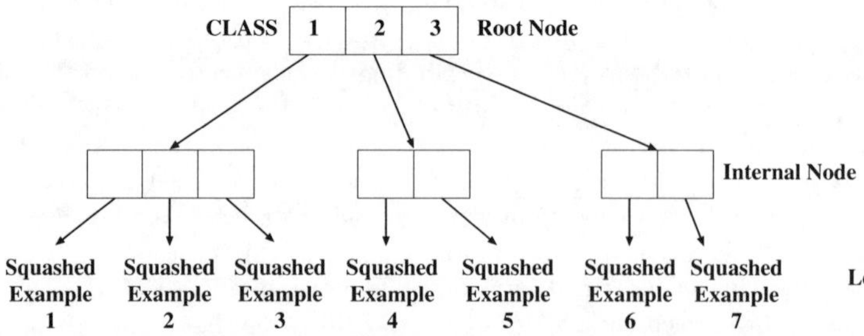

Figure 1: An example of an Squashed-Feature tree

Our approach iteratively squashes the training data set based on the set of example weights which are obtained from a boosting procedure. As we explained in section 3.1, BIRCH employs a threshold L to judge whether an example belongs to a leaf, i.e. a squashed example. In each iteration, a squashed example with a large example weight is typically divided since we employ a smaller threshold for the corresponding leaf node. On the other hand, a squashed example with a small example weight is typically merged with another squashed example since we employ a larger threshold for the corresponding leaf node. Since this squashing and boosting procedure is iterated so that the training data set is squashed appropriately, we call our approach an SB (Squashing-Boosting) loop, which is sketched as follows.

1. Initial data squashing
   Given m examples $(\mathbf{x_1}, y_1), (\mathbf{x_2}, y_2), \cdots, (\mathbf{x}_m, y_m)$, obtain p squashed examples $(\mathbf{x}_{\text{sub }1}, y_{\text{sub }1}), (\mathbf{x}_{\text{sub }2}, y_{\text{sub }2}), \cdots, (\mathbf{x}_{\text{sub }p}, y_{\text{sub }p})$ by constructing an SF tree. In this phase, the threshold L for judging whether an example belongs to a leaf is uniformly settled to L_0.

2. Application of boosting
   Apply AdaBoost.M2 to $(\mathbf{x}_{\text{sub }1}, y_{\text{sub }1}), (\mathbf{x}_{\text{sub }2}, y_{\text{sub }2}), \cdots, (\mathbf{x}_{\text{sub }p}, y_{\text{sub }p})$, and obtain example weights $D_T(1, y_{\text{sub }1}), D_T(2, y_{\text{sub }2}), \cdots, D_T(p, y_{\text{sub }p})$ and a classification model.

3. Update of thresholds
   For a leaf which represents a set of examples $(\mathbf{x}_{\text{sub }i}, y_{\text{sub }i})$, update its threshold $L(t, \mathbf{x}_{\text{sub }i})$ to $L(t+1, \mathbf{x}_{\text{sub }i})$.

$$L(t+1, \mathbf{x}_{\text{sub i}}) = L(t, \mathbf{x}_{\text{sub i}}) \frac{D_1(i, y)}{D_T(i, y)} \log a(t, i) \tag{10}$$

   where $D_1(i, y)$ is given by (1), and $a(t, i)$ represents the number of examples which are squashed into the leaf i.

4. Data Squashing
   Construct a novel SF tree from the training examples. In the construction, if a leaf has a corresponding leaf in the previous SF tree, use $L(t+1, \mathbf{x}_{\text{sub i}})$ as its threshold. Otherwise, use L_0 as its threshold.

Our SB loop iterates phase 2 to 4 Θ times by incrementing the number t of iterations. We show a summary of our SB loop in figure 2.

It should be noted that we do not employ margins instead of example weights since the theory of margins for AdaBoost does not hold for AdaBoost.M2. We have compared the use of margins as well as the use example weights for data squashing in outlier detection, and found that several experiments exhibit similar results [6].

4.2 Projected SVD Distance

BIRCH employs a distance measure such as average cluster distance and Euclidean distance in constructing a CF tree [13]. These distance measures typically fail to represent distribution of examples since they neglect interactions among attributes.

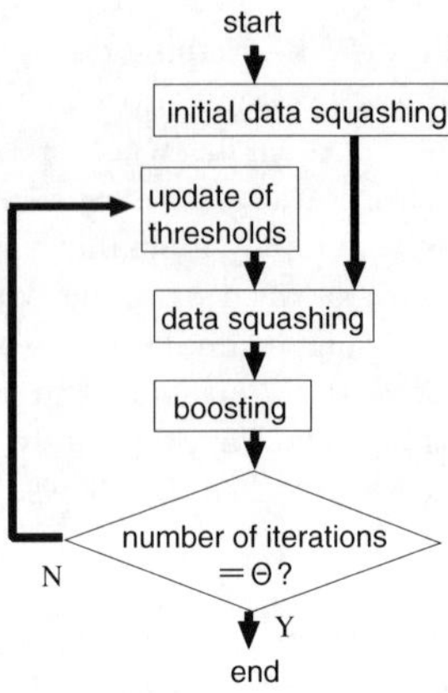

Figure 2: SB (Squashing-Boosting) loop

In order to circumvent this problem, we propose to store the number of examples N_ϕ, an average vector $\boldsymbol{\mu}_\phi$, and a quasi-product-sum matrix W_ϕ in a node ϕ of our SF tree, where $\boldsymbol{\mu}_\phi = \sum_{i=1}^{N_\phi} \mathbf{x}_i$. A quasi-product-sum matrix, which is given by (11), is updated when a novel example is squashed into its corresponding leaf. The update is done by adding the product-sum matrix of the novel example to the quasi-product-sum matrix. A quasi-product-sum matrix of an internal node is given by the add-sum of the quasi-product-sum matrices of its children nodes.

$$W_\phi = \begin{pmatrix} g_{11\phi} & \cdots & g_{1j\phi} & \cdots & g_{1m\phi} \\ \vdots & \ddots & & & \vdots \\ g_{i1\phi} & & g_{ij\phi} & & g_{im\phi} \\ \vdots & & & \ddots & \vdots \\ g_{m1\phi} & \cdots & g_{mj\phi} & \cdots & g_{mm\phi} \end{pmatrix} \tag{11}$$

$$g_{ij\phi} = \begin{cases} \sum_k g_{ijk} \\ \text{(for an internal node, where } k \text{ represents an identifier of its child nodes)} \\ \\ x_{fi}x_{fj} + g'_{ij\phi} \\ \text{(for a novel example, where } x_{fi} \text{ represents an attribute value of an} \\ \text{attribute } i \text{ for an inputted example } f\text{, and } g'_{ij\phi} \text{ represents the original} \\ \text{value of a squashed example } \phi) \end{cases} \tag{12}$$

For instance, we have an example k which is obtained by squashing two examples (1,2) and (5,8). Its quasi-product-sum matrix W_k is given by

$$W_k = \begin{pmatrix} 1^2 + 5^2 & 1 \cdot 2 + 5 \cdot 8 \\ 1 \cdot 2 + 5 \cdot 8 & 2^2 + 8^2 \end{pmatrix} = \begin{pmatrix} 26 & 42 \\ 42 & 68 \end{pmatrix}. \tag{13}$$

Suppose a novel example (6,2) is judged to belong to this example, then the matrix is updated to W'_k, which is given as follows.

$$W'_k = \begin{pmatrix} 6^2 & 6 \cdot 2 \\ 6 \cdot 2 & 2^2 \end{pmatrix} + \begin{pmatrix} 26 & 42 \\ 42 & 68 \end{pmatrix} = \begin{pmatrix} 62 & 54 \\ 54 & 72 \end{pmatrix} \tag{14}$$

Suppose the parent node of the current node has another child node k', and $W_{k'}$ is given by

$$W_{k'} = \begin{pmatrix} 2 & 6 \\ 6 & 26 \end{pmatrix}, \tag{15}$$

then the quasi-product-sum matrix $W_{k''}$ of the parent node k'' is updated as follows.

$$W_{k''} = \begin{pmatrix} 2 & 6 \\ 6 & 26 \end{pmatrix} + \begin{pmatrix} 62 & 54 \\ 54 & 72 \end{pmatrix} = \begin{pmatrix} 64 & 60 \\ 60 & 98 \end{pmatrix} \tag{16}$$

Our projected SVD distance $\Delta(\mathbf{x}_i, k)$ between an example $\mathbf{x}_i$ and a squashed example k is defined as follows.

$$\Delta(\mathbf{x}_i, k) = (\mathbf{x}_i - \boldsymbol{\mu}_k)^t S_k^{-1} (\mathbf{x}_i - \boldsymbol{\mu}_k) \tag{17}$$

where S_k represents the quasi-covariance matrix obtained from the quasi-product-sum matrix W_k.

$$S_k = \begin{pmatrix} Cov(11k) & \cdots & Cov(1jk) & \cdots & Cov(1mk) \\ \vdots & \ddots & & & \vdots \\ Cov(i1k) & & Cov(ijk) & & Cov(imk) \\ \vdots & & & \ddots & \vdots \\ Cov(m1k) & \cdots & Cov(mjk) & \cdots & Cov(mmk) \end{pmatrix} \tag{18}$$

$$\text{where } Cov(ijk) = \frac{g_{ijk}}{N_k} - E(ik)E(jk) \tag{19}$$

$$\text{and } E(ik) \text{ is the } i\text{th element of } \boldsymbol{\mu}_k \tag{20}$$

Our projected SVD distance requires the inverse matrix of S, and we use singular value decomposition [12] for this problem. In the method, S is represented as a product of three matrices as follows.

$$S = U \cdot \begin{pmatrix} z_1 & & & 0 \\ & z_2 & & \\ & & \cdots & \\ 0 & & & z_n \end{pmatrix} \cdot V \tag{21}$$

where U and V are orthogonal matrices. Consider two vectors $\mathbf{x}$, $\mathbf{b}$ which satisfy $S \cdot \mathbf{x} = \mathbf{b}$. If S is singular, there exists a vector $\mathbf{x}'$ which satisfies $S \cdot \mathbf{x}' = 0$. In general, there are an infinite number of $\mathbf{x}$ which satisfies $S \cdot \mathbf{x} = \mathbf{b}$, and we choose the one with minimum $||x^2||$ as a representative. For this we use

$$\mathbf{x} = V \cdot [\text{diag}(1/z_j)] U^T \cdot \mathbf{b}, \tag{22}$$

where we settle $1/z_j = 0$ if $z_j = 0$, and $\text{diag}(1/z_j)$ represents a diagonal matrix of which jth element is $1/z_j$. This is equivalent to obtaining $\mathbf{x}$ which minimizes $||S \cdot \mathbf{x} - \mathbf{b}||$, and corresponds to obtaining an approximate solution for $S \cdot \mathbf{x} = 0$ [12].

5 Experimental Evaluation

5.1 Experimental Condition

We employ artificial data sets as well as real-world data sets in the experiments. Each of our artificial data sets contains, as classes, four normal distributions with equal variances and the covariances are 0. We show means and variances of the classes in table 1. We varied the number of attributes 3, 5, 10. Each class contains 5000 examples.

Table 1: Means and variances of classes in the artificial data sets, where μ_i represents the mean of an attribute i for each class

class	μ_1	μ_2	μ_3	μ_4	μ_5	μ_6	μ_7	μ_8	μ_9	μ_{10}	variance
1	-6	2	-9	3	10	5	-4	-10	2	9	7
2	7	-2	0	10	3	-4	4	-7	3	7	9
3	-2	-3	-9	-6	3	8	8	1	-2	-3	5
4	-5	-5	8	5	1	-7	6	6	7	-6	8

We employed the KDD Cup 1999 data set in the UCI KDD Archive [1] as the source of the real-world data sets. Since it is difficult to introduce a distance measure of data squashing for a nominal attribute, we deleted such attributes before the experiments. As the result, each data set contains 12 attributes instead of 43. We selected the normal-access class and the two most frequent fraudulent-access classes, and defined a 3-class classification problem. We have generated ten data sets by choosing 10000, 20000, $\cdots$ 90000, and 97278 examples from each class.

We measured classification accuracy and computational time using 5-fold cross-validation. For artificial data sets, we have chosen boosting without data squashing and boosting with a single data squashing in order to investigate on effectiveness of our approach. We also evaluated our projected SVD distance by comparing it with average cluster distance and Euclidean distance. The threshold L was settled so that the number of squashed examples becomes approximately 3% of the number of examples, the number of iterations in boosting was settled to $T = 100$, and the number of iterations of data squashing in our approach was $\Theta = 3$. For real-world data sets, we compared our projected SVD distance with average cluster distance. We omitted Euclidean distance due to its poor performance in the artificial data sets.

5.2 Experimental Results and Analysis

5.2.1 Artificial Data Sets

We show the results with our projected SVD distance in figure 3. From the figure, we see that our SB loop, compared with boosting with single data squashing, exhibits higher accuracy (approximately 8 %) though its computational time is 5 to 7 times longer for almost all data sets. These results show that a single data squashing fails to squash data appropriately, while our SB loop succeeds in doing so by iteratively refining the squashed data sets. Moreover, degradation of accuracy for our approach, compared with boosting without data squashing is within 3 % except for the case of 10 attributes, and our method is 5 to 6 times faster. These results show that our data squashing is effective in speeding-up boosting with a little sacrifice in accuracy.

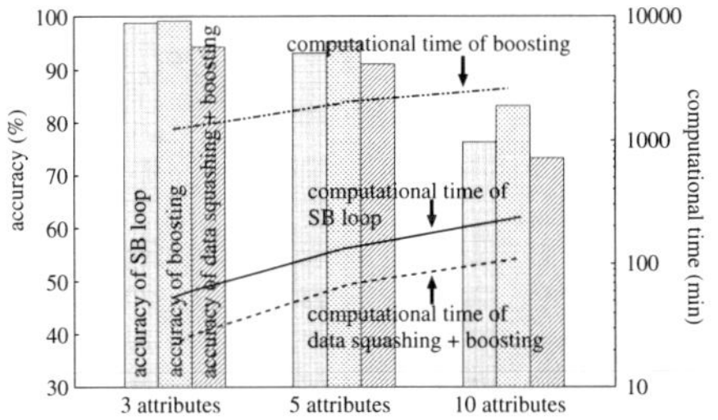

Figure 3: Effect of SB loop with projected SVD distance for the artificial data sets

We also show the results with average cluster distance and Euclidean distance in figure 4. The figure shows that our approach is subject to large degradation of accuracy compared with boosting, especially when Euclidean distance is employed. These results justify our projected SVD distance, which reflects distribution of examples to distance.

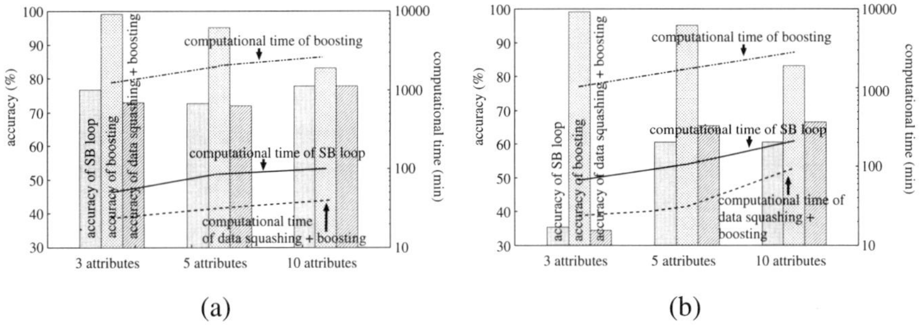

Figure 4: Effect of SB loop for the artificial data sets with average inter-cluster distance (a) and Euclidean distance (b)

5.2.2 Real-World Data Sets

We show experimental results with our projected SVD distance and average cluster distance in figure 5. In both cases, compared with boosting with single data squashing, SB loop exhibits approximately 8 % of improvement in accuracy in average though its computational time is approximately 4 to 6 times longer. Compared with boosting without data squashing, when our projected SVD distance is employed, our approach shortens computational time at most to 1/35 with a small degradation in accuracy. Moreover, the accuracy of our SB loop is no smaller than 92 % when our projected SVD distance is employed.

The good performance of our approach can be explained from characteristics of the data set. In the data set, two attributes have large variances which are more than 10000 times greater than the variances of the other attributes. Therefore, data squashing is practically performed in terms of these attributes, and is relatively easier than the cases with the artificial data sets. Moreover, these attributes are crucial in classification since our approach sometimes improves accuracy of boosting without data squashing.

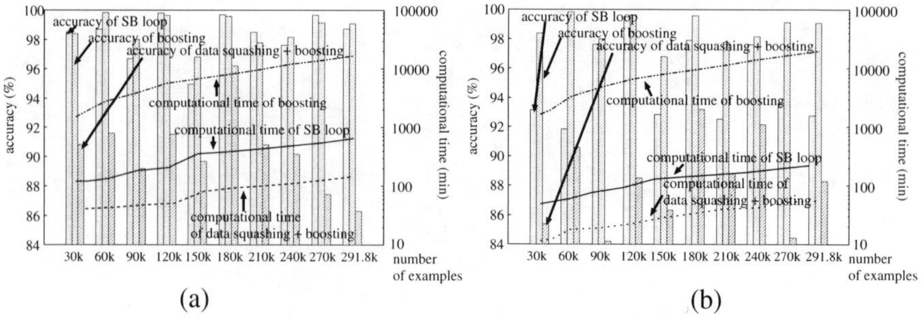

Figure 5: Results of the KDD Cup 1999 data with projected SVD distance (a) and average cluster distance (b)

6 Conclusion

The main stream of conventional data mining research has concerned how to scale up a learning/discovery algorithm to cope with a huge amount of data. Contrary to this approach, data squashing [4] concerns how to scale down such data so that they can be dealt by a conventional algorithm.

Our objective in this paper represents a speed-up of boosting based on data squashing. In realizing the objective, we have proposed a novel method which iteratively squashes a given data set using example weights obtained in boosting. Moreover, we have proposed, in data squashing, the projected SVD distance measure, which tries to reflect distribution of examples to distance.

We experimentally compared our approach with boosting without data squashing and boosting with a single data squashing using both artificial and real-world data sets. Results show that our approach speeds-up boosting 5 to 6 times while its degradation of accuracy was typically less than approximately 3 % for artificial data sets. Compared with boosting with a single data squashing, our approach requires 5 to 7 times of computational time, but improves accuracy approximately 8 % in average. For the real-world data sets from the KDD Cup 1999 data set, our projected SVD distance exhibits approximately 8 % of higher accuracy in average compared with average cluster distance, while the required computational time is almost the same.

Acknowledgement

This work was partially supported by the grant-in-aid for scientific research on priority area "Active Mining" from the Japanese Ministry of Education, Culture, Sports, Science and Technology.

References

[1] S. Bay. *UCI KDD Archive.* Dept. of Information and Computer Sci., Univ. of California Irvine, 1999. http://kdd.ics.uci.edu/.

[2] D. Comer. The Ubiquitous B-Tree. *ACM Computing Surveys*, 11(2):121 – 137, 1979.

[3] T. G. Dietterich. An experimental comparison of three methods for constructing ensembles of decision trees: Bagging boosting and randomization. *Machine Learning*, 40(2):139 – 157, 2000.

[4] W. DuMouchel, C. Volinsky, T. Johnson, C. Cortes, and D. Pregibon. Squashing flat files flatter. In *Proceedings of the Fifth ACM International Conference on Knowledge Discovery and Data Mining*, pages 6 – 15, 1999.

[5] Y. Freund and R. E. Schapire. Experiments with a new boosting algorithm. In *Proceedings of the Thirteenth International Conference on Machine Learning*, pages 148 – 156, 1996.

[6] S. Inatani and E. Suzuki. Data squashing for speeding up boosting-based outlier detection. In *Proceedings of the Thirteenth International Symposium on Methodologies for Intelligent Systems*, 2002. (accepted for publication).

[7] L. Kaufman and P. J. Rousseeuw. *Finding Groups in Data.* Wiley, New York, 1990.

[8] H. Liu and H. Motoda. *Feature Selection.* Kluwer, Norwell, Mass., 1998.

[9] H. Liu and H. Motoda, editors. *Instance Selection and Construction for Data Mining*, Norwell, Mass., 2001. Kluwer.

[10] T. Nakayasu, N. Suematsu, and A. Hayashi. Learning classification rules from large-scale databases. In *Proceedings of the 62th National Conference of Information Processing Society of Japan*, volume 2, pages 23 – 24, 2001. (in Japanese).

[11] D. Pavlov, D. Chudova, and P. Smyth. Towards scalable support vector machines using squashing. In *Proceedings of the Sixth ACM International Conference on Knowledge Discovery and Data Mining*, pages 295 – 299, 2000.

[12] W. H. Press et al. *Numerical Recipes in C - Second Edition.* Cambridge Univ. Press, Cambridge, U.K., 1992.

[13] T. Zhang, R. Ramakrishnan, and M. Livny. BIRCH: An efficient data clustering method for very large databases. In *Proceedings of the 1996 ACM SIGMOD International Conference on Management of Data*, pages 103 – 114, 1996.

Active Mining
H. Motoda (Ed.)
IOS Press, 2002

Reducing Crossovers in Reconciliation Graphs Using the Coupling Cluster Exchange Method with a Genetic Algorithm

Hajime Kitakami and Yasuma Mori
{kitakami,mori}@its.hiroshima-cu.ac.jp
Faculty of Information Sciences
Hiroshima City University
3-4-1 Ozuka-Higashi, Asa-Minami-Ku, Hiroshima 731-3194, JAPAN

Abstract. In molecular evolutionary genetics, reconciliation graphs that represent a kind of n-level hierarchy ($n \geq 2$) are constructed from two ordered trees. In order to achieve effective reconciliation and a result that is easy to use, it would be important to reduce the number of crossovers that are developed in a reconciliation graph. The authors propose a methodology that effectively reduces the number of crossovers in such a graph by coupling a new heuristic search, called cluster exchange, with a genetic algorithm (GA). The authors' heuristic search does not require the construction of an interconnection matrix between each two-level hierarchy. An interconnection matrix must be constructed with a constraint between two leaf layers, and the matrix approaches a unit matrix by means of matrix transformations. The authors define one of the stop conditions as the number of identical candidate solutions in the search. If the number of solutions exceeds the value of a monotonic decreasing function with the depth of the search as the domain, the stop condition is satisfied. When the solution after the heuristic search lacks sufficient quality, our method allows us to find a better solution by applying a GA to the solution. The final solution is better than those obtained using either method separately.

1 Introduction

The recent evolution of the Internet has facilitated access to large amounts of information in the form of text and images. In many instances, it might be desirable to integrate some of the different databases that are available on the Internet. However, even when two different sets of information have similar contents, they almost never use the same semantics or syntax because they have been created in different areas or fields in a loosely coupled environment. This poses a research problem as to how to integrate databases with semantic heterogeneity. A method for integrating and analyzing heterogeneous tree databases using visualization techniques would have great value.

We focus on the reconciliation graphs that are mainly used in the field of molecular evolutionary genetics. Such graphs are constructed from two ordered heterogeneous trees and a kind of n-level hierarchy ($n \geq 2$). In order to achieve both effective reconciliation and results that are easy to use, it is important that a methodology be developed that would reduce the number of crossovers in a reconciliation graph.

Goodman et al. [6] introduced the concept of reconciliation work in the field of molecular evolutionary genetics. Subsequently, Page et al. [13, 15, 16, 17] summarized and applied the concept mathematically. The reconciliation work can be achieved in two ways: (1) by mapping across two trees and (2) by duplicating subtrees in the resultant mapping. A reconciliation graph is constructed by connecting the leaf layers of two ordered heterogeneous trees. Reconciliation includes complicated manual operations since the graph has a large number of crossovers. However, existing reconciliation work has not used automatic processing because reconciliation graphs in past studies have been small.

The computer reconciliation is also useful for improving a taxonomic tree [8] using a gene tree [21], and discovering conceptual differences among different people involving diverse structures [23] and overcoming structural heterogeneities [14, 18] that are constructed from XML and/or relational schemas. Structure heterogeneities are appeared in the integration of enterprise databases, which include hundreds of tables with thousands of attributes in complex and disparate structure. A method for finding the directed graph with minimum-crossovers has applications in many fields and its development is very significant. Moreover, this method is concerned with retrieving data that can be represented by a tree structure [3].

On the other hand, the existing simple heuristic searches require the development of methods to escape locally optimal solutions [12]. In order to escape locally optimal solutions, many researchers have studied genetic algorithms (GAs) [2, 12], which have the capacity for global searches. However, not all GAs are faster than heuristic searches since a GA generates a large number of generations of the population that it constructs.

This paper proposes a methodology that effectively reduces the number of crossovers in reconciliation graphs by coupling a new heuristic search, called cluster exchange, and a GA. Any method that reduces the number of crossovers on a reconciliation graph involves searching for a kind of optimal solution.

2 Data Model

Our data model is used to define the cluster exchange and the GA. This section describes the data model from the viewpoint of both structure and operations. The initial reconciliation graph shown in the left-hand side of Figure 1 is constructed from two ordered heterogeneous trees. First, Section 2.1 defines the structure of the ordered tree included in the reconciliation graph. Second, Section 2.2 defines the interconnection matrix, which is the structure used to define our proposed computational models. Moreover, basic operations for the matrix are described in the section. Finally, Section 2.3 describes two methods for finding maximum clusters and a minimum cluster, respectively.

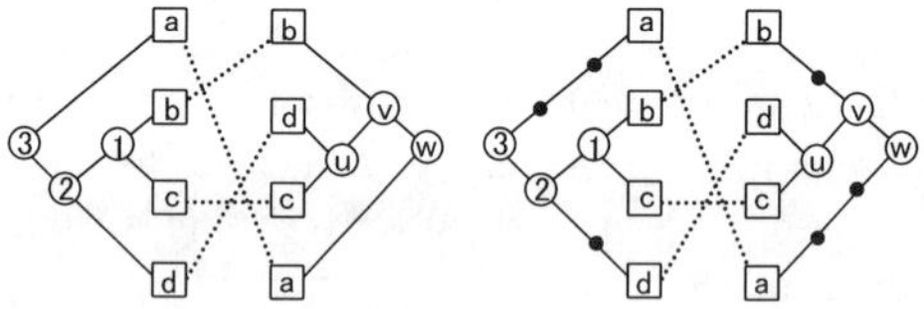

Figure 1: An example of a reconciliation graph

2.1 Tree structure

The tree structure has known properties such as a unique root node, no cycle, no nodes, and no incoming edges other than the root.

For simplicity, we normalize all trees with a height of $n-1$. If the depth of a leaf is less than $n-1$, we add dummy nodes so that the path length from the root to the leaf is $n-1$. For example, when we add dummy nodes marked by "•" for the graph of the left-hand side in Figure 1, the graph is normalized, as shown in the right-hand side of the figure. We call the directed graph, $G(V, R, n, \Sigma)$, an ordered tree T with order Σ and height $n-1$, if the directed graph satisfies zero-crossover and is an n-level hierarchy ($n \geq 2$) that satisfies the following conditions, where V is a set of nodes and R a set of edges:

(1) V is partitioned into n subsets. That is: $V = N_1 \cup N_2 \cup \cdots \cup N_n (N_i \cap N_j = \phi, 1 \leq i < j \leq n)$, where N_i is called the ith level and n is the number of levels in the hierarchy.

(2) N_1 contains one element, which is called the root of T.

(3) R is partitioned into $n-1$ subsets. That is: $R = B_1 \cup B_2 \cup \cdots \cup B_{n-1} (B_i \cap B_j = \phi, i \neq j), B_i \subset N_i \times N_{i+1}, i \leq j \leq n-1$, where any two edges, (p_1, q) and $(p_2, q) \in R$, satisfy $p_1 = p_2$. For every edge, $(p, q) \in R$, p is called the initial (or parent) node, and q is called the final (or child) node.

(4) The set of nodes with zero in-degree consists of N_1 only. The set of nodes with zero out-degree is N_n, and each node is called a leaf.

(5) The order σ_i of N_i is given for each i, where the term "order" indicates the sequence of all nodes of N_i; $\sigma_i = d_1^{(i)}, d_2^{(i)}, \cdots, d_\alpha^{(i)}$, where $d_p^{(i)} \in N_i$, α denotes the number of nodes of N_i. The ordered binary relationship of the sequence is represented by the binary relationships, $d_1^{(i)} < d_2^{(i)}, d_2^{(i)} < d_3^{(i)}, \cdots, d_{\alpha-1}^{(i)} < d_\alpha^{(i)}$. The n-level hierarchy defined by the directed graph is denoted by $G(V, R, n, \Sigma)$, where $\Sigma = (\sigma_1, \sigma_2, \cdots, \sigma_n)$.

	b	d	c	a
a	0	0	0	1
b	1	0	0	0
c	0	0	1	0
d	0	1	0	0

Figure 2: An example of an interconnection matrix

2.2 *Interconnection matrix*

The reconciliation graph can be regarded as a kind of n-level hierarchy ($n = 8$) when it is viewed from the left- to the right-hand side, as shown in Figure 1. An interconnection matrix Mat is not constructed between each two-level hierarchy. Instead, an interconnection matrix is constructed with a constraint between two leaf layers, and the matrix approaches a unit matrix using matrix transformations. The interconnection matrix representing a relationship between two leaf layers, shown in Figure 1, is represented in Figure 2. The left-hand and right-hand trees are respectively represented as $a < b < c < d$ and $b < d < c < a$ in the ordered relation with respect to the set of leaves. These two ordered relations are represented as two ordered leaf lists, $OL_1 = [a, b, c, d]$ and $OL_2 = [b, d, c, a]$, respectively. Leaf c in the left-hand tree and the ordered leaves, $[b, d, c, a]$, in the right-hand tree have a connective relationship that is defined as the vector (0 0 1 0) located in the third row of the matrix. Since leaf c is connected to the third leaf in the left-hand ordered tree, the third element in the vector is unity. The number of crossovers between the ordered leaf lists, $C(Mat)$, is defined as:

$$C(Mat) = \Sigma m_{j,\beta} m_{k,\alpha} [1 \leq j < k \leq n, 1 \leq \alpha < \beta \leq n] \tag{1}$$

where $m_{i,j}$ is the (i,j)-component of the matrix Mat. If the connective relationship is represented as the row number of unity located in each column vector, the number of crossovers can be easily computed. After this, this simple expression is used to represent the relationship. In Figure 2, the simple expression is represented as a list $[2,4,3,1]$. In this example, unity is located in the second row of the first column, the fourth of the second, the third of the third, and the first of the fourth. The number of crossovers in the example is 4. In general, the number of crossovers given by this simple expression is computed by counting the number of times the ascending order of the list is violated. If the simple expression is sorted from left to right in a ascending order, the number of crossovers will be zero.

There are two kinds of basic operations for the previously mentioned interconnection matrix. One of them is an exchange operation for two different leaves on the left-hand tree, and the other is that for two different leaves on the right-hand tree. In order to exchange two leaves, b and a, in the right-hand tree of Figure 1, the vector (0 1 0 0) of the first and the vector (1 0 0 0) of the fourth columns must be exchanged in the matrix, as shown in Figure 2. Extended operations to exchange two sets of rows or columns are needed in order to apply our computational model. A named cluster from each set will be described in detail in a later section. The set $\{b, d, c\}$ in the right-hand tree of Figure 1 corresponds to the set of columns from the first to the third column, shown in Figure 2, and is an example of a cluster. Even if one cluster is exchanged with another $\{a\}$, which corresponds with the fourth column, the exchange operation would never create any crossover among branches of the tree. The exchange operation creates a new ordered relation represented as $a < b < d < c$ in the right-hand tree. Our extended operation for the matrix is thus based on the cluster exchange. The operation is useful for the combination of the cluster exchange method and the GA, as stated in a later section.

2.3 Cluster search

There are two types [9, 10] of clusters in our computational model, the cluster exchange and the GA. This section describes the method for finding each of them. The first one is called a maximum cluster and the other, a minimum cluster. The former is useful because it not only exchanges two clusters through the cluster exchange method but also creates an initial population and applies a mutation in the GA. The latter is useful to generate the recombination in the GA. In order to find the maximum clusters for the matrix, either the row or column side in the matrix must be selected. If the row side is selected, the focus is on the left-hand ordered tree.

2.3.1 Maximum cluster

After selecting which ordered tree to focus on, two maximum clusters that are independent of each other will be found from two different leaves selected randomly. Let us consider the method used to search the two maximum clusters for the tree. The two clusters are useful to distinguish two different leaves, $p, q \in N_n$, where $p \neq q$. The subtrees used to find the two clusters, C_1 and C_2, shown in Figure 3, can be searched using the following procedure:

(1) After finding the branch node, r, for two different leaves using the recursive backward search, $p, q \in N_n$, two subtrees, $subT_1$ and $subT_2$, are selected from all the subtrees connected to the branch node. When $subT_1$ and $subT_2$ include the respective leaves, p and q, the same nodes for $subT_1$ and $subT_2$ do not exist. $SubT_1$ and $subT_2$ are also the maximal trees selected from the possible subtrees to distinguish between two different leaves, $p, q \in N_n$.

(2) The two sets, C_1 and C_2, related to the leaves are computed from the two subtrees, $subT_1$ and $subT_2$, respectively, using the recursive forward search, where $C_1 \cap C_2 = \phi$.

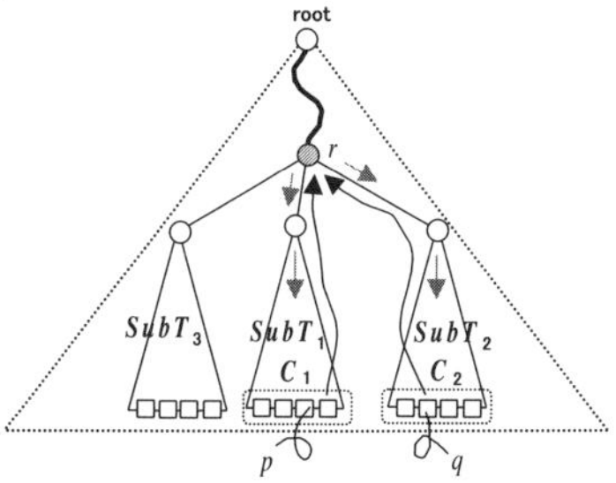

Figure 3: An example of a search of two maximum clusters for two leaves, p and q

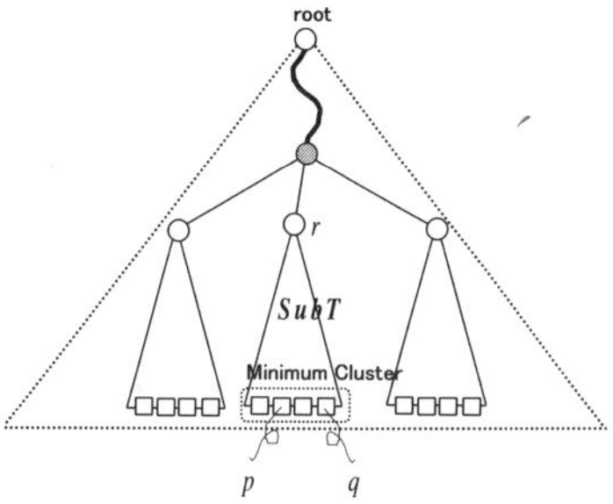

Figure 4: An example of a search of one minimum cluster for two leaves, p and q

2.3.2 Minimum cluster

After deciding which ordered tree to focus on, one minimum cluster will be found from the two different leaves, $p, q \in N_n$, selected randomly. The minimum subtree, $SubT$, shown in Figure 4, includes the two elements, p and q. The root of the minimum subtree is the first shared node, r, for the two elements in the figure. All leaves included in the minimum subtree are called a minimum cluster for the two elements. When the recombination of the GA selects another minimum cluster from an ordered tree in another matrix, the exchange of the two minimum clusters is carried out for each ordered tree, whether it is the left- or the right-hand tree.

3 Computational model

The two search methods, the cluster exchange and the GA, have been combined to overcome their individual disadvantages. Each method will stop the processing if the number of crossovers is reduced to zero. However, if the reconciliation graph given by users does not have an optimal solution with zero crossover, neither method, in general, could stop the processing without a threshold. Our stop condition is defined as the number of occurrences for a current semi-optimal solution. Moreover, when either method stops processing, a user will decide whether to execute the next method in order to improve the quality of the solution. The summarized processing is as follows:

(Step1) Inputting an initial reconciliation graph, the number S of the modification for the crossover of the graph, and a stop condition that is the number of occurrences for a current semi-optimal solution;

(Step2) Modifying the Sth time of the initial graph using a mutation operation for the matrix;
(Step3) While (user decides to execute the next search) do
(Step4) Applying the cluster exchange method until satisfying the stop condition;
(Step5) Outputting a prompter to allow a user to decide whether he needs to execute the next search method;
(Step6) Exiting the while-loop if the user does not decide to execute the next search method;
(Step7) Applying the GA until satisfying the stop condition;
(Step8) Outputting a prompter to allow a user to decide whether he needs to execute the next search method;
(Step9) Ending;
(Step10) Outputting the results;

The previous processing (Step 2) is used to change the initial reconciliation graph before applying the two methods. S generally has a zero value when there is no interest in having the variation of the initial graph affect the quality of the solution. A later section describes each of the cluster exchange method in Step 4 and the GA in Step 7.

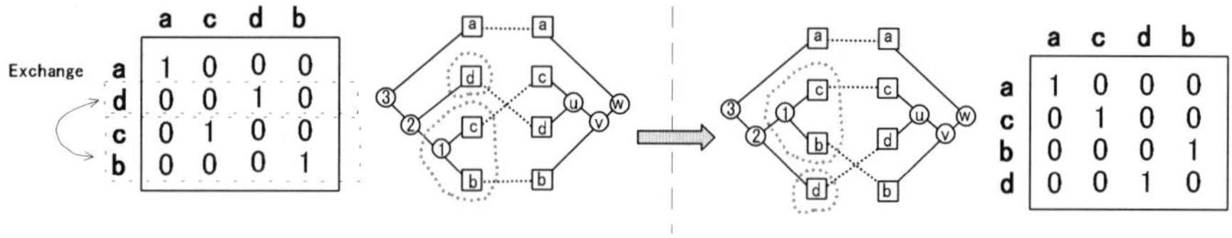

Figure 5: An example of exchanging two clusters

3.1 Cluster exchange method

The cluster exchange method is achieved by exchanging two clusters found from either the row side or the column side of the interconnection matrix. The aim of the method is to make the matrix approach a unit matrix. The two maximum clusters, C_1 and C_2, are found from two leaves, p and q. The leaves, p and q, are selected from the matrix in order to exchange the diagonal element with zero into unity, since each diagonal element of the unit matrix is unity. The cluster exchange method is achieved by moving the focused position of the exchange operation sequentially from the upper-left to the lower-right corner of the matrix along the diagonal elements. The focused position for each exchange operation is called a pivot position. The ith pivot position indicates the focused position of the (i, i)-element. Figure 5 shows an example of the cluster exchange method in order to assign the second pivot position into unity. Each cluster is circled by a dotted line in the figure. In this example, the method exchanges two clusters, $\{d\}$ and $\{c, b\}$, which respectively include two leaves, d and c, in the left-hand ordered tree. After that, the cluster exchange method is applied to the third pivot position in this operation. There are two alternatives for exchanging the two rows, $\{b\}$ and $\{d\}$, or exchanging the two columns, $\{d\}$ and $\{b\}$, to assign the pivot position into unity. In general, there are many alternatives for the combination of rows and/or columns to assign a pivot position to unity. If any pair of two clusters, C_1 and C_2, found at the ith pivot position, includes at least one leaf that has already been exchanged, the method moves to the $(i-1)^{th}$ position in order to resume the previous operation. We can summarize the above sketch as follows:

(Step1) Making an interconnection matrix for two leaves selected respectively from two ordered trees that construct a reconciliation graph given by the user; Assigning the set, S, of processed leaves into empty set; $i := 1$;
(Step2) While (the stop condition is not satisfied) do
(Step3) Executing the next exchange operations at the ith pivot position;
(Step4) While (the execution fails at the position) do
(Step5) If $i = 1$, then exiting the while-loop;
(Step6) Executing the next exchange operations at the $(i-1)^{th}$ pivot position;
(Step7) $i := i - 1$;
(Step8) Ending;
(Step9) Storing the status of the processing at the ith pivot position in the memory;
(Step10) $S := S\cup$ the leaf that was processed by the exchange operations;
(Step11) $i := i + 1$;
(Step12) Ending;
(Step13) Outputting the results

The stop condition in Step 2 is satisfied when the number of crossovers is zero or the number of occurrences for a current semi-optimal solution is greater than the value of the function, $f(p) = \alpha(m-p)$, where α is a constant and m is the size of the matrix. Moreover, by means of Steps 3 and 6, candidates are found for leaves exchanged when the exchange operation moves sequentially from from the jth column to the mth column at the pivot position, where $i \le j \le m$. If the column of jth has unity for k^{th} row and the leaf for the column of jth is one of the candidates, the cluster exchange between the ith column and the jth column is carried out after applying the cluster exchange between the ith row and the kth row. The variable k satisfies the relation, $i \le k \le m$. After that, the candidate of the next operation will be the $(j+1)^{th}$ column. Of course, if the $(j+1)^{th}$ column is not appropriate, the next candidate will be the $(j+2)^{th}$ column.

3.2 GA

In general, a GA [2] is achieved by repeatedly applying such basic operations as selection, recombination, and mutation for an initial population after the population is created. The processing can stop when a candidate solution satisfies one of the thresholds given by user. In solving the problem of reducing the number of crossovers for the reconciliation graph, one of the thresholds will be satisfied if either the number of crossovers for the graph is zero or no more improvements of the candidate solution are observed. No more improvements will probably be observed if the occurrences of the candidate solution are greater than the value of $\beta \times pop \times m$, where β is a constant, pop is the size of the population, and m is the size of the interconnection matrix. The selection operation selects individuals with high probability based on fitness. Higher fitness is defined as a lower number of crossovers for the graph [10]. The processing of the genetic algorithm is carried out as follows:

(Step1) Randomly create an initial population of chromosomes (individuals) with a genotype.
(Step2) Iteratively perform the following substeps on the population of chromosomes until the termination criterion has been satisfied:
(a) Evaluate the fitness of each individual in the population.
(b) Create a new population by mating the current population; apply mutation and recombination as the parent chromosomes mate. The operations are applied to individuals in the population who are chosen because of their probability based on fitness.

(Step3) Return the best individual as the result of the genetic algorithm for the run.

The genotype in our GA is defined as two ordered leaf lists including an interconnection matrix. The genotype can be defined as $[(a, b, c, d), (b, d, c, a)]$, for example, as shown in Figure 1. The first element (a, b, c, d) represents $a < b < c < d$ as the ordered relation. The second element (b, d, c, a) represents $b < d < c < a$ as the ordered relation. The basic operations of the GA are applied to the population with the genotype. Only a basic operation that does not generate any crossover among tree branches is allowed. Not all exchanges are allowed for the population.

In order to prevent the loss of the recombination and/or the mutation, the production of dysfunctional offspring should be avoided as soon as possible, and the parent phenotype should be inherited in the new population. The rest of the section will describe the methods for the recombination and the mutation.

3.2.1 Recombination

The operation of recombination between two individuals generates two child chromosomes by exchanging two parts that are selected from the chromosomes of a different parent. This operation is applied to two leaf sequences included in either the left- or right-hand tree with respect to the two chromosomes and allows exchanging the same subsets included in the leaf sequences. If two different subsets are exchanged with each other, dysfunctional offspring are generated. Therefore, exchanging the same subsets is only allowed in the recombination operation. Each subset has to include at least two leaves since the recombination aims to exchange the order of leaves included in the subsets. The subset can be obtained by finding the two minimum clusters for the two leaves, p and q, from the two ordered trees. Moreover, the operation of the cluster exchange applies to both the left-hand and the right-hand tree.

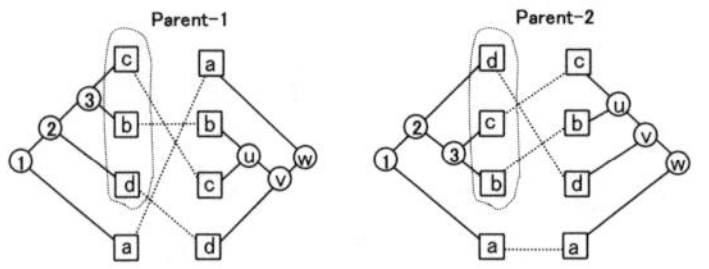

Figure 6: An example of two parent chromosomes

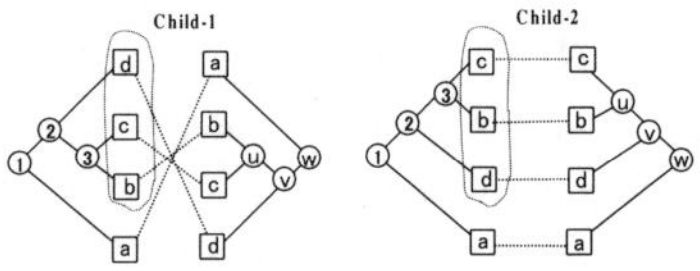

Figure 7: An example of the generation of two child chromosomes

Figure 6 shows an example of two parent chromosomes, parent-1 and parent-2. The chromosomes, parent-1 and parent-2, have interconnection matrices, $(4, 2, 1, 3)$ and $(2, 3, 1, 4)$, respectively, since they have $[(c, b, d, a), (a, b, c, d)]$ and $[(d, c, b, a), (c, b, d, a)]$ as the genotype. The number of crossovers is four for the former and two for the latter.

For the recombination between the two parent chromosomes, parent-1 and parent-2, we will focus on each left-hand tree, as shown in Figure 6. If the system selects two leaves, c and d, it finds the two minimum clusters shown in the dotted circles of the figure. When the system exchanges the two minimum clusters, two child chromosomes can be obtained, as shown in Figure 7. The child chromosome, child-1, represented as $[(d, c, b, a), (a, b, c, d)]$, has the interconnection matrix $(4, 3, 2, 1)$. Therefore, the number of crossovers is six in the left-hand graph. The right-hand graph has zero-crossovers since the child chromosome, child-2, represented as $[(c, b, d, a), (c, b, d, a)]$, has the matrix $(1, 2, 3, 4)$. By the same method, the recombination can be carried out while focusing on each right-hand tree.

3.2.2 Mutation

Consideration will now be given to an exchange of two leaves included in a leaf sequence in order to execute a mutation for a chromosome. The two leaf nodes are randomly selected from nodes included in the leaf sequence that is included in either the left- or the right-hand tree. If a exchange for two leaves selected randomly is executed, the exchange may generate a new crossover within the branches of the tree. In order to avoid the generation, the two maximum clusters are exchanged with respect to the two leaves.

4 Performance evaluation

We have implemented a prototype system in Quintus Prolog [19] on a Sun workstation ULTRA80 (450MHz) in order to verify the proposed method. The graph consists of 40 records or leaves in each tree and is constructed from a real taxonomic tree [8] and a phylogenetic tree [7]. The graph has 380 records that are branch nodes, and the minimum number of crossovers is four. The performance of the following three points are investigated and evaluated:

(1) Whether the cluster exchange method has a better solution than the GA;

(2) Whether the monotonic decreasing function with the depth of the search as the domain is more useful for one of the stop conditions; and

(3) Whether the coupling of the cluster exchange method and the GA provides better solutions than either one alone.

In order to investigate the above (1), the cluster exchange method and the GA in CPU time are compared as they are used to solve the same problem, where α of the monotonic decreasing function described in 3.1 is assigned to 2 and β described in 3.2 is 1. After the CPU time of the cluster exchange method is translated into the number of generations, they are regarded as the horizontal line on the graph, as shown in Figure 8, and the number of crossovers is plotted as the vertical line on the graph. In executing the cluster exchange method, the number of crossovers is changed non-monotonically. For simplicity, only the number of crossovers for improved solutions is plotted. The results indicated that the cluster exchange method is faster than the GA. In the comparison, an elite strategy was used for the selection operation, and a value of 10 was used as the population size. The other strategies did not yield much difference in the performance evaluation.

The above (2) is investigated by comparing both the monotonic decreasing function and a constant in CPU time. Both are used to stop the processing. Three kinds of initial reconciliation graphs are provided, as shown in Table 1, since the computational time depends on the graph. The number of crossovers for the first one, initial status A, is 411. The second one, initial status B, is given 100 mutations for the first one. The third one, initial status C, is given 1,000 mutations for the first one. The monotonic decreasing function is more effective for both initial status A and B than for the constant. The initial status C provides the same result, which is not quite satisfactory, by using both the function and the constant. The result by using the GA shown is also unsatisfactory shown in Figure 9.

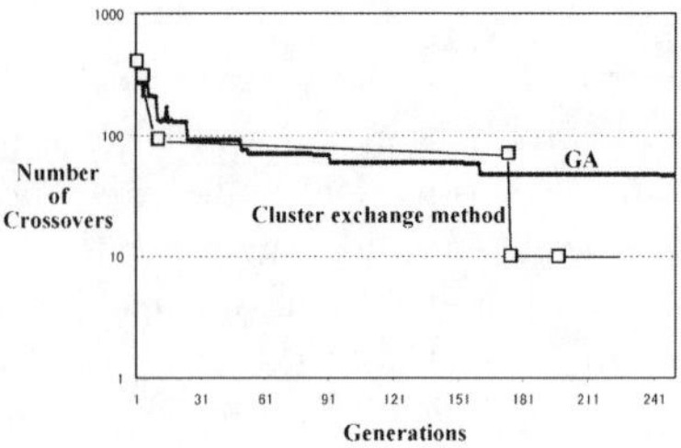

Figure 8: Performance evaluation of the cluster exchange method and the GA

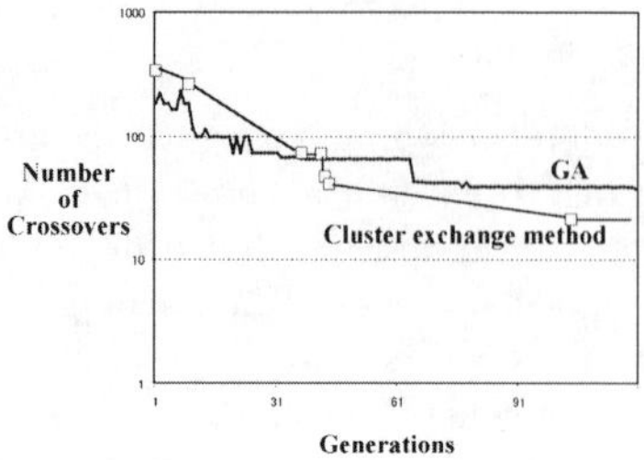

Figure 9: Processing time for an initial status, C

In order to improve the result, we propose applying the GA after finishing the cluster exchange method. This improvement corresponds to investigating the above (3). Table 2 shows the CPU time and the number of final crossovers applied to their coupling. The coupling is useful for each initial status of three. The final solution using the coupling is of higher quality than that obtained using either method separately.

Table 1: CPU time (sec) evaluated in different initial status

Initial status \ Number of ocurrences	4	5	23	50	Value of monotonic function
A	1.12 (88)	1.28 (88)	11.91 (10)	15.42 (10)	10.92 (10)
B	1.21 (23.60)	4.40 (9)	6.32 (9)	9.29 (9)	5.1 (9)
C	1.72 (21)	1.76 (21)	4.15 (21)	10.46 (21)	4.29 (21)

※ The value in ( ) is the number of crossovers when the stop condition is satisfied

Table 2: CPU time (sec) evaluated from coupling the cluster exchange method and GA

Applied procedure \ Population size	4	6	10	15
Cluster exchange	4.29 (21)	4.31 (21)	4.33 (21)	4.30 (21)
GA	6.45 (18)	55.21 (15)	114.35 (16)	830.94 (13)
Total time	10.74 (18)	59.52 (15)	118.68 (16)	835.24 (13)

※ The value in ( ) is the number of crossovers when the stop condition is satisfied

5 Related work

The reconciliation graph is a kind of directed graph with n-level hierarchies. In general, the problem of n-level hierarchies can be interpreted by extending the problem to two-level hierarchies. Unfortunately, the two-level hierarchy problem is an NP-hard problem [4, 5, 22], which must be solved using effective heuristic searches [4, 20] invoking either the barycentric or median method. The barycentric method reorders the row (and/or column) barycenters using sorting. The median method is similar to the barycentric method.

Applying either the median method or the barycentric method to each interconnection matrix that is defined between two layers solves the problem of n-level hierarchies. However, all the matrices interact with each other in the computational process. In order to avoid the interaction, the reconciliation graph is regarded as a graph that is constructed from two ordered trees; then, the cluster exchange method applied to the graph defines an interconnection matrix between only two leaf layers and does not define any matrix between the other layers. The other layers indicate non-leaf layers in each ordered tree. The operation of the cluster exchange for the matrix has a function to avoid the generation of crossovers among tree branches. The method does not allow a direct exchange of either two rows or two columns; however, it does allow the exchange of two clusters found in either two rows or two columns. A monotonic decreasing function with the depth of the search as the domain is used as one of the stop conditions. Moreover, when the final solution after the cluster exchange method lacks sufficient quality, our system allows finding a better solution by applying our GA.

The chromosome defined as one-dimensional array constructed by values, 0 and 1, is extended to two-dimensional array in our GA. It seems that this extension is similar to the concept of a GP (Genetic Programming) [1, 11] defined as the natural extension of GA. However, there are lots of differences between our GA and the GP. The GP has the chromosome defined as a tree structure, and applies such operators as mutation, permutation, recombination and so forth to the tree structure. On the other hand, the application of our GA operators is restricted to the two leaf layers that are represented by two-dimensional array. Moreover, our GA does not need the modification that any node name is changed to another node name in the mutation of GP.

6 Conclusions

This paper proposes a methodology to reduce the number of crossovers in reconciliation graphs that use both a cluster exchange method and a GA. The cluster exchange method defines one interconnection matrix between the leaf layers of two ordered trees and exchanges two maximum clusters so that the matrix approaches a unit matrix. The maximum clusters, C_1 and C_2, include leaves p and q, respectively, and are found in the ordered trees included in the reconciliation graph. After terminating the cluster exchange method, the GA repeatedly applies operations such as selection, recombination, and mutation to the computed result. Creation of an initial population and application of mutation involve exchanging the two maximum clusters so that the matrix approaches a unit matrix. Recombination exchanges the two minimum clusters on two chromosomes. Each minimum cluster is a part of a chromosome and is found from a minimum subtree, which includes leaves p and q, given randomly by the system.

In order to verify our methodology, we evaluated the performance of a real reconciliation graph constructed from 380 nodes. This demonstrated that the cluster exchange method is faster than the GA. Moreover, when the GA was executed after using the cluster exchange method, higher quality solutions were found than when using either method separately.

We envision the following future projects:

(1) Developing a method of distributed parallel processing for reconciliation graphs.

(2) Developing an extended method to solve more general computational problems.

Acknowledgments

We thank the staff of the international DNA data banks for their help in obtaining the taxonomy databases. This work was supported in part by a Grant-in-Aid of Scientific Research (C) (2) from the Japanese Society for the Promotion of Science and a Hiroshima City University Grant for Special Academic Research.

References

[1] W. Banzhaf, P. Nordin, R. E. Keller, and F. D. Francone. *Genetic Programming - An Introduction on the Automatic Evolution of Computer Programs and Its Applications.* Morgan Kaufman Publishers, Inc., 1998.

[2] L. Davis, editor. *Handbook of Genetic Algorithms.* Van Nostrand Reinhold, 1991.

[3] M. De Marsicoi et al. Indexing pictorial documents by their content, a survey of current techniques. 15:119–141, 1997.

[4] P. Eades, B. D. McKay, and N. C. Wormald. On an edge crossing problem. In *Proc. 9th Australian Computer Science Conference*, pages 327–334, 1986.

[5] M. R. Garey and D. S. Johnson. Crossing number is np-complete. *SIAM J. of Algebraic and Discrete Methods*, 4(3):312–316, 1983.

[6] M. Goodman, J. Czelusniak, A. E. Romero-Herrera, and G. Matsuda. Fitting the gene lineage into its species lineage: A parsimony strategy illustrated by cladograms constructed from globin sequences. *Systematic Zoology*, 28:132–168, 1979.

[7] T. Hashimoto, Y. Nakamura, T. Kamaishi, and M. Hasegawa. Early evolution of eukaryotes inferred from protein phylogenies of translation elongation factors 1α and 2. *Archiv fur Protistenkunde*, 148:287–295, 1997 in German.

[8] H. Kitakami, Y. Mori, M. Arikawa, and A. Sato. An integration methodology for autonomous taxonomy databases using priorities. In *Proc. of the 5th International Conference on Database Systems for Advanced Applications (DASFAA'97)*, pages 243–251, World Scientific Publishing Co. Pte. Ltd., April 1997.

[9] H. Kitakami and M. Nishimoto. Constraint satisfaction for reconciling heterogeneous tree databases. In *Proc. of the 11th International Conference on Database and Expert Systems Applications (DEXA 2000)*, 1873:624–633, Lecture Notes in Computer Science, Springer-Verlag, September 2000.

[10] H. Kitakami, H. Maruyama, and Y. Mori. A web-based system to support a part of visual reconciliation work. In *Proceedings of the International Conference on Artificial Intelligence (IC-AI'2001)*, pages 210–215. CSREA Press, June 2001.

[11] J. R. Koza. *Genetic Programming: On the Programming of Computers by Means of Natural Selection.* The MIT Press, 1992.

[12] D. L. Kreher and D. R. Stinson. *Combinatorial Algorithms, The CRC Press Series on Discrete Mathematics and Its Applications.* CRC Press, 1999.

[13] B. Ma, M. Li, and L. Zhang. From gene tree to species trees. *SIAM J. Computing, Society for Industrial and Applied Mathematics*, 30(3):729–752, 2000.

[14] R. J. Miller, L. M. Haas, and M. A. Hemandes. Schema Mapping as Query Discovery. In *Proc. of 26th International Conference on Very Large Databases*, pages 77–88, Morgan Kaufmann, 2000.

[15] R. D. M. Page. Maps between trees and cladistic analysis of historical associations among genes, organisms, and areas. *Systematics Biology*, 43(1):58–77, 1994.

[16] R. D. M. Page and M. A. Charleston. Reconciled trees and incongruent gene and species trees, in: B. mirkin, f. r. mcmorris, f. s. roberts and a. rzhetsky (eds.), mathematical hierarchies in biology, dimacs series in discrete mathematics and theoretical computer science. *American Mathenatical Society*, 37:57–70, 1997.

[17] R. D. M. Page and M. A. Charleston. From gene to organismal phylogeny: Reconciled trees and the gene/species tree problem. *Molecular Phylogenetics and Evolution*, 7:231–240, 1997.

[18] Lucian Popa, Mauricio A. Hernamdes, Yannis Velegrakis and Renee J. Miller. Mapping XML and Relational Schemes with Clio. In *Proc. of 18th International Conference on Data Engineering*, pages 498–499, IEEE Computer Society Press, 2002.

[19] Quintus Prolog, Swedish Institute of Computer Science, http://www.sics.se/isl/quintus/, 2001.

[20] K. Sugiyama, S. Tanaka, and M. Toda. Methods for visual understanding of hierarchical system structures. *IEEE Transaction on Systems, Man, and Cybernetics*, pages 109–125, 1981.

[21] Virginia Morell. Web-crawling up the tree of life, news & comment. *Science*, 273, 1996.

[22] J. N. Warfield. Crossing theory and hierarchy mapping. *IEEE Transaction on Systems, Man, and Cybernetics*, pages 505–523, 1977.

[23] T. Yoshida, T. Kondo, and S.Nishida. Discovering Conceptual Differences among Different People via Diverse Structures. In *Proc. of the Third Pacific-Asia Conference on Knowledge Discovery and Data Mining*, 1574:494–498, Springer-Verlag, 1999.

Active Mining
H. Motoda (Ed.)
IOS Press, 2002

Outlier Detection using Cluster Discriminant Analysis

Arata Sato, Takashi Suenaga and Hitoshi Sakano
{ara,suenaga,sakano}@rd.nttdata.co.jp
NTT Data Corporation
Kayabacho Tower Bldg.,21–2, Shinkawa 1-chome, Chuo-ku, Tokyo 104–0033, JAPAN

Abstract. In this paper, we propose novel outlier detection method using *cluster discriminant analysis* (CDA). It is difficult or sometimes impossible to detect outliers automatically. Because one cannot define objective criteria from the view point of statistics. As a result, in practical applications, outliers have to be detected subjective by looking for isolated data in scatter plot. However, data mining is often used for the analysis of multidimensional data that has hundred of attributes and is very difficult to represent in the form of scatter plot. To overcome this problem, we already proposed novel low dimensional mapping method named CDA. We demonstrate effectiveness of CDA through the experimental result to detect outliers.

1 Introduction

An important requirement of data mining systems is the ability to detect values that are excessively large or small (i.e., outliers). If outliers are simply caused by noise in the source data, they adversely effect the analysis of this data. Various methods have therefore been proposed for detecting outliers.

However, it is difficult to detect outliers automatically because it is impossible to provide objective statistical criteria for doing so [1]. As a result, in practical applications such as customer relationship management (CRM) systems, outliers have to be detected subjectively by looking for isolated data in scatter plots. However, data mining is often used for the analysis of multidimensional data that has tens or hundreds of attributes and is very difficult to represent in the form of a scatter plot. It is thus necessary to find some way of mapping such data into two- or three-dimensional spaces that can be intuitively understood by humans. There are various ways of implementing this dimensional reduction, including statistical methods such as principal component analysis [2] and multi-dimensional scaling [3], and neural network methods such as self-organized feature mapping [4]. However, these techniques are mostly ineffective for the detection of outliers.

We have proposed cluster discriminant analysis [5] as a method for visualizing multivariate data, and we have shown that it is capable of producing useful scatter plots. This method analyzes the cluster structures in multidimensional space, and then looks for a way of mapping these clusters to a lower-dimensional space while preserving these cluster structures, thereby allowing them to be visualized. Since this method detects outliers as clusters in their own right, is it a powerful tool for the detection of outliers.

Section 2 of this paper clarifies the problems of the conventional visualization techniques and outlines our proposed approach. Based on this approach, we describe the derivation of cluster discriminant analysis as a visualization algorithm for cluster structures, and we discuss the feasibility of detecting outliers based on cluster discriminant analysis. We also discuss the possibility of using cluster discriminant analysis to observe discriminant hyper-surfaces. In section 3, we experimentally demonstrate the effectiveness of cluster discriminant analysis for detecting outliers and observing discriminant hyper-surfaces. Section 4 is conclusion of this paper.

2 Detection of outliers by cluster discriminant analysis

2.1 Cluster discriminant analysis

The purpose of visualizing distributions in data mining is to ascertain the distribution structure. This can only be achieved if the distribution structures can be preserved in a lower dimensional (2- or 3-dimensional) space.

However, in methods based on conventional approaches (e.g., principal component analysis) that preserve distance relationships, one has to preserve as many as possible of the distance relationships between data having tens or even hundreds of dimensions in a space of just a few dimensions. This is of course impossible in principle.

To avoid this problem, we employ a new structure-preserving approach whereby the structures are analyzed in a higher-dimensional space before they are destroyed, and then a mapping to a lower-dimensional space that preserves these structures is determined. There are various types of structures worth preserving, such as cluster structures and curve structures; in this paper we consider a visualization algorithm that focuses on cluster structures.

To implement a structure-preserving visualization algorithm, it is first necessary to analyze the cluster structures of the undistorted data distribution in a higher-dimensional space. This process is referred to as clustering. Various algorithms have been prepared for this purpose, such as the k-means method [6], and these can all be used.

The next step is to find a mapping to a lower-dimensional space that leaves these cluster structures intact. To produces favorable visualization results, a mapping of this sort should preserve the distance relationships between the clusters in a higher-dimensional space without making them overlap each other. This can be expressed in statistical terms by saying that the mapping should minimize the intra-cluster variance and maximize the inter-cluster variance. This sort of mapping is generally implemented by an algorithm known as discriminant analysis [7].

It thus follows that a visualization algorithm that preserves cluster structures can be implemented by using discriminant analysis to map the results of clustering in a higher-dimensional space to a lower-dimensional space. We call this visualization algorithm cluster discriminant analysis.

2.2 Detection outliers by cluster discriminant analysis

Cluster discriminant analysis is a visualization method in which clustering is first performed in a higher-dimensional space, and then discriminant analysis is used to find a way of mapping the clusters that have been found to a lower-dimensional space. Since it uses this two-step approach, we considered that cluster discriminant analysis should be able to detect outliers easily. Outliers are data that lie outside the normal region of most of the data. Cluster discriminant analysis thus regards outliers as separate clusters at the clustering stage, and as a result it maps them to positions separate from the normal data.

In conventional statistical methods—such as the method proposed by Knorr et al. [1] and methods where threshold values are provided for Mahalanobis distances and principal component scores—the problem of how to set the threshold values remains unresolved. As a result, humans have to set the threshold values based on experience. This makes such methods impractical for use in commercial data mining, where scatter plots are used to perform manual detection instead. The detection of outliers by cluster discriminant analysis should therefore prove to be a useful technique for commercial applications.

2.3 The observation of discriminant hyper-surfaces

To address the problem of deriving identification rules, it is essential to study the properties of discriminant hyper-surfaces. In this study, we attempted the visualization of discriminant hyper-surfaces as well as the detection of outliers.

To observe discriminant hyper-surfaces, we used a slightly modified version of the algorithm used in the original cluster discriminant analysis. First of all, the data is divided into categories based on the target variables, and then clustering is performed on the data in each category. The categorized data is then recombined, and visualization is performed by mapping the clusters into a lower-dimensional space found by performing discriminant analysis with the cluster labels regarded as categories.

With this method, visualization should still be possible when the discriminant hyper-surfaces degenerate in the lower-dimensional space. Moreover, it should be possible to visualize non-linear hyper planes as well as linear hyper planes.

However, it is important to bear in mind that the selection of an algorithm using this technique is only a sufficient condition. In other words, when a discriminant hyper-surface can be observed with this method, then it is clear that a discriminant hyper-surface exists in the higher-dimensional space and a rule extraction algorithm can be selected based on this information. But when no discriminant hyper-surfaces can be observed, it is impossible to determine whether it is essentially impossible to discriminate the clusters or whether the structure of the identification hyper planes is incapable of being visualized by this method.

3 Application to the analysis of meningitis data

In this section we describe the use of cluster discriminant analysis to data relating to cases of meningitis, and we show experimentally that it is effective at extracting outliers that have an adverse effect on the analysis.

For this experiment we used medical data relating to the differential diagnosis of meningitis published in the KDD Challenge 2000 [8]. This data consists of the records for 140 patients. These records contain 38 attributes, ranging from age and sex to leukocyte counts and cerebrospinal fluid cell counts. The actual data consisted of two types of variable-numerical variables and categorical variables. The presence of categorical variables resulted in cases where rank defects occurred in the covariance matrices, making it impossible to apply discriminant analysis. We therefore decided to exclude the attributes of categorical variables from the data, and for the actual analysis we used 16 numerical variables as the attributes.

The objective of analyzing the meningitis example data was to extract identification rules relating to the following three target variables [8]:

- Important factors for diagnosis (Diag):
Meningitis can be broadly divided into bacterial and viral forms. It is basically diagnosed according to the ratio of coenocytes (cells with multiple nuclei) to monocytes (cells with a single nucleus) in the cerebrospinal fluid. A large number of coenocytes indicate bacterial meningitis, and a large number of monocytes indicate viral meningitis.

- Important factors for finding the strain (CULT_FIND):
Since different strains of meningitis are treated differently, the discovery of factors that have an important bearing on finding out the strain of the disease is of great significance.

- Important factors for sequelae (COURSE):
The presence or absence of sequelae (lasting effects of the disease) depends on whether or not treatment is given at a suitable period. However, it is not known which factors are related to convalescence and sequelae. Consequently, the discovery of factors that determine the presence or absence of sequelae is also of great significance.

In the following, these three target variables are referred to as factors.

In our experiments, we used the k-means method as the clustering algorithm, and we used Fisher's canonical discriminant analysis as the discriminant analysis algorithm.

3.1 Detection of outliers

Figure 1(a) shows the results of visualization by cluster discriminant analysis with the number of clusters set to 3. In the following figures, a single point in the scatter plots corresponds to the data for a single patient, and points indicated by the same symbols indicate that they belong to the same cluster. The encircled data in the figures exists separately from the other data and is thought to correspond to outliers, at least in a statistical sense.

Figure 1(b) shows the results of visualization with the number of clusters set to 3 after this outlier data has been excluded. In this figure, the encircled data is situated further away from the other data. These outliers appeared to differ from the other data in terms of attributes relating to the cerebrospinal fluid.

Finally, Figure 1(c) shows the results of visualization after excluding the data detected in Figure 1 (b) (which seem to correspond to abnormal values). In this figure, none of the data is distributed separately from the rest, and it can thus be assumed that the detection of outliers is complete.

3.2 Observing discriminant hyper-surfaces

To extract identification rules, we observed the discriminant hyper-surfaces.

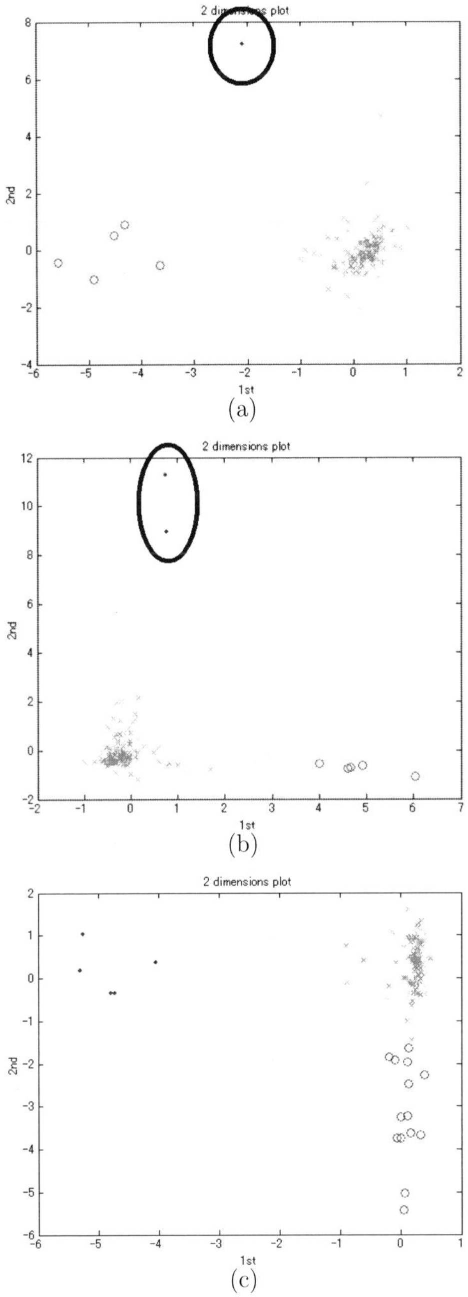

Figure 1: The process of detecting outliers from the meningitis data

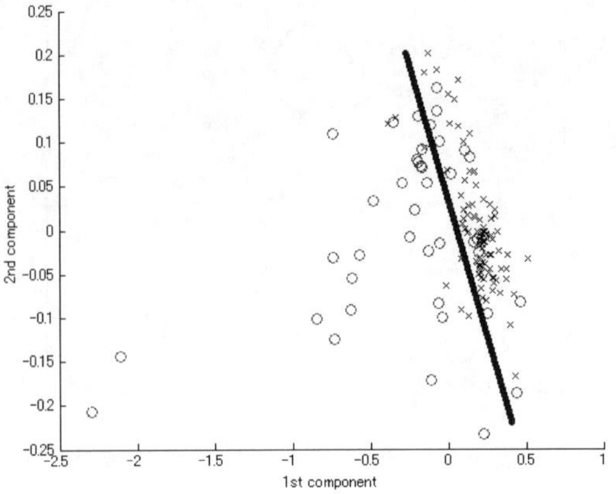

Figure 2: Results of observing discriminant hyper-surfaces relating to the identification of important factors for diagnosis (Diag)

Important factors for diagnosis (Diag)

Figure 2 shows the results of applying the visualization algorithm to the meningitis data from which outliers had been excluded. In this figure, open circles represent cases of bacterial meningitis, and crosses represent cases of viral meningitis.

In this figure, there is no complex interweaving between the data, and the discriminant hyper-surface looks linear. It thus seems that it is possible to extract rules adequately with a linear algorithm such as linear discriminant analysis. For example, it seems that a discriminant hyper-surface such as the straight line drawn in the figure exists in a higher-dimensional space, indicating the possibility of forming an effective rule for identification.

Important factors for finding the strain (CULT_FIND)

Figure 3 shows the results of visualizing the discriminant hyper-surfaces relating to important factors for finding the strain. In this figure, open circles represent cases where the strain could be found, and crosses represent cases where the strain could not be found. In this figure there appears to be an overlap between some of the data from different categories. However, since the part where the overlapping occurs does not constitute the entire data set, it is difficult to judge whether this shows that the data is basically impossible to discriminate or whether this is due to limitations of the visualization algorithm.

Important factors for sequelae (COURSE)

Figure 4 shows the results of visualizing the important factors for sequelae. In this figure, open circles represent cases where there were no sequelae, and crosses represent cases where there were sequelae. Like Figure 3, the data from different categories also appears to overlap in this figure. However, unlike Figure 3, there are many overlapping parts between the data from each category, and it thus appears to be very difficult to extract rules by using a linear algorithm.

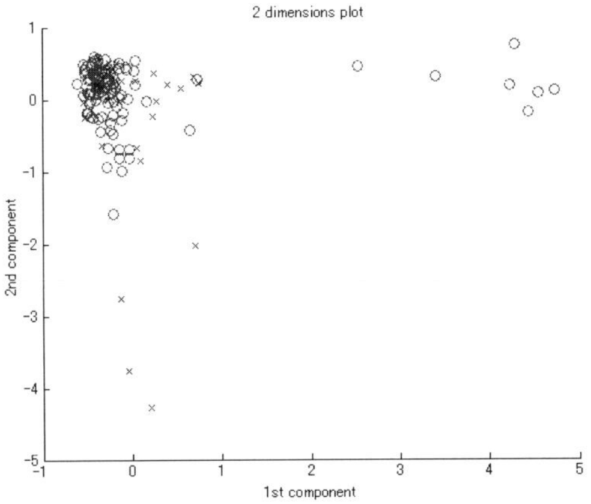

Figure 3: Results of observing discriminant hyper-surfaces relating to the identification of important factors for finding the strain (CULT_FIND)

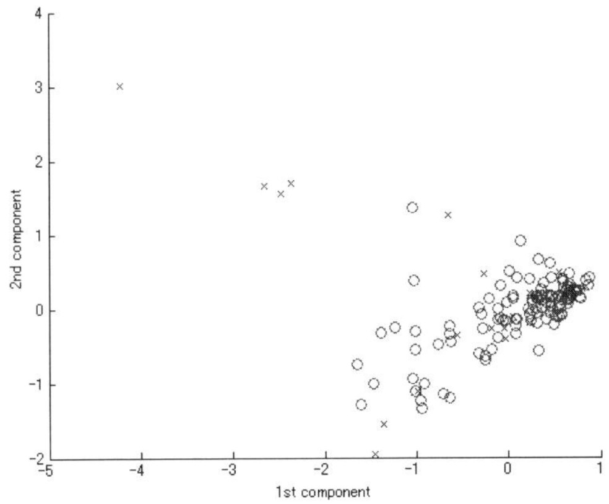

Figure 4: Results of observing discriminant hyper-surfaces relating to the identification of important factors for sequelae (COURSE)

Table 1: Identification accuracy relating to Diag (16 attributes)

	Learning data		Test data	
Category	vir	bac	vir	bac
No. of data items	20	20	20	77
Identification accuracy (%)	100	95	79	70

Table 2: Identification accuracy relating to Diag (3 attributes)

	Learning data		Test data	
Category	vir	bac	vir	bac
No. of data items	20	20	20	77
Identification accuracy (%)	100	75	92	75

3.3 Rule extraction

Finally, we tried to extract identification rules from these results. Although many different rule extraction algorithms can be used for the identification problem, the meningitis data used here contains a very small amount of data per attribute so we decided to employ a rule extraction algorithm based on linear discriminant analysis and sequential backward selection, and to derive rules by solving identification problems with small numbers of attributes.

In the following, we will show that the results obtained using the above algorithms are not inconsistent with the results of measurements relating to the validity of the linear identification algorithm.

Important factors for diagnosis (Diag)

Table 1 shows the identification accuracy of discriminant analysis when all the attributes are used.

The identification accuracy achieved with the test data represents the reliability of the extracted algorithm. That is, the reliability of the rules was found to be 70-80%, which shows that the discriminant hyper-surface obtained by cluster discriminant analysis does not contradict the observed results.

$$\text{Discrimination function} = -1.4 \times X_1 - 1.4 \times X_2 - 11.6 \times X_3 + 0.67 \tag{1}$$

$$\left(\begin{array}{ll} X_1: & \text{The number of days since the patient started to get a headache} \\ X_2: & \text{Irritant proteins} \\ X_3: & \text{Number of coenocytes in the cerebrospinal fluid} \end{array} \right)$$

Next, we selected variables by using the sequential backward selection. As a result, we selected three variables and extracted the rule shown in Equation (1). Table 2 shows the identification accuracy of this rule.

These results show that it is possible to solve this identification problem with a considerably high level of reliability based on three attributes.

Table 3: Identification accuracy relating to CULT_FIND (16 attributes)

	Learning data		Test data	
Category	T	F	T	F
No. of data items	20	20	11	84
Identification accuracy (%)	80	85	81	47

Table 4: Identification accuracy relating to COURSE (16 attributes)

	Learn data		Test data	
Category	n	p	n	p
No. of data items	15	15	7	98
Identification accuracy (%)	93	100	71	40

The extracted rule shows that X_3 (the number of coenocytes in the cerebrospinal fluid) affects the identification about 8 times as much as X_1 (other irritant proteins) and X_2 (the number of days since the patient started to get a headache). Since doctors also first focus on the number of coenocytes in the cerebrospinal fluid in order to judge whether a case of meningitis is bacterial or viral during diagnosis, it can be judged that the rule obtained here is valid. However, the identification accuracy becomes extremely low when either of the other two attributes is disregarded, resulting in a meaningless rule. This demonstrates that the judgment of whether a case of meningitis is bacterial or viral (an essential part of the diagnosis performed by a doctor) cannot be made solely on the basis of the number of coenocytes, and must include other attributes.

Important factors for finding the strain (CULT_FIND)

Table 3 shows the identification accuracy of the discriminant analysis when all the attributes are used. In this table, T indicates cases where the strain could be found, and F indicates cases where the strain could not be found.

These results show that the use of linear discriminant analysis does not facilitate the extraction of rules for identifying important factors for finding the strain. This is consistent with the above finding that it is difficult to extract rules with a linear algorithm from the results of observing the discriminant hyper-surfaces.

However, the results do not show that it is impossible to extract identification rules by non-linear algorithms.

Important factors for sequelae (COURSE)

Table 4 shows the identification accuracy of discriminant analysis when all the attributes are used. In this table, n indicates cases where there were no sequelae, and p indicates cases where there were sequelae.

The discriminant hyper-surface observation results indicated overlapping between clusters, and the identification accuracy was low in experiments using linear discriminant analysis, so there is no contradiction with the visualization results.

4 Conclusion

In this paper we have introduced a new visualization algorithm called cluster discriminant analysis, and we have experimentally demonstrated the validity of this algorithm in terms of detecting outliers and observing discriminant hyper-surfaces.

In the future, we will check to see whether or not the outliers detected here correspond to abnormal values that are meaningful from a medical viewpoint, and we will verify the extracted rules. Furthermore, by applying this technique to other types of data [5] we will examine the relationship between the outliers extracted by cluster discriminant analysis to values that are considered to be abnormal in the relevant field.

In this study, the only valid experimental result we obtained is that there is no inconsistency between linear discriminant analysis and the results of observing linear discriminant hyper-surfaces. However, the extent to which practical results can be obtained in relation to non-linear identification algorithms is another interesting area for further study.

References

[1] E. M. Knorr and R. T. Ng. Algorithms for Mining Distance-Based Outliers in Large Datasets. *Proceedings of the 24th VLDB Conference*, pages 392–403, 1998.

[2] H. Hotelling. Analysis of a complex statistical variables into principal components. *Journal of Educational Psychology*, pages 417–441, pages 498–520, 1933.

[3] W. S. Torgerson. Multidimensional scaling: I. *Theory and method*, pages 401–419, 1952.

[4] T. Kohonen. Self-organized formation of topologically correct feature maps. *Biological Cybernetics*, pages 59–69, 1982.

[5] T. Suenaga, A. Sato, and H. Sakano. KES2002(to be submitted).

[6] J. MacQueen. Some methods of classification and analysis of multivariate observations *Proceedings of the Fifth Berkeley Symposium on Mathemtical Statistics and Probability.*, pages 281–297, 1967.

[7] R. A. Fisher. The use of multiple measurements in taxonomic problems. *Annals of Eugenics.*, pages 179–188, 1936.

[8] E. Suzuki, (ed.) Proc. Int'l Workshop of KDD Challenge on Real World Date, 2000. http://www.slab.dnj.ynu.ac.jp/challenge2000.

III

USER REACTION AND INTERACTION

Active Mining
H. Motoda (Ed.)
IOS Press, 2002

Evidence-Based Medicine and Data Mining

Developing a Causal Model via Meta-Learning Methodology

Masanori Inada*[1] and Takao Terano*[2]
m-inada@mac.email.ne.jp, terano@gssm.otsuka.tsukuba.ac.jp
*[1]Department of Clinical Laboratory, Toranomon Hospital
2-2-2 Toranomon, Minatoku, Tokyo 105-8470, JAPAN
*[2]Graduate School of Systems Management, Tsukuba University
3-29-1, Otsuka, Bunkyoku, Tokyo 112-0012, JAPAN

Abstract. This paper discusses the applicability of data mining for Evidence-Based Medicine (EBM), which means integrating individual clinical expertise with the best available external clinical evidence from systematic researches. The objectives of data mining in medicine include the derivation of valuable knowledge which will be able to provide new comprehension beyond conventional medical experience. Although data mining is useful to explore latent knowledge, the outcome usually just has low-grade evidence because of the non-controlled bias and confounding. In the practice of EBM, we emphasize the point that experimental studies such as randomized controlled trials provide strong evidence and observational retrospective studies are hard to contribute for generating strict clinical evidence. Hence, the mined hypothesis must be refined through the validation process designed as prospective studies or experimental studies independent of primary data mining tasks. As an alternative process, meta-learning methodology would provide a new manner that evaluates the validity of mining results to refine mined knowledge and integrates knowledge obtained at several research groups. The meta-learning methodology which includes applying several data mining algorithms with cross-validation can contribute to generating evidence by supporting construction of credible models such as causal models. We propose a framework to develop causal models via meta-learning methodology.

1 Introduction

In this paper, we discuss the applicability of data mining for Evidence-Based Medicine (EBM) [11] and propose a framework to develop causal models via meta-learning methodology [2, 16]. The objectives of data mining in medicine include the derivation of valuable knowledge, which will be able to provide new comprehension beyond conventional medical experience. The meta-learning methodology, which includes applying several data mining algorithms with cross-validation, can contribute to generating evidence by supporting the construction of credible models which integrate expert knowledge and the mined knowledge.

Recently the concern of EBM is increasing in the medical field. The EBM contributes to realization of reliable medicine. On the basis of outcomes shown in clinically relevant researches, the EBM provides standardized medical diagnoses and treatments. The standardization of medicine is expected for the prevention of medical mistakes and repression of medical expenses.

EBM is realized by three stages: generating evidence, being accessible to evidence and utilizing evidence. Information technologies such as data analysis, creating database and retrieving information contribute to EBM at each stage. Data mining, which is a new concept of data analysis, in the interdisciplinary field of KDD (knowledge discovery in databases), encompassing statistical, pattern recognition, machine learning and visualization tools, can contribute to generating evidence, because it derives the useful pattern from abundant data. Modern hospitals are well equipped with information systems, which provide relatively inexpensive means to collect and store the medical data. It is worthy to use stored data for EBM from the viewpoints of expense and efficiency.

Although data mining is useful to generate evidence from medical field data, the outcomes usually just are low-grade evidence because of the non-controlled bias and confounding. In the practice of EBM, we emphasize the point that experimental studies such as randomized controlled trials provide strong evidence and observational retrospective studies are hard to contribute for generating strict clinical evidence. Hence, the mined hypothesis must be refined through the validation process designed as prospective studies or experimental studies independent of primary data mining tasks. As an alternative process, the meta-learning methodology would provide a new manner that evaluates the validity of mining results to refine mined knowledge and integrates knowledge obtained at several research groups.

To acquire the practical causal models, it is important that experts participate in model implementation. To build the models by expert knowledge using modeling tools such as structural equation modeling [1] is worthy. In order to make the objective knowledge based on data reflect to the models, the meta-learning methodology provides a way to understand the given data set deeply because several data mining algorithms are applied to same data set. Obtained insights are useful to build practical causal models. This procedure is a new methodology for evidence generation.

The paper is organized as follows. The second section describes EBM briefly. The third section discusses the applicability of data mining for EBM after clarifying the problem of observational study. The fourth section provides a framework to develop causal models as new mining methodology to build the practical models that integrate expert's subjective knowledge and objective knowledge based on data.

2 Evidence-Based Medicine

According to Sackett [11], EBM is the conscientious, explicit and judicious use of current best evidence in making decisions about the care of individual patients. The practice of evidence-based medicine means integrating individual clinical expertise with the best available external clinical evidence from systematic research. The concrete procedure is shown as follows.

1. Formulate the problem of the patient
2. Collect medical information effectively
3. Appraise the information critically
4. Adapt the information to the patient
5. Evaluate posteriorly

In this procedure, the critical appraisal of the information is very important. When using the information as evidence, the quality of the information must be strictly evaluated.

The level of Evidence depends on the design of the clinical research. Several organizations have shown the grade of evidence. Table1 shows evidence levels defined by AHCPR (the Agency for Healthcare Policy and Research, now the Agency for Healthcare Research and Quality). The evidence obtained from meta-analysis [7] of several randomized controlled trials is best grade. The evidence from at least one randomized controlled trial is better level. The evidence of well-designed controlled study without randomization and quasi-experimental study are good evidence. The evidence obtained from the observational study such as the case control study and from expert committee reports or opinions and/or clinical experiences of respected authorities are low-grade evidence. The outcomes of experimental study are classified to high-grade evidence. Although the outcomes of observational study are classified to low-grade evidence, it is important for clinical researches, because experiments such as the risk evaluation by harmful material are sometimes impossible ethically.

The digital libraries, which provide the environment for the EBM practice, have been built. The cochrane library [4] provides pieces of evidence reviewed by specialists. In addition, AHRQ (the Agency for Healthcare Research and Quality) , NGC (the National Guideline Clearinghouse) [8] and PubMed [10] provide services related to the EBM.

Table 1: Classification of Evidence Levels for Interventions

Ia	Evidence obtained from meta-analysis of randomized controlled trials.
Ib	Evidence obtained from at least one randomized controlled trial.
IIa	Evidence obtained from at least one well-designed controlled study without randomization.
IIb	Evidence obtained from at least one other type of well-designed quasi-experimental study.
III	Evidence obtained from well-designed non-experimental descriptive studies, such as comparative studies, correlation studies and case studies.
IV	Evidence obtained from expert committee reports or opinions and/or clinical experiences of respected authorities.

3 Applicability of Data Mining for EBM

In order to realize EBM, we must develop the practical EBM methodology and collect many pieces of evidence. In this section we first clarify positioning of data mining on the tasks of generating evidence, and describe the applicability of data mining for EBM. We second refer to meta-analysis which generates strong evidence and third discuss the methodology of meta-learning to evaluate the validity of mined results.

3.1 Generating Evidence and Data Mining

Data mining is recognized as observational study at the point that it treats the data obtained without the clear analysis purpose. As shown by Table1, results of observational study are low-grade evidence and are hard to use as evidence. Here it is necessary to clarify the role of observational study.

Although experimental study such as randomized controlled trials produces high-grade evidence, there are ethical and economical restrictions. In epidemiology investigations, exposure experiments that evaluate risk factors could not be conducted ethically.

Economical restrictions often interfere with execution of the experimental study. In the aspects of research expense and efficiency of data gathering, observational study is better than experimental one. We must select observational study at the study that contains an ethical problem. The role of observational study is to complement experimental study.

Observational study can be classified as follows [9].

Case-series is a report on a series of patients with an outcome of interest. No control group is involved.

Case-control study involves identifying patients who have the outcome of interest (cases) and control patients without the same outcome, and looking back to see if they had the exposure of interest.

Cross-sectional study is the observation of a defined population at a single point in time or time interval. Exposure and outcome are determined simultaneously.

Cohort study involves identification of two groups (cohorts) of patients, one which did receive the exposure of interest, and one which did not, and following these cohorts forward for the outcome of interest.

If study designs of observational study are classified from the viewpoint of time sequences, case-control study is retrospective, cross-sectional study is prevalence and cohort study is prospective. Usually, the study design of data mining is retrospective or prevalence study. From the reason that confounding factor and bias never controlled, the pieces of knowledge derived from these designs of study are mere hypotheses. The process of the hypothesis verification is necessary to refine it as evidence. For example, it is useful to add prospective study or experimental study. This subject is an important point to emphasize in the medical application of data mining. At least data mining is useful for medicine data analysis as exploratory data analysis tool.

In the conventional medical data analysis, statistical techniques have been used to verify the hypothesis contemplated by researchers. Statistical techniques mainly treat simple hypothesis or simple knowledge structure. So it is useful to apply the data mining for more complex knowledge structure. Human body is characterized by homeostasis and fluctuation. Even if measurement is accurate, it is difficult to grip true phenomena. Whereas a specific pattern appearing in typical diseases enables clinicians to diagnose, interpretation of clinical laboratory data is not easy because interactions between many factors cause diseases complexly. We point out that it is necessary to recognize disease models as complicated models. Approach to the medical data analysis by data mining provides a starting point for complex modeling of medical data.

In business fields the novelty or the remarkableness of knowledge are appreciated. On the contrary, in the medical fields the credible knowledge is required. There is a demand for the certain knowledge discovery from large-scale data analysis to acquire the complex knowledge structure and reevaluate known medical knowledge. Construction of the credible knowledge base is needed to realize the EBM.

To acquire reliable knowledge the interpretation/evaluation process of the knowledge is important. For example, the interpretation of the model obtained using the neural network is very difficult because of black-box approach. By only evaluation of the classification accuracy, from the ethical aspect in the medical application the obtained model will be difficult to utilize. Although, for example, trying the interpretation of the knowledge using decision tree analysis about the cluster obtained by the self-organization mapping is worth. Meanwhile techniques that focus the evaluation process of knowledge have been developed. Ishino and Terano have developed the SIBILE [15] which makes users evaluate the mined knowledge interactively during mining process. We are developing the SIBILE [15] enhanced toward the medical application using the criteria of sensitivity (true positive rate) and specificity (true negative rate).

3.2 Meta-Analysis and Meta-Learning

The previous section described the applicability of data mining about primary analysis to raw data for generating evidence. This section discusses the applicability of data mining about secondary analysis to outcomes obtained by primary analyses.

The special report of Science in 1995 [14] raises the problem that epidemiology researches will be denied shortly after being announced one after another. Causes were searched by the interview to the epidemiology researchers and the medical statisticians. The article concluded a cause is difference of bias size of each study by the fact that confounding factors are not sufficiently controlled because of the insufficient randomization. Because of the fact that any researches cannot avoid the bias to some extent, evaluating more than 30 similar studies is required to avoid the risk. Meta-analysis [7] is used to analyze the several independent research results.

Meta-analysis can find out the whole direction of conclusions, integrating the statistics obtained on each study. Meta-analysis is an important methodology in the EBM and provides strong evidence. There is a crucial problem called publication bias that negative results are hard to be published. The world scale of research registration system needs to be built for fundamental solution.

Meta-analysis, which integrates statistics on the outcomes of researches, cannot treat complicated knowledge structure. So, the meta-learning methodology, which is an element technology of the distributed or parallel KDD is expected. Meta-learning is defined to learn from the learned knowledge loosely. Several research groups have proposed various methodologies of meta-learning [16]. Many research groups present the goal of the meta-learning as the improvement of the classification accuracy, the appropriate selection of the learning algorithms and the selection of the appropriate bias.

Chan and his colleagues in Columbia University have proposed the meta-learning methodology that builds meta-classifier using meta-level training data set consist of the predictions obtained by base-classifiers learned at some data sites [3]. Stolfo et al. [12] urged the meta-classifier shows the high classification accuracy than base-classifiers based on the empirical results of frauds detection problem of the credit card. METAL projects [6] have developed the tools to guide the appropriate data mining technique for inexperienced user and to guide the appropriate preprocessing techniques. Suyama and Yamaguchi have developed the CAMLET that is a platform for automatic composition of inductive applications with method repositories [13]. The problem of the appropriate bias selection was discussed in the special issue of Machine Learning in 1995 [5] as the framework to guide the design and the development of the new machine learning system. The expression bias as the selection problem in the appropriate hypothesis space and the procedure bias as the problem of the model selection are discussed. The SIBILE is a meta-learning tool that deals with feature selection problems as expression bias selection problem.

We position the meta-learning methodology proposed by Chan [2] as a knowledge acquisition phase to support construction of the practical model. The methodology includes the comparison of the predictive performance and integration of learned models. The meta-learning methodology provides an opportunity of validity evaluation of mined models or knowledge. In the next section we discuss the validity evaluation.

3.3 Validity Evaluation and Model Building Support

In order to avoid unsuitable results influenced by bias or confounding, we introduce the meta-learning methodology and carry out comparative evaluation of learning results. The role of generalization of knowledge is expected of the meta-learning that carries out comparison and integration of the models obtained by data mining.

We here define the internal validity and external validity as follows: the internal validity is how much appropriately the given data set is explained, the external validity is how much appropriately the models or knowledge is applicable to unknown data set. The internal validity can be evaluated by the comparison between results of various mined models from the same data set by data mining algorithms. The external validity limited in a specific data source can be evaluated by the cross-validation technique. To generalize the model or knowledge beyond the specific data source, comparing and integrating the model or knowledge across several research groups is necessary. As discussion above, the meta-learning methodology is effective for validity evaluation.

The protocol about data collection and data analysis has been necessary as a premise of application of the statistical technique. It is the misuse of the statistical technique to apply the techniques that are not incorporated in the protocol after data collection. This criticism which avoids the problem that researchers change the meaning of information to what they seeks is important in the medical fields.

However, to perform only limited analysis shown in the protocol makes lose an opportunity to extract many pieces of knowledge from data. In the data mining, a strong demand about data collection is not required. Many methods such as feature selection and the data preprocessing are proposed on the premise that the extensive abundant stored data is used effectively. In the meta-learning methodology, different data mining algorithms are applied to the same data set, and the trials that effectively acquire many types of knowledge from data set are performed. To understand the given data set deeply, and to explore more complicated knowledge structure widely, the rough exploratory data analysis mentioned above is useful.

4 Proposed Mining Methodology

In this section we present a novel data mining methodology (Figure 1), which includes a knowledge evaluation method with meta-learning about the performance of predicted values. The objective of the proposed methodology is to obtain high credible knowledge expressed as a causal model for medical data analysis. The methodology consists of the following phases: model exploration, model implementation, model validation and model integration.

The model exploration is m $\times$ n parallel KDD processes containing m data mining algorithms and n-fold learning. We evaluate each learned model by comparing its predictive values against unknown test data. Simultaneously we get many types of knowledge, which is useful to construction of practical models, from results of several mining algorithms. In the model implementation phase we generate practical models from both background knowledge and acquired knowledge from the data. We build the

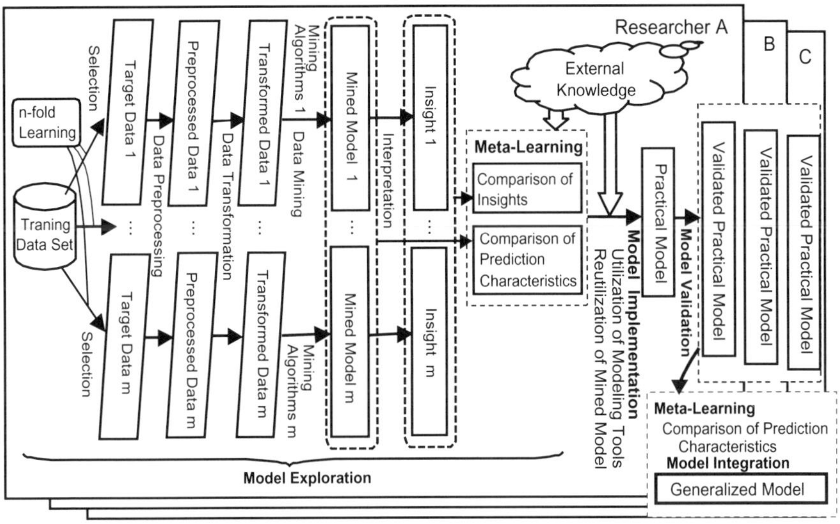

Figure 1: Proposed Mining Methodology

models using model-building tools such as structural equation modeling [1] or select a suitable mined model among model exploration phase. Next we verify the prediction performance of the practical models with test data set.

In the model integration phase, we compare the prediction performance of practical models obtained by independent different data sources to generalize the acquired knowledge. This phase is to generate a generalized model integrated from individual practical models.

The expected effects according to this methodology are summarized as follows.

- Many types of knowledge, which contribute building practical models, are extracted from given data set by application of several data mining algorithms.
- Suitable modeling is conducted by results of the model exploration phase.
- Relative criteria for model evaluation are obtained, because the predictive characteristics of various data mining models are compared.
- Practical models including background knowledge are built, when the user oneself builds models using model-building tools.
- The internal validity can be evaluated by comparing the results of several data mining algorithms, external validity can be evaluated by cross-validation composed of n-fold learning and use of unknown test data and generalization of acquired knowledge is enabled by model integration phase.

5 Conclusion

In this paper we have discussed the applicability of data mining to EBM and have proposed a method to develop causal models. We summarize as follows.

- To refine mined hypothesis as evidence, it is necessary to add another verification research. Generally the study design of data mining is retrospective observational study. Because there are non-controlled bias and confounding in stored data, the acquired knowledge is mere hypothesis.

- Data-mining technology contributes to reevaluate the present medical knowledge by discovering complicated knowledge structure. Data mining is an exploratory data analysis tool. To model biomedical data, the viewpoint as the complex systems is necessary.

- Because of the ethical aspect in medical fields, the process of interpretation/evaluation in KDD is important for medical application. We must develop techniques that support this process.

- The meta-learning methodology provides useful knowledge in the practical model construction by users, and it provides the opportunity of validity evaluation.

EBM is a still immature domain. We need to develop mining technologies to medical application aggressively.

References

[1] K.A. Bollen. Structural Equations with Latent Variables. John Willy & Sons, 1989.

[2] P.K. Chan. An Extensible Meta-Learning Approach for Scalable and Accurate Inductive Learning. PhD Thesis, Graduate School of arts and Sciences at Columbia University ,1996.

[3] P.K. Chan and S.J. Stolfo. On the Accuracy of Meta-learning for Scalable Data Mining. Jornal of Intelligent Information Systems, 8(1): 5-28, 1997.

[4] The Cochrane Collaboration homepage, 2002. http://hiru.mcmaster.ca/cochrane/

[5] M. desJardins and D. Gordon, editors. Special issue on Bias evaluation and selection, Machine Learning. 20(1/2), 1995.

[6] METAL Project homepage, 2002. http://www.metal-kdd.org/

[7] B. Mullen. Advanced BASIC Meta-Analysis. Lawrence Erlbaum Associates, 1989.

[8] National Guideline Clearinghouse homepage, 2002. http://www.guideline.gov/index.asp

[9] NHS Research and Development. Centre for Evidence-Based Medicine, Evidence-Based Medicine Glossary, 2002. http://cebm.jr2.ox.ac.uk/docs/glossary.html

[10] PubMed homepage, 2002. http://www.ncbi.nlm.nih.gov/PubMed/

[11] D.L. Sackett, W.M.C. Rosenberg, J.A.M. Gray, R.B. Haynes and W.S. Richardson. Evidence-based Medicine: what it is and what it isn't (editorial). British Medical Journal, 312: 71-72, 1996.

[12] S.J. Stolfo, D.W. Fan, W. Lee, A. Prodromidis and P. K. Chan. Credit Card Fraud Detection Using Meta-Learning: Issues and Initial Results, In Proc. AAAI-97 Workshop on AI Methods in Fraud and Risk Management, 1997

[13] A. Suyama and T. Yamaguchi. Automatic Composition of Inductive Applications Using Ontologies. Journal of Japanese Society for Artificial Intelligence, 15(1): 155-161, 2000.

[14] G.Taubes. Epidemiology Faces Its Limits. Science, 269(14): 164-169, 1995.

[15] T. Terano and Y. Ishino. Interactive Knowledge Discovery from Marketing Questionnaire Using Simulated Breeding and Inductive Learning Methods. In Proc. 2nd Int. Conf. on Knowledge Discovery & Data Mining (KDD ' 96), pages 279-282, Portland, OR, 1996.

[16] R. Vilalta and Y. Drissi. Research Directions in Meta-Learning, In Proc. of the 2001 International Conference on Artificial Intelligence, Ed. H.R. Arabnia, Las Vegas, 2001.

Active Mining
H. Motoda (Ed.)
IOS Press, 2002

KeyGraph for Classifying Web Communities

Yukio Ohsawa*,**, Yutaka Matsuo*, Naohiro Natsumura*
Hirotaka Soma**, Masaki Usui**
{osawa,matsuo,matumura}@miv.t.u-tokyo.ac.jp
soma@mfj.or.jp,usuim@jn.nittobo.co.jp
**Japan Science and Technology Corporation
**The University of Tsukuba
3-29-1 Otsuka, Bunkyo-ku Tokyo 112-0012, JAPAN

Abstract. Textual communication in message boards is analyzed for classifying Web communities. We present a communication-content based generalization of an existing business-oriented classification of Web communities, using *KeyGraph* a method for visualizing the co-occurrence relations between words and word clusters in text. Here, the text in a message board is analyzed with *KeyGraph*, and the structure obtained is shown to reflect the essence of the content-flow. The relation of this content-flow with participants' interests is then formalized. Structural features of relations between participants and words, determining the type of the community, are shown to be computed and visualized.

1 INTRODUCTION

Communities are catching social attentions because the decision of a human relies on information available in one's belonging group of people [1, 2, 3]. This trend existed from long time ago, as long as the history of human culture. Meetings for group-wise decision making and conversations with coworkers have been the information source for each participant. The recent outbreak of community-wares come from the growth of the variety of powerful communities such as enterprise decision teams, customer groups etc., hand in hand with the growth of communities on the Internet, e.g., mailing lists, message boards, and chat rooms. These trends made various community-based human activities for people with various value criteria.

As a result, various range of Web communities appeared - match (friends or mail-pals) making, problem solving, clubs, etc. We can classify them into various types of communities for various aims. In one of match-making type, frictions have been caused in the real world by partners who came to know each other in the virtual world, i.e., the Internet. On the other hand, in a problem solving community, the content flow rather than the participants' characters are the main focus of attentions for most participants. The latter type may enable the creation of useful knowledge, rather than new human relations. However, in a problem solving community, new comers may feel hard to take part in the communication because the discussions are highly specialized and the value criteria shared are strongly biased, so people with average knowledge seems to be excluded. This sometimes disturbs the growth of community. Between these two extreme types, chats about easy topics allowing the entry of people with various values exist, where the excitement is remarkable but collisions due to gaps between value criteria can lead to flaming and the disorder of topics.

Thus, the aims and problems underlying each type of community have great impact onto business issues, e.g., whether the community creates useful knowledge, how easy it is to join the community, or why the community is good for making advertisement for commerce [4]. In this paper, we analyze the text of communication in message boards, which we take as an example of Web communities. The types of communities in existing classification as in [4] are generalized systematically, by featuring each type of community on three dimensions we introduced. These dimensions correspond to topological features of the output graph of *KeyGraph* [5], a co-occurrence graph of words and word-clusters for textual information.

2 SIX TYPES OF COMMUNITIES

The classification of communities in [4] is being accepted as a model of Web-based societies as business target. Because the book is in Japanese, let us show its classification of Web communities as follows.

A. Topic based community High-quality and up-to-date discussions for solving difficult problems appear. Experts of the corresponding area organize or lead the communication. This tends to focus the participants to a closed group of people with leading opinions.
Ex) http://www.cnn.com/COMMUNITY/

B. Problem solving community People with similar interests exchange ideas and knowledge, not of as high quality as in A, for solving the shared problem. The community is often hosted by ones particular about relevant areas, and one(expert)-to-many communications are likely to occur. Ex) e.g., http://www.about.com/

C. Product/service evaluation by users Products/services in the market are evaluated by users and the evaluations are circulated by the community of users. Reports and questions on experiences are prevalent. Ex) http://www.epinions.com/

D. Mutual supporting forum of users Cooperative exchanging of knowledge about products/services users already use, formed by a number of question askers and a few leaders.
Ex) http://supportforum.sun.com/

E. Community for friend-making or leisure The user- and content-management is the least considered, and free communication is desired. The quality of discussion may be low, but the group grows easily. Ex) http://www.yahoo.com

F. Club Private community organized by user(s).
Ex) http://clubs.yahoo.com/

3 THREE DIMENSIONS OF COMMUNITIES

The discussion above leads to the classification of communities on the u,v, and w as follows.
u : The centralization strength of contexts, defined as the topics or people: u =2: centralized strongly, u =1: weakly centralized, u =0: not centralized

v: The coherence of communication context: v=2: strongly coherent context(s), v=1: various (weakly coherent) contexts, v=0: not sharing contexts
w: Orientation to creative decisions: w=2: create ideas for new decisions, w=1: apply someone's knowledge to make decisions, w=0: do not make decisions new to oneself

Introducing these attributes, we can put the community classes **A** through **F** above as follows.

A: u=2, v=2, w=2.
B: u=2, v=1, w=1.
C: u=0, v=1, w=1.
D: u=1, v=1, w=1.
E: u=0, v=0, w=0 to 2.
F: u=1, v=1, w=0 to 2.

This classification implies the existence of other types of communities because we should have 27 (3*3*3) types. In the case of (u, v, w) = (0,1,0), e.g., ones in the poor-mannered community site, the communication spreading to various contexts by anonymous participants leads not to decisions, but to quarreling with words as "die" "stupid" etc.

In this paper, we propose a method to compute u, v and$w$, given a community message board and its text of messages and message writers. Realizing this method, we can guess what sense a community with seemingly complex communications is making and anticipate what sense it will make. For example, if we obtain (u, v, w) = (2,1,1), we can guess there is an expert leading the community to solve problems in a certain domain, as in a community of type **B**. This leader can be considered a good teacher for ones interested in the domain. On the other hand, if we obtain (u, v, w) = (1,1,0), we can guess the community is a club-like gathering and can join without hesitation. Further, if the community is creating useful knowledge, the communication content can be a textbook for business.

4 COMMUNICATION ANALYSIS WITH KeyGraph

4.1 Threads and Sub-community Relations

We compute u, v and w based on thread-based co-occurrence of words in a message board. If there is a message M not a response to a previous message and there are responses to M, those responses and M are called a thread as a set. If a pair of words co-occurs frequently in the same thread, we regard the pair as of high co-occurrence. For example, Figure 1 is a part of a message board talking about ecological issues. The part in Figure 1 focuses attention onto how to make a hybrid car - known as gentle to the atmosphere - prevalent. A group of participants who frequently co-occur in the same threads can be regarded as sharing the (at least temporary) context of interest.

Suppose there are two or more such groups of participants. A participant who stays in one group can be regarded as concentrated in a narrow specific context. On the other hand, one talking to many groups can play a significant role as a messenger helping those groups exchange information and consider new topics, or sending commands to multiple sections, although she may otherwise have just an unstable interest drifting among groups. In either case, she has a potential to develop her ideas to prevail to a wide part of the community, which may lead to innovating ideas. Generally, a leader can be:

a. An innovator thinking of new ideas, or
b. A messenger circulating new ideas to various people.

These people are followed by people in each local group or sub-community [1]. People who only organize or manage a community, e.g., an administrator of a mailing list, is not usually called a leader because such one does not show directions of communication. However, if this organizer gives new topics to talk about, he can be seen as a leader or:
c. Topic starter and coordinator as a role in the community.

That is, a leader is a participant from whom the community comes to grow in a radiating manner. In a centralized Web community, such a member catches attentions of other members, more strongly than mutual attentions between non-leader members. The stronger the centralization, the more the communication between the leader and others overwhelm distributed (among non-leaders) communication. As a result, the value of u in Section 3 can be expressed by the extent the community forms a one-to-many radiation structure.

On the other hand, a community where new ideas and leaders often occur has multiple groups among which information is exchanged directly via weak ties and stimulates changes [6, 7], rather than centralization. New ideas are sometimes imported to a certain group X from other groups, and people in X can create new knowledge by talking on the new information. As pointed in [8], new idea-combinations trigger innovations.

4.2 KeyGraph for Seeing Sub-community Relations from Threads

In order to catch the characteristics of community types, we construct a graph representing human-human and topic-human relations based on the text in message boards. From the discussion above, this graph is desired to satisfy the following conditions:

Condition 1: People in the same group, i.e., sharing the communication context are included in the same cluster of people or words in the graph.

Condition 2: People touching multiple clusters of words or people are depicted in the graph.

As a graph satisfying both conditions, we apply *KeyGraph* [5]. In *KeyGraph*, the input data D of a1, a2, ... occurring sequentially is dealt in the form as:

$D =$ a1, a2, ... an. b1, b2, b3 ..., bm. c1, c2, ... cp ...

We call each datum "a1", "a2", ...or "cp" a word, and the sequence between two nearest periods a sentence. Accordingly, D is called a document. The algorithm of *KeyGraph* is summarized as follows, with the metaphor of building (make basis, columns, and then the roofs)

The Process of *KeyGraph* [5]

Step 1: Take frequent words in D, and connect a pair of words with an edge called a stick if the number of sentences including the pair is larger than a preset threshold. Here some connected graphs come out, and we call each of them a basic cluster.

Step 2: If a word co-occur (appear in the same sentence) with words in a certain basic cluster more than a preset threshold frequency, the word and the cluster is connected by an edge called a column (dotted line in Figure 2).

Brazil to Spend 25BilliononRenewableEnergy
– Elizabeth7/25/99
Re : BraziltoSpend25 Billion on Renewable Energy
- Sam 9/06/99
Solar Cells Thinner Than Human Hair
- Elizabeth 7/25/99
Re: Solar Cells Thinner Than Human Hair
- Paul Wilcoson 6/11/00
Re: Solar Cells Thinner Than Human Hair
- Marlan Cowley 10/10/00

Figure 1: A message board with two threads.

Step 3: If a word is outside of basic clusters, where columns more than a certain threshold number come across, we call the word a roof (the double circles in Figure 2).

In the case of Figure 1, by assigning each participant name or each word in communication to a word and each thread to a sentence in *KeyGraph*, we obtain document
D = "Elizabeth Sam. Elizabeth Paul-Wilcoson Marlan-..."
from the communication history. The result of *KeyGraph* is as in Figure 2, where the lower part corresponds to the part of communication shown in Figure 1. Generally, a roof comes to be of low frequency but a significant position in the graph. In a community, sticks in clusters mean strong ties between people, whereas roofs and columns can be a messenger connecting multiple groups or an innovator touching various participants, and words realizing weak ties between clusters [6, 7]. Nodes and sticks in basic clusters were shown in black and columns and roofs were in red in *KeyGraph*, but let us focus attention to the topological structure of each graph if the paper is printed in black and white.

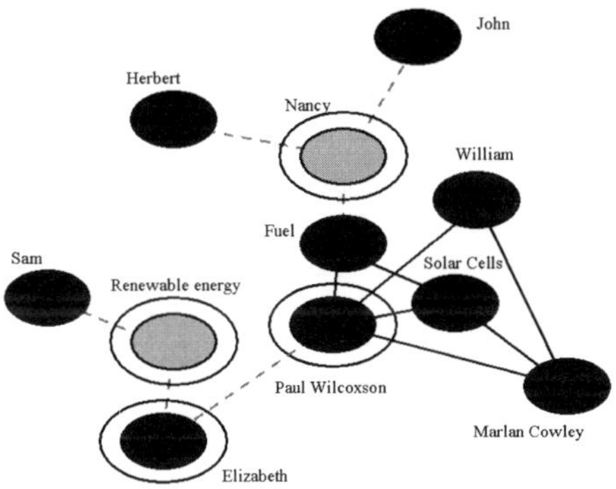

Figure 2: *KeyGraph* for a community of Figure 1

4.3 Relational Structure Representing Features of Community

Based on the output graph of *KeyGraph*, let us first consider u, the strength of centralization of a community: That is, the extent the graph looks like few-to-many radiation

Table 1: (Cx, Lx) for node x vs the position of x

	Large Cx	Small Cx
Large Lx	x is in a peripheral cluster	x is far from an enhanced center
Small Lx	x is in a central cluster	x is the center, not clustered

means the strength of leader if the few nodes represent participant-names, and the leading topic if the few nodes represent words in the communication.

A function appropriate for expressing the degree of centralization is thus desired. A small world has been introduced as a typical form of community in the nature, e.g.,a group of creatures, human, and Web pages [9, 10, 11]. The extent a community looks like a small world, i.e., small-worldliness S is defined as:

$$S = C/L. \tag{1}$$

A community of large S is called a small world. Here, C is the relative density of graph G, defined as the average rate of edges existing among nodes surrounding each node in G, (this rate is 2/3 for "Solar Cells" in Figure 2, because the number of edges connecting "Fuel" "Paul Wilcoson" and "Marlan Cowley" surrounding "Solar Cells" is 2, among all 3 possible edges). L is the average value of the shortest distance from each node in G to other nodes in G. Both C and L are normalized with dividing by their values for a random graph of the same number of nodes and edges as G. Differing from [12], here we define the distance between two nodes without a connecting path in G as extremely long - 100 times longer than other nodes.

If human relations in a community are represented by a graph, C means the effect of daily local communications of participants or basic concepts usually talked about. On the other hand, L means the difficulty of information propagation between participants linked via various relations including strong (usual or daily) and weak (unusually communicating) ties [11]. A small world of large S can be regarded as a community where weakly-tied pairs of strongly-tied local sub-communities exchange information to distill new ideas. That is, these variables represent the role of participants and concepts to other participants.

We extend the idea of small world, to formalize u and v as well as creativity w. Here, we introduce Cx and Lx separately for each node x and consider the role of nodes in the overall community. This leads to featuring the community, because the role of each particle, i.e., participant or a concept/idea expressed by words, to the community can be reflected to the role of the community to participants' thoughts and concept developments from communication.

Common facts about community formation are shown in Table 1. For example, a leader in a centralized community contacts other members, but those non-leaders do not make communications with each other often. If the non-leader members communicate and co-work to make decisions, we do not call it a centralized community, because the community comes to be a decentralized problem solving system. Thus both Cx and Lx take small values if x represents a leader. On Table 1, we formalize the role of a node in a graph as:

$$\alpha = 1/(CxLx). \tag{2}$$

$$\beta = Lx. \tag{3}$$
$$\gamma = Cx/Lx. \tag{4}$$

Here, each LHS symbol denotes α : The strength of leadership of x, β : The isolation of x, and γ : Casting (receiving) ideas distilled in clusters including (around) x.

Then, we express the extent of the centralization by υ, the coherence of communication context by ζ, and the application of various information to idea distillation and innovation in local communities by ω, of the community represented by the overall graph as in Eq.(5).

$$\upsilon = Max\alpha, \ \zeta = avr1/\beta, \ \omega = avr\gamma, \tag{5}$$

Here, Max and avr respectively mean the maximum and the average values for all nodes. We can then define the values of u, v and w in Section 2 as in Eq.(6), corresponding to Section 3.

$$\begin{aligned}
u &= 2 \quad (if\ \upsilon \geq \theta_1), u = 1(if\ \theta_1 > \upsilon \geq \theta_2), \\
u &= 0 \quad (if\ \theta_2 > \upsilon). \\
v &= 2 \quad (if\ \zeta \geq \theta_3), v = 1(if\ \theta_3 > \zeta \geq \theta_4), \\
v &= 0 \quad (if\ \theta_4 > \zeta). \\
w &= 2 \quad (if\ \omega \geq \theta_5), w = 1(if\ \theta_5 > \omega \geq \theta_6), \\
w &= 0 \quad (if\ \theta_6 > \omega).
\end{aligned} \tag{6}$$

Here, θ_1 to θ_6 are the given thresholds for variables. Based on experiences with various communities, we set θ_1=3.0, θ_2=1.5, θ_3=.3, θ_4=.1, θ_5=3.0, and θ_6=1.0 where communities could be classified to **A** to **F** with the keenest fitting to our impression of exiting Web communications.

5 THE RESULTS OF KeyGraph FOR WEB COMMUNITIES

In *KeyGraph* applied to a community as explained above, we can see co-occurrences of (a) participant-participant, (b) participant-word, and (c) word-word. These three types of co-occurrence mean (a) context-sharing group of people, (b) the interest of people in concepts, and (c) hidden context or concept underlying co-occurring set of words. Because the contexts, interests and concepts maybe hidden, i.e., do not appear explicitly in communications, let us look at the output figures of *KeyGraph*, and make numeric evaluations of the implication of each output.

Example 1: A community of type D, for distributed (cooperative but not centralized) solution of sub-problems relevant to a given problem
Figure 3 shows the result of *KeyGraph* for a community talking about energy-saving methods, in a community-collection site of ecology. Many opinions here are concentrated in the recent popular topic "hybrid cars." The value of u is 1 reflecting "Honda" as a (not strongly) leading topic, so we can see not a human leader but a leading topic organizing some part of the community. The graph is scattered to connected sub-graphs, and the value of v is 0. This means the community is of low consistency of context, e.g. aiming at solving multiple problems. w is 1, meaning the community is somehow dedicated to creative decisions relevant to hybrid cars. (u,v,w)=(1,0,1) does not correspond to any of **A** to **F** defined in Section 3. In fact, little new idea was observed to come out of this community - they exchanged existing experiences and knowledge.

However, if we take a part of the graph, we can find features of creative solving of coherent problems: The left-hand cluster of Figure 3 shows a structure in the order of "car brand (Honda) → first order functions (fuel, efficiency) → second order functions (solar, wind)." Here, the first order functions are functions of which average car users are aware, whereas the second order functions are recognized by car users with advanced consciousness of ecology. This connected graph, representing the relations between interests in these functions, have been formed by the communications of these different users. Thus, the left-hand side cluster has multiple contexts relevant to each other, relevant to car functions. If you trace the communication including words "Honda" "solar power" etc., in the order of node-connection, you will understand the meaning of second-order functions with smooth shifts from familiar (beginner's) to unfamiliar (advanced) contexts. This cluster takes (1,1,1) as the values of (u, v, w) supporting the fact this community is of type **D**, i.e., weakly centralized (around cars from Honda) for solving sub-problems relevant to a share problem (introduction of solar and hybrid cards to roads).

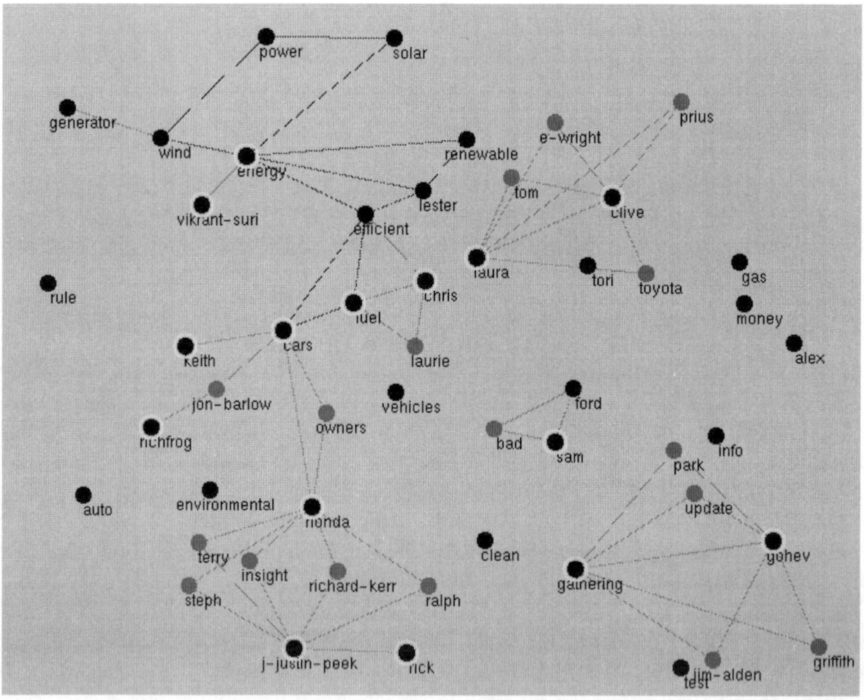

Figure 3: *KeyGraph* for a community on ecology.

Example 2: Type F as weakly centralized communications on freely selected topics under a shared context

Web communities about wine came to developed in various directions, and here we deal with one where users organize the message board for themselves. Figure 4 shows the result of *KeyGraph* for a community of wine collectors. The participants have specific knowledge about wine, and there is no sommelier much more particular about wine than other participants. Thus the community is decentralized by nature, although it has some participants with strong opinions as "tablewine_01", arrowed in the upper half of the graph. In this weakly centralized community with some local clusters (representing local communications) linked to each other as in a small world, participants discover new interests. (u, v, w) is (1, 1, 2) in this case, to form type **F** i.e., weakly centralized communications on freely selected topics in a shared context. This result matches with the fact people feel they are making a club here. In fact, the community is called a club on the Web, where participants sometimes create knowledge or serve others with new satisfactory information about wine bottles to collect.

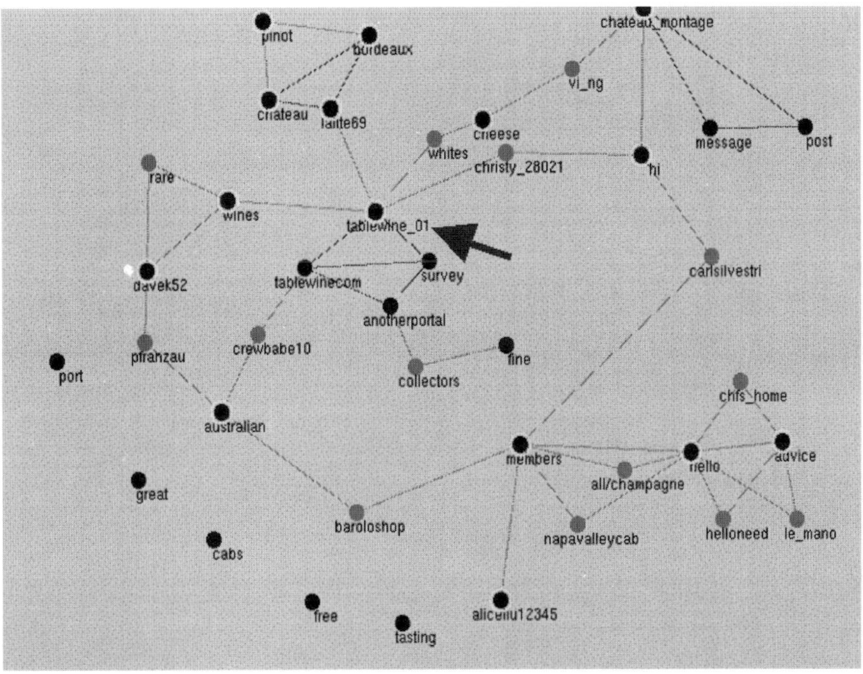

Figure 4: *KeyGraph* for "FINE WINE COLLECTORS" without a leading sommelier.

6 CONCLUSIONS AND FUTURE WORK

We made a three-dimensional space for featuring the essence of communities. This helps in surveying the essence of a community, e.g. whether the community creates useful knowledge, how easy it is to joint the community, or whether/why the community is good for making advertisement for commerce. In contrast with previous analysis of human networks on computer mediated communities (CMC) [13], we showed the raw and shallow textual information in communication-sites reflects the dynamics of various communities, across wide types from business aspect.

References

[1] Rogers, E.M., *Diffusion of Innovations*, Free Press (1962)

[2] Bordia, P. and Rosnow, R.L., Rumor as Group Problem Solving : Development Patterns in Informal Computer-Mediated Groups, *Small Group Research*, Vol.30 pp.8 - 28. (1999)

[3] Nonaka, I., and Takeuchi, H., *The Knowledge Creation Company* , Oxford University Press (1995)

[4] Ishikawa, N., *Competitive Advantage Community Strategy*, Soft Bank Publishing (2000) In Japanese

[5] Ohsawa, Y., et al, *KeyGraph*: Automatic Indexing by Co-occurrence based on Building Construction Metaphor, *Proc. Advanced Digital Library Conference (IEEE ADL'98)* pp.12-18 (1998)

[6] Granovetter, M.S., The Strength of Weak Tie, *The American Journal of Sociology* Vol. 78 pp.1360 - 1380 (1973)

[7] McPherson, J.M., Popielarz, P.A., and Drobnic, S., Social Networks and Organizational Dynamics.*American Sociological Review*, Vol.57 pp.153-170 (1992)

[8] Goldberg, D.E. The design of innovation: Lessons from genetic algorithms, lessons for the real world. Navigating Complexity, IlliGAL Report 98004 (1998).

[9] Watts, D.: *Small World: the Dynamics of Networks between Order and Randomness*, Princeton (1999)

[10] Watts, D. and Strogatz, S. Collective Dynamics of Small World Networks, *Nature*, 398 (1998)

[11] Albert, R., et al, The Diameter of the World Wide Web, *Nature*, 401 (1999)

[12] Matsuo, Y., Ohsawa, Y., and Ishizuka, M., A Document as a Small World, *New Frontiers in Artificial Intelligence, LNAI 2253*, (Springer Verlag), pp. 444 - 448 (2001)

[13] Garton, L., and Wellman, B., Studying On-Line Social Networks, *Doing Internet Research*, edited by Steve Jones, Thousand Oaks, CA (1999)

Active Mining
H. Motoda (Ed.)
IOS Press, 2002

Case Generation Method for Constructing an RDR Knowledge Base

Keisei Fujiwara, Tetsuya Yoshida, Hiroshi Motoda and Takashi Washio
{fujiwara,yoshida,motoda,washio}@ar.sanken.osaka-u.ac.jp
Institute of Scientific and Industrial Research,
Osaka University
Mihogaoka, Ibaraki, Osaka 567-0047, JAPAN

Abstract. Ripple Down Rules (RDR) Method is an incremental Knowledge Acquisition (KA) approach that is able to capture human expertise efficiently. The expert's KA tasks in RDR are 1) to identify the correct class label of each misclassified case and 2) to select important attributes that distinguish the misclassified case from the previous correctly classified case. The latter task is more difficult than the former one since it requires much thought on human expert. This paper proposes a method for reducing the task on human expert by generating context-bounded cases and utilizing them to replace the latter task with the former one. Experiments on the datasets from UCI were carried out to evaluate the proposed method and the result confirmed that it is effective and as good as the standard RDR method on most datasets.

1 Introduction

Ripple Down Rules (RDR) method[2] is one of the methodologies for realizing efficient knowledge acquision and maintenance. It is effective since it allows the incremental acquisition of knowledge. In addition, no knowledge engineers are required since the consistency of knowledge base is also maintained during the knowledge acquision phase.

The following two tasks are required for human experts during the knowledge acquision process in the standard RDR system:

task1: identify the correct class label of a case

taks2: induce the condition which distinguishes between two cases with different class labels based on the difference in the attribute values.

Compared with **task1**, **task2** is much harder for experts since there can be many candidates for the condition with different generalization capability. Thus, selecting an approporiate one can be very difficult for experts.

This paper proposes an efficient knowledge acquision method in which **task2** is replaced with **task1**. In the proposed method the expert is required only to carry out **task1** on cases which are generated by the method to induce the condition in **task2**. Since the expert does not need to carry out **task2**, it is expected that his/her cognitive load can be reduced.

Experiments were carried out to compare the proposed method with the standard RDR method which requires the expert to carry out **task2** in addition to **task1**.

2 Ripple Down Rules

The characteristics of RDR are summarized as follows:

- incremental and consistent: allowing incremental addition of exceptional rules contributes to maintaing the consistency of knowledge base.
- direct interface: knowledge is acquired through the interaction between experts and computers and no knowledge engineers are required.

Structure of knowledge base: a binary tree with Yes and No branches is utilized to represent the knowledge base in RDR as shown in Fig.1. Each node stores an If-Then rule and a Cornerstone Case which triggered the addition of the node into the knowledge base.

Inference: suppose the class label of a case (which is called Current Case in RDR) is asked for the knowledge base. The inference starts from the root node and repeats the following process. At each node when the condition in the If-Then rule of the node is satisfied, Yes branch is followed; if not, No branch is followed. The conclusion for the current case is given by the conclusion part of the node in the inference path for the case whose condition part is lastly satisfied.

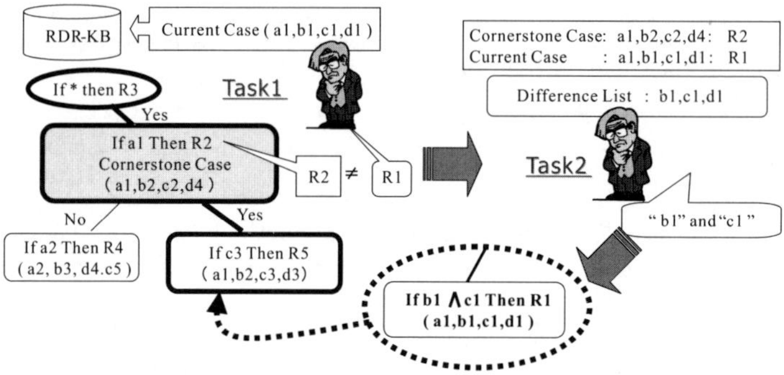

Figure 1: Knowledge acquision in RDR.

Knowledge acquision from experts: knowledge acquision is carried out when the expert does not agree with the result of the RDR system. First, the system returns the cornerstone case for the invoked If-Then rule. Second, the system creates a Difference List, which enumerates all the attributes with different values between the current case and cornerstone case. Third, the expert is required to select a set of attributes which best distinguish the current case from the cornerstone case. Note that any element in the list satisfies the current case but does not satisfy the cornerstone case. Fourth, the selected attributes and their values in the current case are utilized to create the condition part for the If-Then rule. Fifth, by treating the class label from the expert as the conclusion, a new node is created with the If-Then rule and the current case as its cornerstone case. Finally, the new node is attached as a child node to the end node, which is the last node in the inference process.

The difference between 2 cases is represented as the Difference List in Fig. 1. In the RDR system, knowledge constrained within the context of Difference List is acquired efficiently in the form of If-Then rule with **task2** [6]. However, deciding the appropriate pairs of attribute and its value can be difficult and thus **task2** can be hard for human experts.

3 Concept for Case Generation

In our approach cases to be asked for experts in **task1** are generated along with the work for the approximate construction of case base [5]. In this approach a case base for unknown logic function can be constructed approximately with the minimum number of cases. From a negative case and its nearest neighbor positive case, an item lattice shown in Fig.2 is constructed based on their difference. The cases in the lattice are generated and experts are asked to identify their class labels. The case base is refined gradually based on the class labels while its consistency being maintained. Furthermore, the uppor bound for the number of necessary questions (i.e., cases) is estimated analytically to determine the boundary for distinguishing between the positive and negative cases. In

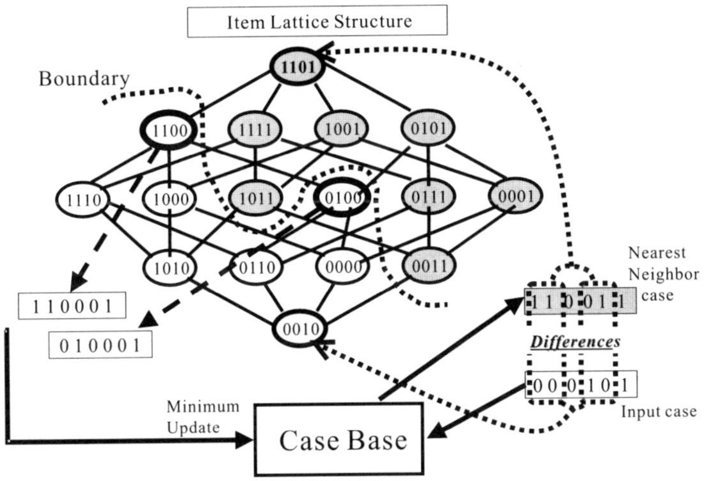

Figure 2: Estimation of the boundary in item lattice.

the context of RDR the current case for which the RDR system induced the wrong class label corresponds to the negative case. The cornerstone case for the induced class label corresponds to the nearest neighbor positive case. Thus, based on the difference list between these cases, the cases which lie in the item lattice for the difference list are generated and the boundary to distinguish between the cornerstone case and current case in the lattice is estimated based on the class label from experts through **task1**.

4 Six Algorithms for Generating Cases

This subsection describes 6 algorithms for generating cases based on the concept in Section 3 in order to replace **task2** with **task1**. The algorithms for generating cases in the item lattice are designed with respect to how many and what kind of cases should be generated to estimate the boundary in the lattice.

4.1 Knowledge Aquisition in RDR

This section describes the algorithm for knowledge acquision from experts in the conventional RDR system. Notations in the algorithm are as follows:

C_{cu} : Current Case

C_{co} : Cornerstone Case

CB : the set of cases which are already encountered
$class(C)$: class label of case C
$DiffList(C_1, C_2)$: difference list between cases C_1 and C_2 [1]
$RdrClassify(C)$: class label of case C from RDR
$ExpertClassify(C)$: class label of case C from an expert (**task1**)
$ExpertCondition(DiffList(C1, C2))$: condition specified by the expert (**task2**) to distinguish between C_1 and C_2. The condition must satisify C_1.

Algorithm for knowledge acquision

Step 1 : If no case is available to acquire knowlege, terminate the knowledge acquision process. If not, classify a case C_{cu} with the RDR system. If $C_{cu} \in CB$, return $class(C_{cu})$. If $C_{cu} \notin CB$, get the class label through $ExpertClassify(C_{cu})$ (**task1**) and add C_{cu} into CB.

Step 2 : If $class(C_{cu}) = RdrInference(C_{cu})$, go to Step 1, since knowledge acquision is not necessary.

Step 3 : (a) calculate $DiffList(C_{cu}, C_{co})$ and show it to the expert, (b) get the condition *Conditon* from the expert through $ExpertCondition(DiffList(C_{cu}, C_{co}))$ (**task2**), (c) if *Condion* is empty, go to Step 1.

Step 4 : Create If-Then rule with *Condion* as the condition and $class(C_{cu})$ as its conclusion (i.e., a rule $Condition \Rightarrow class(C_{cu})$). Add a new node with that rule and C_{cu} as its cornerstone case into the RDR binary tree. Go to Step 1.

4.2 Six Algorithms for Estimating the Condition

Knowledge acquision in RDR with the proposed algorithms is carried out by replacing **task2** (Step 3(b)) in Subsection 4.1 with **case generation** and **task1** as:

- Input : C_{cu},C_{co}
- In the item lattice based on $DiffList(C_{cu}, C_{co})$ (see Fig.3)
  1. Generate a case C_x based on the search strategy. Ask an expert to perform $ExpertClassify(C_x)$ and receive $class(C_x)$ (**task1**).
  2. When the class boundary is found, create the *Condition* which represents the boundary.
  3. Exit when the search space is fully explored. If not, go to 1.
- Output : *Condition*

Different search strategies can be used for generating cases in the item lattice to estimate the condition for the if-then rule which is stored in the newly added node in RDR. Different boundaries are estimated in the item lattice depending on how cases are generated in the search for the boundary. Six algorithms are proposed. They differ

[1] In this paper the difference list is calculated as the list of values in C_1 which are different from those in C_2. For instance, for the current case C_{cu} = (0,0,0,1,0,1) and cornerstone case C_{co} =(1,1,0,0,1,1) in Fig.1, the difference list $DiffList(C_{cu}, C_{co})$ = (0,0,1,0).

in from which case the search is started and how the lattice is searched. Search is started either from the current case C_{cu} or the cornerstone case C_{co}. The lattice is searched as breadth-first like, depth-first like, or just for 1 step. For instance, in Fig. 2 algorithm 1FCU searches the lattice just 1 step From the CUrrent case, BFCO in Breadth-first From the COrnerstone case, and DFCU in Depth-first From the CUrrent case. Search is terminated when the exact boundary is found, or when an approximate one is estimated. Here, the latter is appoximate since it does not guarantee that the cases which satisfy *Condition* necessarily have the same class label with the current case. The characteristics for the proposed 6 algorithms are summarized in Table 1.

Table 1: Characters of 6 Algotithms

Algorithm	Start Case	Search Method	Worst Q num
1FCU	C_{cu}	1step	N
1FCO	C_{co}	1step	N
BFCU	C_{cu}	BFS	2^N
BFCO	C_{co}	BFS	2^N
DFCU	C_{cu}	DFS	N^2
DFCO	C_{co}	DFS	N^2

N: number of different attributes, Worst Q num: the maximum number of questions asked for an expert in the worst case

The characteristics of the proposed 6 algorithms are illustrated in Figs. 3, 4 and 5. Each node in the figures represents the possible case to be generated and the search is carried out in the order of the number on the node. The bold dotted curve in each figure represents the estimated class boundary which is implied by the determined condition through the case generation and task1 on the generated cases.

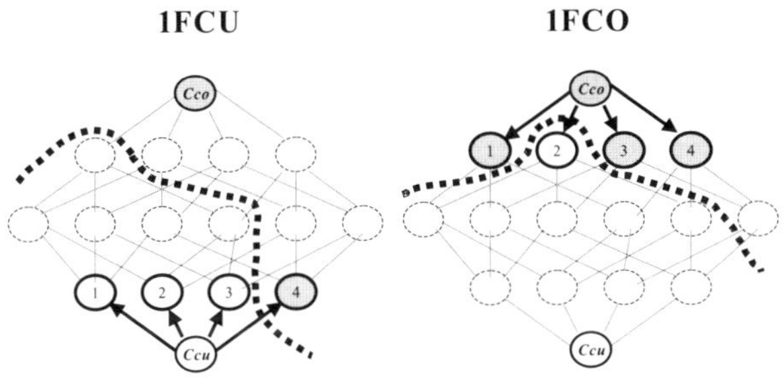

Figure 3: 1 step search from the current case and cornerstone case.

5 Experiment and Discussion

5.1 Experiment

Datasets

Experiments were carried out on one artificial dataset and 13 datasets from UCI repository[1]. On each dataset randomly selected 75% cases were used as the training cases and the remaining 25% cases were used as the test cases. Note that the cases

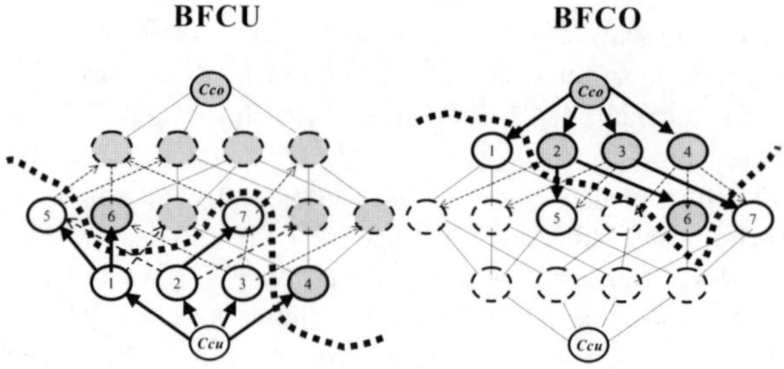

Figure 4: BFS: Breadth-First Search from the current case and cornerstone case.

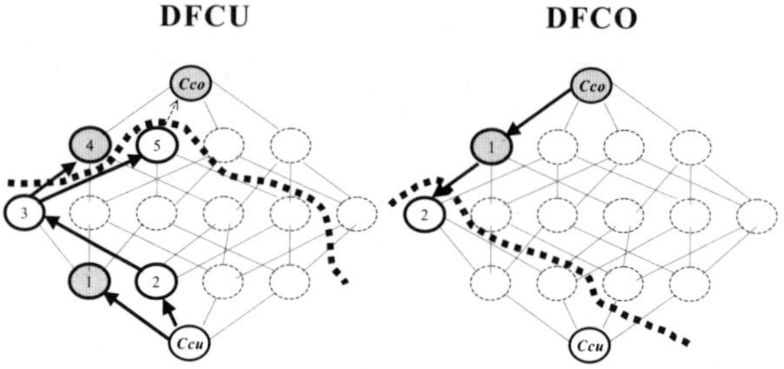

Figure 5: DFS: Depth-First Search from the current case and cornerstone case.

which were encountered in training were removed from the test cases. The above process is repeated 10 times for each dataset and the average is taken for the evaluation.

Simulated Expert

A simulated expert which is constructed by C4.5 was utilized in the experiment for each dataset. C4.5 is used to construct the decision tree for each dataset. Then the decision tree is transformed into a set of If-Then rules (C4.5rules), which is treated as a simulated expert. For each case **task2** is simulated by taking the intersecion of the diffrecence list for the case and the If-Then rule which is used to judge its class label [3].

Experiments were carried out for the RDR method with the proposed case generation algorithms and the standard RDR method in which **task2** is carried out by the simulated expert. The results were analyzed with regard to:
Error Rate: the rate of misclassified cases in the test cases, and **Size** : the number of nodes in the constructed RDR binary tree.

Knowledge acquisition from the simulated expert for the training cases was terminated when the number of interactions reached the predefined number[2]. Here, the number of interactions is defined as the sum of the number of $ExpertClassify$ (**task1**) and that of $ExpertCondition$ (**task2**). Note that in the proposed method the number is equal to that of $ExpertClassify$ since $ExpertCondition$ is not utilized.

5.2 Results and Discussions

Results are summarized in Table. 3.

[2]In the experiment the number is set to the number of 75% cases in each dataset.

Table 2: Datasets Used in the Experiment

dataset	# cases	# classes	# attributes	dataset	# cases	# classes	# attributes
Samp+	4096	2	Nom.*12	Pen-Digits	10992	10	Num. 16
Car Evaluation	1728	4	Nom. 6	Pima Indians	768	2	Num. 6
Nursery	12960	5	Nom. 8	Shut	14500	7	Num. 9
Tic Tac Toe	956	2	Nom. 9	Yeast	1484	10	Num. 8
Voting Records	435	2	Nom. 16	Ann-thyroid	7200	3	Mixed.*** 15/6
Page Blocks	5473	5	Num.** 9	Cmc	1473	3	Mixed. 7/2
Iris	150	3	Nom.**4	German	1000	2	Mixed. 13/7

+Artificial dataset, *nominal attributes, **numerical attributes, ***nominal and numerical attributes : #nominal/#numerical

Comparison with the standard RDR method

Error Rate: In 12 datasets the error rate for the proposed RDR method is equivalent to or smaller than that for the standard one.
Size: In 9 datasets the size for the proposed RDR method is equivalent to that for the standard one.
Number of interactions with the expert: Note that the number of **task1** and that of **task2** are equally counted as 1 interaction, albeit the latter is much harder for the expert. This means that the proposed method enables the construction of an equivalent RDR Knowledge Base with the standard one with respect to the error rate and size by reducing the load on the expert.
Effect of the number of classes: Unfortuntaly the proposed method did not work well for PenDigit and Yeast, both of which has 10 classes. Since the proposed method tries to find the boundary under which the the class label is the same with the current case, the estimated condition might be too specific when the number of class gets large.

Comparison of Six Algorithms

From which case the search should be started?: From Table 3 the current case should be used as the starting point in search since the error rate tends to be low (in 34 combinations out of 40). For instance, DFCU is the best and DFCO is the worst as a whole. Since the current case is misclassified by the RDR KB, it would be better to carry out knowledge acquision so that the KB can be refined around that case. Generating cases which are similar to the current case playes the role of the above process. In the following discussion the current case is used as the starting point in search.
Search strategy: As for 1FCU (maximum N cases are generated) or DFCU (maximum N^2 cases are generated), the error rate and size are equivalent to the standard RDR method. On the other hand, BFCU(maximum 2^N cases are generated) was not effective and did not even work for Pen Digit(16 attributes) and German (21 attributes). The number of interactions required for adding just one node to the RDR binary tree exceeded the predefined threshold in these datasets. When the number of attributes are large as in these datasets, the search space in the item lattice becomes huge and thus is difficult to find out the boundary. As a future direction we plan to utilize the feature selection method [4] to reduce the size of item lattice.

Table 3: Results of experiment.

Data set	Error rate(%)						
	RDR	1FCU	1FCO	BFCU	BFCO	DFCU	DFCO
Samp	0.17	0.10	0.23	0.14	0.04	0.10	1.19
CarEvaluation	4.00	0.78	1.18	0.62	0.94	0.71	5.20
Nursery	2.45	1.29	1.29	1.30	1.44	1.53	4.76
TitTacToe	5.46	7.42	12.00	12.84	13.72	11.49	25.70
Voting Records	0.26	1.03	0.00	4.29	3.43	0.39	4.36
Page Blocks	0.87	1.94	1.98	5.59	5.83	1.83	3.03
Iris	0.00	0.00	0.81	1.08	1.60	0.00	5.38
Pen Digit	2.66	7.07	13.50	-	-	6.97	44.96
Pima Indians	7.29	10.31	7.86	12.50	13.33	7.71	15.94
Shut	0.11	0.09	2.18	0.09	0.96	0.15	0.28
Yeast	11.55	22.36	23.05	26.88	25.67	23.24	39.87
Ann-thyroid	0.46	0.46	0.51	0.50	0.62	0.44	2.28
Cmc	16.92	14.79	18.19	16.96	19.57	17.53	22.80
German	9.76	10.84	9.24	-	-	10.00	13.00

Data set	KB size						
	RDR	1FCU	1FCO	BFCU	BFCO	DFCU	DFCO
Samp	79	60	80	69	75	84	182
CarEvaluation	96	67	72	66	73	74	161
Nursery	478	372	406	370	434	406	1063
TitTacToe	63	64	74	54	61	77	121
Voting Records	6	5	5	4	4	6	11
Page Blocks	35	79	60	13	11	87	143
Iris	4	7	6	6	6	7	15
Pen Digit	282	293	144	-	-	251	539
Pima Indians	17	29	28	8	9	30	47
Shut	21	35	25	31	31	34	95
Yeast	92	118	89	53	49	109	166
Ann-thyroid	14	40	38	27	36	42	147
Cmc	84	116	123	89	91	134	175
German	36	40	38	-	-	34	51

Table 4: Comparation of the proposal method with the standard RDR.

Data set	Property				Comparison					Rating
					RDR		Best of Algorithms			
	CaseNum	Nominal	Numerical	Class	error	size	error	size	Algorithm	
Samp	4096	12		2	0.17	79	0.04	75	BFCO	◎
Car Evaluation	1728	6		4	4.00	96	0.62	66	BFCU	◎
Nursery	12960	8		5	2.45	478	1.29	372	1FCU	◎
TicTacToe	956	9		3	5.46	63	7.42	64	1FCU	○
Voting Records	435	16		2	0.26	6	0.00	5	1FCO	◎
Page Blocks	5473		10	5	0.87	35	1.83	87	DFCU	○
Iris	150		4	3	0.00	4	0.00	7	1FCU,DFCU	◎
Pen Digit	10992		16	10	2.66	282	6.97	251	DFCU	×
Pima Indians	768		8	2	7.29	17	7.71	30	DFCU	○
Shut	14500		9	7	0.11	21	0.09	31	BFCU	◎
Yeast	1484		8	10	11.55	92	22.36	118	1FCU	×
Ann-thyroid	7200	15	6	3	0.46	14	0.44	42	DFCU	◎
Cmc	1473	7	2	3	16.92	84	14.79	116	1FCU	◎
German	1000	13	7	2	9.76	36	9.24	38	1FCO	◎

◎ means the proposed method is better than, ○ as equivalent to, and × as worse than the standard RDR method.

Difference in the nominal and numerical attributes: Although the proposed algorithms worked well for the datasets with nominal attributes, they did not work well for some datasets with numerical attributes. With numerical attributes the number of questions per node tend to increase since it is possible to consider a lot of cases with subtle difference in the values. For instance, for a nominal attribute "traffic light" the number of candidates for the value is usually quite limited as "red, yellow, yello". On the other hand, for a numerical attribute as "weight of car" the number of candidates for possible value tends to increase, such as "650, 660,870,...". Thus, many cases can be generated in the item lattice, which results in the increase in the number of interactions to reduce the error rate. DFCU seems better for numerical attributes from Table 4. The difference between DFCU and 1FCU, is due to the different characteristics of the boundary found in these algorithms. In DFCU the boundary is determined based on two cases with different class labels. On the other hand, in 1FCU the boundary is determined based on the neighboring cases for the current case. It is conjectured that such a difference in the boundaries might bring about the different result for the datasets with numerical attributes.

Which is best?: 1FCU and DFCU are the two best algorithms from the experiment, both of which utilize the current case as the start point in search. These are equivalent with respect to the error rate and size. Details of 1FCU and DFCU are described in Appendix. Since Table 4 indicates that DFCU is better than 1FCU for datasets with numerical attributes, DFCU is the best. However, since the difference is quite small, it is necessary to carry out more experiments to draw more definite conclusion.

6 Conclusion

This paper has proposed a case generation method for constructing an RDR Knowledge Base only through the classification of the generated cases by the expert. The experiment showed that the proposed method can construct an RDR knowledge base which is as good as the one with the standard RDR method that requires an expert to induce the condition for distinguishing cases. Our immediate future plan is to utilize the feature selection [4] to scale up the number of attributes.

Acknowledgement

This work was partially supported by the grant-in-aid for scientific research on priority area "Active Mining" funded by the Japanese Ministry of Education, Culture, Sport, Science and Technology.

References

[1] C.L. Blake and C.J. Merz. UCI repository of machine learning databases, 1998. http :// www.ics.uci.edu / ~mlearn / MLRepository.html.

[2] P. Compton, G. Edwards, B. H. Kang, and et al. Ripple down rules: Possibilities and limitations. In *Proc. of the 5th Knowledge Acquisition for Knowledge Based Systems Workshop*, 1991.

[3] P. Compton, P. Preston, and B.H. Kang. The use of simulated experts in evaluating knowledge acquisition. In *Proc. of the 9th Knowledge Acquisition for Knowledge Based Systems Workshop*, 1995.

[4] Huan Liu and Hiroshi Motoda, editors. *FEATURE SELECTION FOR KNOWLEDGE DISCOVERY AND DATA MINING.* Kluwer academic publishers, 1998.

[5] Ken Satoh and Ryuichi Nakagawa. Discovering critical cases in case-based reasoning. In *Online Proceedings of the Sixth International Symposium on Artificial Intelligence and Mathematics*, Florida, 2000.

[6] Takuya Wada, Tadashi Horiuchi, Hiroshi Motoda, and Takashi Washio. A description length based decision criterion for default knowledge in the ripple down rules method (accepted for publication). *Knowledge and Information Systems*, 3(2):146–167, 2001.

Appendix

A Details of the Algorithms for constructing the Condition

This section describes the details of how the condition for the newly created If-Then rule is constructed in 1FCU and DFCU, which are the two best algorithms from the experiment.

A.1 1FCU

Step1 Create a list of conditions *cond*, which is initialized to an empty list.

Step2 For each attribute A_i which corresponds to the value in $DiffList(C_{cu}, C_{co})$,

- Step2.1 Copy the current case C_{cu} to C_{tmp}. In C_{tmp}, set the value for A_i to $v_i^{C_{co}}$, which is the value in the cornerstone case C_{co} for A_i.
- Step2.2 Receive the class label for C_{tmp} through **task1**.
- Step2.3 If the class label for C_{tmp} is different from that for C_{cu}, add $v_i^{C_{cu}}$, which is the value for A_i in the current case C_{cu}, into *cond*.

Step3 If *cond* is not empty, return the conjunction of all elements in *cond*; otherwise, return *cond*.

The behavior of 1FCU algorithm is illustrated in Fig. 6. For instance, for the current case and cornerstone case which are shown in Fig. 6, 4 cases are generated in Step2.1. Since the class labels for the cases which are denoted as $(a2, b1, d1, f1)$ and $(a1, b1, d2, f1)$ are different from that for C_{co}, $b1$ and $f1$ are inserted into *cond* in Step2.3, respectively. Finally, the condition $b1 \wedge f1$ is returned from the algorithm.

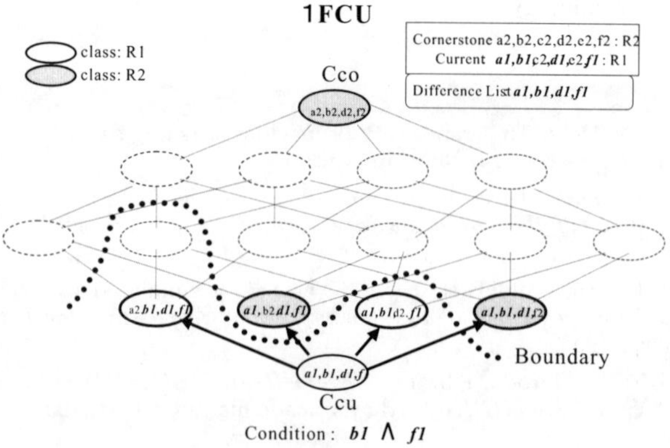

Figure 6: 1FCU: 1 step search From CUrrent case

A.2 DFCU

Step1 Create a list of conditions *cond* and a list of cadidates $cadidateList$, both of which are initialized to empty lists. Insert C_{cu} into C_{now}.

Step2 For each attribute A_i which corresponds to the value in $DiffList(C_{cu}, C_{co})$,

- Step2.1 Generate a case by copying C_{now} and setting the value for A_i to $v_i^{C_{co}}$, which is the value for A_i in the cornerstone case C_{co}.
- Step2.2 Add the created case into $candidateList$.

Step3 If $candidateList$ is empty, go to Step4. Otherwise, remove the first case from $candidateList$ and insert it into C_{tmp}.

- Step3.1 Receive the class label for C_{tmp} through **task1**.
- Step3.2 If the class label for C_{tmp} is different from that for C_{cu}, go to Step3.
- Step3.3 Insert C_{tmp} to C_{now}. Discard $candidateList$ and insert an empty list into $candidateList$. Go to Step2.

Step4 If $DiffList(C_{now}, C_{co})$ is not empty, return the conjunction of all elements in $DiffList(C_{now}, C_{co})$; otherwise, return $DiffList(C_{now}, C_{co})$.

The behavior of DFCU algorithm is illustrated in Fig. 7. For instance, for the current case and cornerstone case which are shown in Fig. 7, C_{now} is set to C_{cu} in Step1 and 4 cases are generated in Step2. Since the class label for the secondly generated case, which is denoted as (a1,b2,d1,f1), is not different from that for C_{cu}, C_{now} is updated to it in Step3.3 and search is continued.

When C_{now} is set to the case which is denoted as (a1,b2,d1,f1), since the class labels for all the generated candidate cases are different from that for C_{cu}, search is terminated. Finally, by taking $DiffList(C_{now}, C_{co})$, the condition $d1 \wedge f1$ is returned from the algorithm.

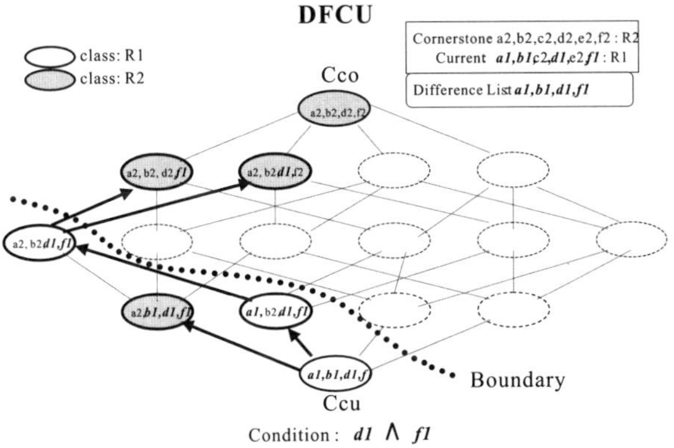

Figure 7: DFCU: Depth-first search From CUrrent case

Active Mining
H. Motoda (Ed.)
IOS Press, 2002

Acquiring Knowledge from Both Human Experts and Accumulated Data in an Unstable Environment

Takuya Wada, Tetsuya Yoshida, Hiroshi Motoda and Takashi Washio
{wada,yoshida,motoda,washio}@ar.sanken.osaka-u.ac.jp
Institute of Scientific and Industrial Research,
Osaka University
Mihogaoka, Ibaraki, Osaka 567-0047, JAPAN

Abstract. A Knowledge Acquisition method "Ripple Down Rules" can directly acquire and encode knowledge from human experts. It is an incremental acquisition method and each new piece of knowledge is added as an exception to the existing knowledge base. This knowledge base takes the form of a binary tree. Since RDR acquires new piece of knowledge only when the decision made by the so far built tree is wrong, there is no clear distinction between the initial KBS building phase and the later KBS management/refinement phase. There is another type of knowledge acquisition method that learns directly from data. Induction of decision tree is one such representative example. Noting that more data are stored in the database in this digital era, use of both expertise of humans and these stored data becomes even more important. Further, it is not appropriate to assume that the knowledge is stable and maintains its usefulness. Things change over time. It is not good to keep old useless knowledge in the knowledge base when such change happens. This paper attempts to integrate inductive learning and knowledge acquisition under a situation in which we can't assume a stable environment. We show that using the minimum description length principle (MDLP), the knowledge base of Ripple Down Rules is automatically and incrementally constructed from data. We, thus, can use both human expertise and data simultaneously. When it is found that some change takes place, useless knowledge is automatically deleted based on MDLP, still keeping the consistency of knowledge base. Experiments are carefully designed and tested to verify that the proposed method indeed works for many data sets having different natures.

1 Introduction

With the recent development of computer network, huge amount of information in various forms is communicated through the network, and it is required to provide a methodology to construct a reliable and adaptive knowledge-based system (KBS), which is accessible to both experts and users over the network. Two functions are required to realize such a reliable and adaptive KBS. One is the capability to reconstruct the internal structure of the KBS so that it can adapt to environmental changes in the domain while maintaining its performance. The other is to acquire knowledge from both human experts and data concurrently or alternately by incorporating the recent development in research on machine learning. "Ripple Down Rules [2] (RDR)" method has the capability to realize the above two functions.

In RDR knowledge is directly acquired and encoded from human experts without requiring high-level models of knowledge. Since it is an incremental KA method, there is no clear distinguish between knowledge acquisition and knowledge maintenance.

This paper proposes to incorporate the concept of the Minimum Description Length Principle [7] (MDLP) into the RDR method. The proposal consists of two parts: (1) extension of RDR to cope with changes in class distribution and (2) integration of both inductive learning method (a machine learning method that builds an RDR knowledge base incrementally from data) and the standard RDR method (a knowledge acquisition (KA) method that captures expertise incrementally from a human expert). Data sets from UCI repository [1] are utilized to evaluate the proposed method with respect to acquiring knowledge from both data and experts and to coping with the changes in the class distribution.

2 Ripple Down Rules Revisited

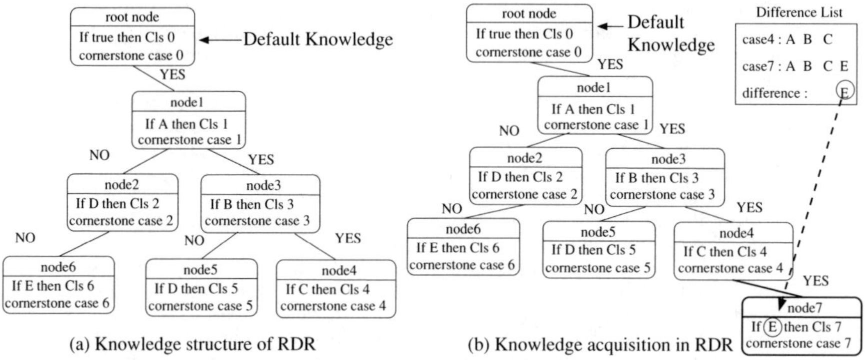

Figure 1: Knowledge structure of the Ripple Down Rules method

The basis of this method is the maintenance and retrieval of cases. When a case is incorrectly retrieved by an RDR system, the KA (maintenance) process requires the expert to identify how a case stored in a KBS differs from the present case. The structure of an RDR knowledge base is shown in Figure 1(a). Each node in the binary tree is a rule with a desired conclusion (If-Then rule). Each node has a "cornerstone case (CS-case)" associated with it, that is, the case that prompted the inclusion of the rule. An inference process for an incoming case starts from the root node of the binary tree. Then the process moves to the YES branch of the present node if the case satisfies the condition part of the node, and if it doesn't, the process moves to the NO branch. This process continues until there is no branch to move on. The conclusion for the incoming case is given by the conclusion part of the node in the inference path for the case whose condition part is lastly satisfied. Note that this node which has induced the conclusion for the case is called "last satisfied node".

If the conclusion is different from the one which an expert judges the case to be, knowledge (new rule) must be acquired from the human expert, and this rule must be added to the existing binary tree. The KA process in RDR is illustrated in Fig. 1(b). When the expert wants to add a new rule, there must be a case that is misclassified by a rule in RDR. The system asks him/her to select conditions for the rule from the "difference list (D-list)" between these two cases: the misclassified case and the CS-case. Then the misclassified one is stored as the refinement case (new CS-case) with the new rule whose condition part distinguishes these two cases. Depending on whether the last satisfied node is the same as the end node (the last node in the inference path), the new rule and its CS-case are added at the end of YES or NO branch of the end node. Knowledge is never removed or changed, simply modified by the addition of exception rules.

3 The Minimum Description Length Principle

This principle is the normal practice for selecting the most plausible probabilistic model from many alternatives, based on individual observational data for those alternatives. "Description length (DL)" can measure the complexity of the hypothesis. When the hypothesis is a classifier of some representation (decision tree [6] or neural network [4]), given some appropriate encoding method, the value can be the sum of (1) a DL for encoding the hypothesis itself and (2) a DL for encoding the misclassified cases by the hypothesis. According to the MDLP, the model to be selected is the one with the smallest total DL.

One of the differences between the proposed method and the standard RDR is that each node in the former keeps not only the CS-case but also those cases whose last satisfied node is that node. Let P be a set consisting of m cases that has passed a node α in the inference process, and let O be a subset of P ($O \subseteq P$), consisting of r cases for which the node α is the last satisfied node. In our encoding the DL of the tree is calculated first and then the one for the misclassified cases is calculated. This means that the tree information can be used to calculate the DL for the misclassified cases. In other word, the DL for the misclassified cases depends on the DL for the tree. The DL for the tree is calculated based on the pairs of an attribute and its value in P for the knowledge base. On the other hand, the one for the misclassified cases is based on the class information in O.

3.1 The DL of a Binary Tree

Two kinds of information need to be encoded at the node α: the branch information and the If-Then rule information. The DL for the former is mentioned in [8]. The information for the rule [1] consists of 4 components: (**1**) {the number of attributes used in the condition part}, (**2**) {attributes used in the part}, (**3**) {the attribute value for each attribute in (**2**)} and (**4**) {the class in the conclusion part}. Before calculating the DL of (**1**), we need to specify the "attribute-space" of each node in the binary tree. From m cases in P, we obtain the frequency distribution of each attribute-value. The corresponding attribute-space consists of a set of attributes each having at least 2 different attribute-values with each frequency of at least 1 case.

Suppose the space for node α has n attributes $\{A_i | i = 1, 2, ..., n\}$, resulting from P. The way to calculate the DLs for information (**1**), (**2**) and (**4**) are same as in [8], and the paper explains how to calculate the DL of (**3**) in case of nominal attribute only. Thus, we explain here how to calculate the DL for numerical attribute. For numerical attribute A_i, the condition can be $\{? < A_i\}$, $\{A_i \leq ?\}$ or $\{? < A_i \leq ?\}$. Thus, $\log_2 {}_3C_1$ bits are necessary to identify which one to use. Suppose that m_i is the number of candidates for a cut-off value for the attribute A_i. When the condition is $\{? < A_i\}$ or $\{A_i \leq ?\}$, another $\log_2 {}_{m_i}C_1$ bits are necessary. On the other hand, when it is $\{? < A_i \leq ?\}$, $\log_2 {}_2C_1$ bits are needed to indicate which one is encoded first, lower

[1] In RDR, the rule consists of multiple attribute-value pairs and one class-value pair.

bound or upper bound. In the former case, the upper one is encoded with $\log_2 {}_{m_i-k_i}C_1$ bits after the lower one is done with $\log_2 {}_{m_i}C_1$ bits. In the latter, the lower one is encoded with $\log_2 {}_{m_i-l_i}C_1$ bits after the upper one is encoded with $\log_2 {}_{m_i}C_1$ bits. Here, k_i (l_i) means that the lower (upper) cut-off value is the k_i-th (l_i-th) one from the left edge (right edge).

The sum of DLs to encode (**1**), (**2**), (**3**) and (**4**) is the DL necessary to encode the If-Then rule information for the node α. The sum of the DL for the branch-information and the one for the rule-information is the DL for the node α. The whole knowledge base can be encoded by encoding every node in the tree from the root node downward.

3.2 The DL for Class Labels of Misclassified Cases

Suppose that k cases in O has the same class label with the consequence part of the rule in node α[2] . First, $\log_2 {}_rC_1$ bits are necessary to express that $r-k$ cases have different classes from the consequence part. If $k=r$, there is no misclassified case and no further encoding is required. If $k<r$, it is necessary to represent which class the remaining $r-k$ cases have. Suppose that the number of classes which are different from that for the CS-case is s and the number of cases for each class is $p_i(i=1,2,...,s)$. The class labels are sorted in descending order with p_i i.e. $p_s \geq p_{s-1} \geq ... \geq p_2 \geq p_1$. The DL for misclassified cases is calculated using the algorithm shown in Table 1. The function **ceil()** in the table returns the least greatest integer for the argument.

Table 1: Algorithm to calculate DL to specify true classes of misclassified cases

initialize DL **to** 0,	: reset
all_num **to** r,	: the # of O
$right\_num$ **to** k,	: the # of right cases
j **to** $class\_num-1$,	: the # of class candidates
i **to** s, and	: the # of different classes
max_i **to** ∞.	

repeat while ($all\_num \neq right\_num$)
  if $all\_num - right\_num < max_i$,
    then $candi = all\_num - right\_num$,
    else $candi = max_i$,
  if $all\_num - right\_num > j$
    then $candi = candi -$**ceil**$((all\_num - right\_num) \div j)$.
  $DL = DL + log_2(j) + \log_2(candi) + \log_2({}_{all\_num}C_{p_i})$.
  $all\_num = all\_num - p_i$.
  $max_i = p_i$.
  decrement j and i.

The DL for the i-th different class is calculated at the line "$DL = DL + \log_2(j) + ...$". The first term is for specifying which class label the case has, the second one is for the number of cases p_i with the class, and the last one is to encode the locations for p_i cases. With this encoding it is possible to identify the true class labels for $r-k$ cases

[2]Note that we now employ the MLDP. Thus nodes are deleted or new nodes are not added if doing so results in a lower DL, and it is not necessarily true that the RDR classifies correctly all the cases that it has seen in the past

if the encoded bits are decoded. Encoding the entire binary tree in top-down produces the bit string for the class labels which are attached to each misclassified case in the tree.

3.3 The MDLP for the RDR method

The bit string with the total DL (the sum of the DLs in Sections 3.1 and 3.2) encodes the class labels for all cases stored in the KBS. Based on the MDLP for the RDR method, the binary tree with the smallest total DL should be most accurate for prediction. However, it is empirically known that most encoding methods tend to overestimate the DL for the knowledge base compared with the one for the class labels for the misclassified cases [5]. Thus, in general the following weighted sum is used to estimate the total DL:

$$Total\ DL = (DL\ of\ Subsection\ 3.2) + W \times (DL\ of\ Subsection\ 3.1) \qquad (1)$$

W is a weight and usually set to less than 1. In our approach W is set to 0.3 based on our experience [8].

4 Knowledge Acquisition in Three Situations

4.1 Knowledge Acquisition from Human Experts

This situation is the same with the standard RDR method. A human expert is required to select one or more elements from the D-list [3] and the selected ones are treated as the condition part in the newly created node.

4.2 Knowledge Acquisition from Data Alone

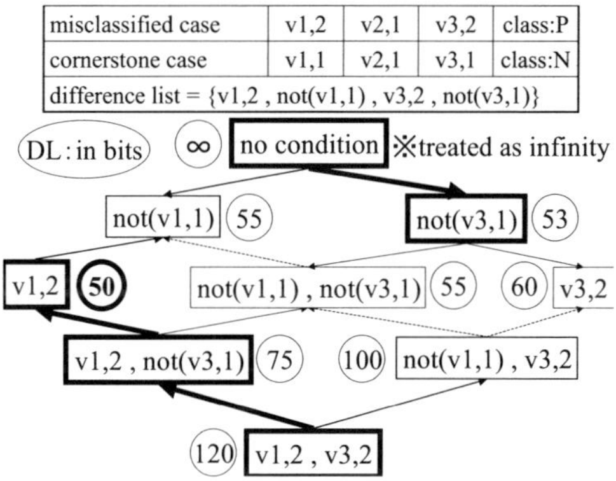

Figure 2: Search by data alone

In this method, based on the MDLP, we want to search all possible sets of elements from the D-list for a set with the minimum total DL.

The objective in our approach is to search for the set of conditions from the D-list so that the total DL is minimized when it is utilized as the condition part in

[3]The list holds a set of conditions such that none of them is satisfied with the CS-case and that all of them are satisfied with the misclassified incoming case.

the newly created node. In our proposed search strategy, the greedy search is carried out both from the most specialized condition to the misclassified case and from the most general condition to it. The search process is terminated when the total DL is captured in a local minimum. The proposed method enables to construct a knowledge base from data inductively without human experts. Admittedly exhaustive search will enable the construction of more accurate knowledge bases at the cost of much more computation. However, greedy search is employed to narrow the search space and to open the possibility for human experts to intervene with their high-level information management capability.

Figure 2 is an example in which an input case misclassified by the so far grown RDR tree has the attributes values $\{v_{1,2}, v_{2,1}, v_{3,2}\}$ and a CS-case whose node has derived the false conclusion has the values $\{v_{1,1}, v_{2,1}, v_{3,1}\}$. A detail of the search algorithm in the lattice of elements in the D-list is omitted due to the space limitation, but it is a greedy search. The search starts with a condition $\{v_{1,2}\&v_{3,2}\}$ which is most specialized to the input case, and it finds a condition $\{v_{1,2}\}$ that falls in a minimum total DL. The search restarts with a condition $\{no\ condition\}$ which is most general for the input case, and it finds another condition $\{not(v_{3,1})\}$ that falls in a local minimum total DL. The condition $\{v_{1,2}\}$ found by the search from the most specialized condition has smaller total DL than the condition $\{not(v_{3,1})\}$ by the search from the most general condition. So finally the condition $\{v_{1,2}\}$ is selected as the condition part of a new node for the incoming case.

4.3 Knowledge Acquisition from Both Data and Human Experts

We can integrate both IL and KA to jointly construct an RDR knowledge base. For example, during the initial phase of KBS development, there is not enough data available and a human expert is the sole source of knowledge, but at a later stage we can switch the source of knowledge to accumulated data without rebuilding the tree from scratch. For another example, when both human experts and data are available, it is possible to use both knowledge sources to construct a knowledge base. One way is to start the search for condition part which is selected from the one selected from the D-list by an expert so that both knowledge sources can be utilized effectively. This can lead to finding a better condition from the viewpoint of MLDP, compared with the one selected by the expert. Notice that we estimate only this latter integration setting in Section 6.

5 The Knowledge Deletion for the Environmental Changes

Part of knowledge acquired previously may become useless when the class distribution for the domain or the noise rate for the cases changes over time. Moreover, such invalid knowledge may hinder efficient acquisition of new pieces of knowledge. A naive way to cope with this issue is to discard the constructed knowledge base completely when such a change is detected and to reconstruct a new knowledge base under new environment. However, when the change is slow, it is difficult to detect. Moreover, since some part of the knowledge base may still be valid for the new environment, it would be reasonable to reuse such knowledge as much as possible. This section proposes the criterion to judge if a node is worth to be kept and if not, how to delete the node from the knowledge base.

5.1 The Criterion to Estimate the Value of Knowledge

When the class distribution changes, the D-list may have no element even if the input case is misclassified by the current KBS. In such a case, the standard RDR method can't add a new node. Even if the list is not empty, it does not make sense to add a node when no element in the list is judged as important by the expert. It may be reasonable to treat the node that induced the misclassification as useless. However, if the misclassification is brought about due to some noise in the input case, the knowledge stored at that node can still be valid and the deletion might lead to inconsistency of the knowledge base. Since the policy of the RDR method is to acquire a new piece of exceptional knowledge based on the inconsistency of the input case with the current knowledge base, the deletion of node should be carried out with caution: otherwise, acquisition of exceptional knowledge cannot be carried out if such inconsistency always triggers the deletion.

The proposed criterion in our approach is based on the assumption that a new node is not to be added even if the input case is misclassified, when adding the node does not decrease the DL, and is carried out as follows. First, tentatively delete the node which induces the wrong conclusion. The cases of the same class as the conclusion part of the deleted node are also deleted (These cases are in O for the node α in Section 3). Other cases of different classes, which were in the deleted node, are restored and redistributed in the tree to their new last satisfied nodes. If the normalized [4] total DL for the knowledge base after deletion is smaller than that of the current one, accept the deletion. Otherwise, recover the current knowledge base by retracting the deletion process.

5.2 Deletion of Node from a Binary Tree

The deletion of the node is carried out as follows.

(0) If the node to be deleted has a child below Yes branch, assign α to the child. If it has a child below NO branch, assign β to the child.

(1) Delete the node from the binary tree.

(2) If there is a node labeled α, attach it below the node γ as a child with the branch label i and go to (3). If there is a node labeled β, attach it below the node γ with the branch label i and terminate. If there is no nodes labeled α and β, terminate. Here, the node γ is the parent node of the node to be deleted, and i is the label (YES or NO) of the branch from the node γ to the node to be deleted.

(3) Add the condition part of the deleted node to that of node α. If the node α has a child below its No branch, reassign α to the child and go back to (3); if not, go to (4).

(4) If a node is labeled β, attach it below the node α as a child with the branch label NO and terminate. If there is no node labeled β, terminate.

---

[4]Because the DL monotonically increases in proportion to the number of cases, comparing the total DLs for knowledge bases with different number of cases makes no sense. Since the number of cases stored in the knowledge base is different before and after the deletion, the total DLs to be compared are normalized as $\mathrm{DL}_\alpha/\mathrm{DL}'_\alpha$ and $\mathrm{DL}_\beta/\mathrm{DL}'_\beta$. Here, DL' denotes the DL for encoding the true class information for the whole cases in the current binary tree without using the tree information, i.e. using the root node information alone.

The above algorithm is illustrated in Fig. 4. Suppose that node No.2 is judged as contradictory. First, the node is deleted from the tree. Then, node No.4, which is the child node of the Yes branch, is connected to the Yes branch of node No.1. At the same time, the condition "a", which is the premise of the If-Then rule in node No.2, is added to node No.4. Next, the subtree below node No.3 is connected to the No branch of node No.7. Finally, the condition "a" is added to node No.7. It is easy to confirm that no consistency arises for the stored CS-cases by this reconstruction of the tree. The illustrated process achieves the deletion of the CS-cases in a KBS and the cases which support the If-Then rule attached to the case simultaneously.

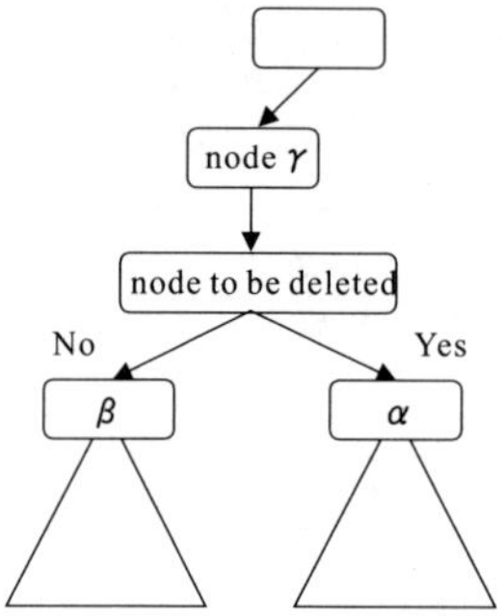

Figure 3: A tree for explaining deletion algorithm

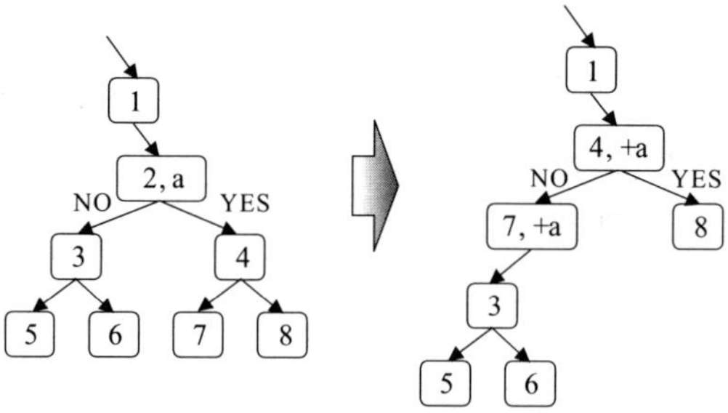

Figure 4: Deletion algorithm

6 Experiment

Experiments were carried out to investigate the effectiveness of the proposed method, using 15 databases from University of California Irvine Data Repository [1] (see Table 2). The class distribution of the problem domain were changed abruptly twice during a simulation for the RDR system to acquire knowledge from the data and a human expert incrementally. The accuracy of the knowledge base and the ratio of total DL for the knowledge base are reported for comparison.

[Generation of change in class distribution] A set of cases X_{chg} with different class distribution from the original dataset X_{org} were generated [5] . Then, they are

[5]To make X_{chg}, first, all cases in X_{org} are sorted with respect to values in lexical order for nominal

Table 2: Summery of data sets

Data Set Name	#of Case	#of Class	#of Attribute	Data Set Name	#of Case	#of Class	#of Attribute
Car	1728	4	Nom.* 6	PageBlocks	5473	5	Num. 10
Nursery	12960	5	Nom. 8	PenDigits	10992	10	Num. 16
Mushrooms	8124	2	Nom. 22	Yeast	1484	10	Num. 8
Krvkp	3196	2	Nom. 36	PimaIndians	768	2	Num. 6
VotingRecord	435	2	Nom. 16	GermanCredit	1000	2	Mix.*** 13/7
BreastCancer	699	2	Nom. 9	Cmc	1473	3	Mix. 7/2
Splice	3190	3	Nom. 60	AnnThyroid	7200	3	Mix. 15/6
Image	2310	7	Num.** 19				

*Nominal attribute, **Numerical attribute, ***This database has two kinds of attributes: nominal attribute / numerical attribute

individually divided into the 75% training data (X_{org}^{train},X_{chg}^{train}) and the 25% test data (X_{org}^{test},X_{chg}^{test}). First, by treating X_{org} as the original population, input cases which are selected randomly from X_{org}^{train} are passed to the RDR system. When the total number of cases passed to the system becomes equal to three times as large as that of the original population, the population is changed to X_{chg}. After that the system receives the cases drawn from the X_{chg}^{train}. When the total number of cases drawn from X_{chg}^{train} becomes three times as large as that of the population, it is changed to X_{org} again.

[Simulated Expert] Simulated Expert [3] (SE) is usually used instead of a human expert for the reproduction of experiments and consistent performance estimation in the RDR research community. Therefore, this paper follows this tradition. Note that when $X_{org}^{train}(X_{chg}^{train})$ is the population, we make a If-Then rule set derived from a decision tree constructed by standard C4.5 [5] using $X_{org}(X_{chg})$ to be the SE. This means that the SE is a really good expert, and he/she can change his/her knowledge according to the changes of an environment on the problem domain. A set of elements selected from the D-list by the SE is defined as the intersection between the list and the condition part of the If-Then rule in the SE which predicts correctly the case misclassified by the RDR system at the KA stage. For a numeric attribute, if there is an element in the D-list that satisfies the inequality condition of the SE rule, this inequality is interpreted as an element in the intersection. Thus, the condition which is nearest to the set selected by SE out of the candidates in the lattice space illustrated in Fig. 2 is the starting condition for searching. Note that no negative expression (not) is treated in the If-Then rule set induced by C4.5. In order to be able to select negative conditions also from the D-list, we binarized attribute-values and force the rule set to have negative ones.

[Accuracy of the knowledge base] We examine the error rate of misclassified case for the test data using the knowledge base at prespecified time points. Note that we use the $X_{org}^{test}(X_{chg}^{test})$ as the test data when the population is the $X_{org}^{train}(X_{chg}^{train})$. The RDR method is incremental, and a different ordering of in-

---

attributes and in ascending order for numerical attributes. Then, they are sorted in lexical order for class label. Finally, the labels for (#of all cases ÷ #of classes ÷ 10) cases are changed by shifting them so that the class label for about 10% in X_{org} is changed to neighboring class.

put cases results in a different knowledge base of RDR [8]. Therefore, we repeated the simulation 10 times, changing the parameter of random sampling for the input case from the population at each simulation.

6.1 Results and Discussions

Figures 5 and 6 show one result out of 10 simulations for the dataset "PenDigits". For each simulation the RDR system received 75000 cases. Two curves with marks in Fig. 5 indicate the change of accuracy of the proposed methods described in Subsections 4.3 and 4.2. Deletion of node was carried out based on the description in Section 5. The other two curves with no marks are for the proposed methods without the deletion of nodes. By comparing two curves for "SE&Data" and "Data" ("SE&Data&No_Deletion" and "Data&No_Deletion") the error rate for the former method is fewer than that for the latter up to the 25000th input case. The result shows that the condition which is selected by SE is a good starting condition for the lattice search.

The class distribution was changed abruptly at the 25000th case and 50000th case in the simulation. Such changes are reflected as the sharp increase in the error rate in the figure, since the rate increases with the change of class distribution. The figures also show that deleting the inconsistent knowledge from the knowledge base contributes to reducing the rate.

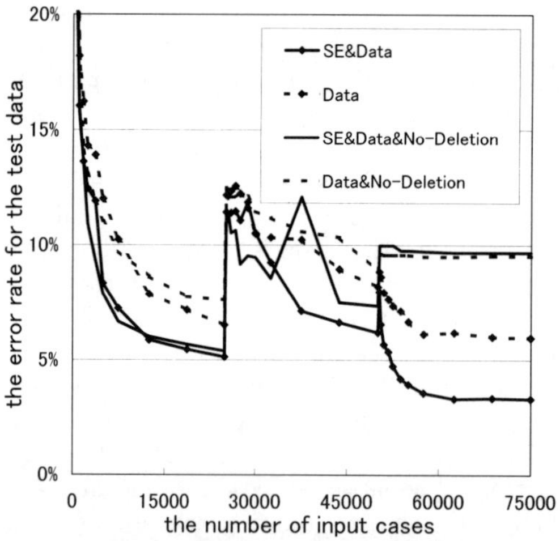

Figure 5: One of the 10 simulations for "PenDigits" (error)

The change in the ratio of DL for each RDR knowledge base is shown as the four curves in Fig. 6. The two curves without marks show that the ratio decreases monotonously through KA process while the class distribution does not change. Although the ratio increases when the distribution is changed, however, it decreases after the 25000th case in two curves: "DL for se&data" and "DL for data". This also suggests that our deletion algorithm works well to keep the size of knowledge base concise. Note that the method with the lowest ratio of DL has the lowest error rate, which confirms the validity of the MDLP in RDR.

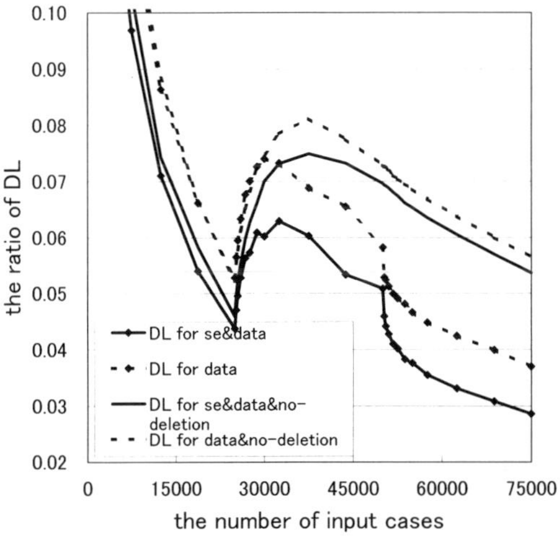

Figure 6: One of the 10 simulations for "PenDigits" (DL)

Table 3 summarizes the results for 15 data sets at the end of each simulation when the RDR system received all the cases. Out of the four methods, only the results for two with deletion are shown here. The columns for **% of cases** represent the ratio of cases which were kept inside the knowledge base with respect to the whole cases. The ones for **RDR** represent the error rate of the knowledge base for the test data. The decision trees were constructed by C4.5 using the cases held in the knowledge bases at the end of simulation. The columns for **C4.5** represent the error rate of such decision trees. Moreover, the column for **C4.5 with whole cases** shows the error rate of a decision trees using all input cases. For instance, the result for the dataset "Mushrooms" suggests that knowledge from SE is effective since the method **RDR from SE and data** shows the lower error rate than **RDR from DATA only**. Note that all the values are the average of 10 simulations according to the reason explained in [**Accuracy of the knowledge base**].

It is interesting to see whether cases matching to the current class distribution are held in the knowledge base when all cases are input to the RDR system. For the dataset "AnnThyroid", more accurate knowledge bases were constructed for **RDR from SE and data** compared with **C4.5 with whole cases**. This probably is the result that the deletion algorithm works well, and can delete the worthless knowledge with node holding it. Other data sets where such a tendency is shown are "Nursery", "Krvkp" and "Image".

Unfortunately, the error rate for **RDR** was high for some data sets with relatively small number of cases (e.g., "Cmc" and "Yeast") compared with **C4.5 with whole cases**. Since the KA and deletion in our approach are based on the MDLP, if only small amount of cases are available, it is difficult to construct a knowledge base with high predictive accuracy. Thus, our current conjecture is that the deletion algorithm tends to delete too many cases, especially when the size of the original datasets is relatively small. For instance, only 44.6% of the original cases were held in the knowledge base for "Cmc" after deletion.

Table 3: Summary of experimental results

Data Set	RDR from SE and data			RDR from data only			C4.5 with whole cases
	% of cases	RDR	C4.5	% of cases	RDR	C4.5	
Car	95.3%	4.2%	6.3%	95.2%	4.3%	6.4%	6.8%
Nursery	94.1%	3.0%	2.7%	93.9%	2.9%	2.7%	3.2%
Mushrooms	93.0%	3.7%	2.3%	93.9%	5.0%	2.4%	4.5%
Krvkp	93.5%	2.3%	3.1%	92.7%	2.0%	2.2%	3.9%
VotingRecord	95.6%	6.3%	3.7%	95.3%	6.1%	3.5%	3.2%
BreastCancer	95.1%	4.6%	5.3%	93.5%	4.9%	4.7%	5.6%
Splice	90.4%	7.9%	6.2%	74.9%	8.8%	10.4%	9.1%
Image	93.9%	2.6%	3.3%	94.7%	5.4%	3.8%	8.0%
PageBlocks	93.0%	4.4%	5.1%	90.9%	4.9%	5.2%	5.8%
PenDigits	93.9%	3.5%	3.8%	94.2%	5.0%	4.5%	8.4%
Yeast	56.6%	42.9%	40.0%	62.2%	41.5%	37.3%	24.2%
PimaIndians	62.4%	26.0%	23.0%	41.7%	33.3%	24.5%	15.2%
GermanCredit	72.2%	22.5%	19.4%	73.5%	20.6%	17.1%	14.4%
Cmc	44.6%	47.9%	43.5%	46.3%	47.4%	42.5%	29.1%
AnnThyroid	95.9%	1.2%	2.2%	95.2%	1.8%	2.7%	5.0%

7 Conclusion

This paper has proposed a KA method which can adapt to the change in class distribution. The proposed method can be used to acquire knowledge either from both human experts and data or from data alone. Experiments with artificial data showed the effectiveness of the method. However, for some dataset with small number of cases the results were actually bad. In addition, the experimental results suggest that deletion of knowledge (nodes in the binary tree) contributes to holding the necessary cases in the RDR system. As an immediate future plan, we intend to evaluate how RDR algorithm performs when KA from a human expert and data is interleaved as described in Subsection 4.3.

References

[1] C.L. Blake and C.J. Merz. UCI repository of machine learning databases, 1998. http :// www.ics.uci.edu / ~mlearn / MLRepository.html.

[2] P. Compton, G. Edwards, G. Srinivasan, R. Malor, P. Preston, B. Kang, and L. Lazarus. Ripple down rules: Turning knowledge acquisition into knowledge maintenance. *Artificial Intelligence in Medicine*, 4:47–59, 1992.

[3] P. Compton, P. Preston, and B.H. Kang. The use of simulated experts in evaluating knowledge acquisition. In *Proc. of the 9th Knowledge Acquisition for Knowledge Based Systems Workshop*, 1995.

[4] D.K. Gary and J.H. Trevor. Optimal network construction by minimum description length. *Neural Computation*, pages 210–212, 1993.

[5] J.R. Quinlan, editor. *C4.5: Programs for Machine Learning*. Morgan Kaufmann, 1993.

[6] J.R. Quinlan and R.L. Rivest. Inferring decision trees using the minimum description length principle. *Information and Computation*, pages 227–248, 1989.

[7] J. Rissanen. Modeling by shortest data description. *Automatica*, pages 465–471, 1978.

[8] T. Wada, H. Motoda, and T. Washio. Knowledge acquisition from both human expert and data. In *Proc. of the Fifth Pacific-Asia Conference on Knowledge Discovery and Data Mining*, pages 550–561, HongKong China, April 2001. Springer-Verlag.

Active Mining
H. Motoda (Ed.)
IOS Press, 2002

Active Participation of Users with Visualization Tools in the Knowledge Discovery Process

Tu Bao Ho, Trong Dung Nguyen, Duc Dung Nguyen, Saori Kawasai
{bao,nguyen,dungduc,skawasa}@jaist.ac.jp
Graduate School of Knowledge Science,
Japan Advanced Institute of Science and Technology
Tatsunokuchi, Ishikawa 923-1292, JAPAN

Abstract. Visualization has proven its effectiveness in exploratory data analysis and a high potential in knowledge discovery in databases. However, although visualization techniques have progressed dramatically in the last decade, it can be seen that visual knowledge discovery still remains in its infancy. While most visual KDD system focuses on visualization of data and/or knowledge at the beginning and at the end of the knowledge discovery process, there is currently very few systems providing interactive visual data mining, or visualization of the knowledge discovery process, e.g., that are much more related to the active role of the user in knowledge discovery. This paper describes our attempt in this research line with a synergistic visualization of data and knowledge in the knowledge discovery process.

1 Introduction

It is well known that there is no inherent superior method/model in terms of generalization performance. The No Free Lunch theorem states that in the absence of prior information about the problem there are no reasons to prefer one learning algorithm or classifier model to another. The problem of *model selection*—choosing appropriate discovered models or algorithms and their settings for obtaining such models in a given application—is difficult and non-trivial because it requires empirical comparative ebyvaluation of discovered models and meta-knowledge on models/algorithms. Unlike the major research tendency that aims to provide the user with meta-knowledge for an automatic model selection as described in the next section, in our view, model selection should be a user-centered process, i.e., semiautomatic and it requires an effective collaboration between the user and the discovery system. In such collaboration, visualization has an indispensable role because it can give a deep understanding of complicated models that the user cannot have if using only performance metrics.

Visualization has proven its effectiveness in exploratory data analysis and a high potential in KDD [3, 5]. However, although visualization techniques have progressed dramatically in the last decade, it can be seen that visual knowledge discovery still remains in its infancy. While most visual KDD system focuses on visualization of data and/or knowledge at the beginning and at the end of the knowledge discovery process, there is currently very few systems providing interactive visual data mining or visualization of the knowledge discovery process that are much more related to the active role of the user in knowledge discovery. Visual knowledge discovery can be viewed as an integration of two disciplines: information visualization and knowledge discovery. Information visual techniques can be divided into three groups of data visualization techniques, distortion techniques, and dynamic/interaction techniques [5]. Each group has a large number of different subgroups of techniques, for example, data visualization techniques include geometric techniques, icon-based techniques, pixel-based techniques, hierarchical techniques, graph-based techniques, 3D techniques, dynamic techniques, hybrid techniques; the distortion techniques include simple distortions such as fish-eyes and complex distortions such as hyperbolic browser; and dynamic/interaction techniques include data-to-visualization mapping, projections, filtering, linking and brushing, zooming, detail on demand [2].

Many KDD systems provide visualization of data and knowledge, and different visualization techniques have been developed or adapted to the KDD process. CART of Salford Systems has a 2D tree visualizer associated with a tree map that provides an overview of the tree. Another 2D tree browser having good features of multi-level dynamic queries and pruning is developed. System MineSet [1] provides several 3D visualizers, in particular a 3D Tree Visualizer. In [6] the authors developed an interactive visualization in decision tree construction for supporting an effective cooperation of the user and the computer in classification. D2MS shares many features with WinViz [5] and Cviz [4] that both use parallel coordinates. WinViz allows the user to visually examine a tabular database and to formulate query interactively and visually. Cviz is an attempt to integrate visualization into the KDD process.

The goal of this work is to develop a research system for knowledge discovery with visualization support for model selection. The system called D2MS (Data Mining with Model Selection) provides the user with the ability of trying various alternatives of algorithm combinations and their settings. The quantitative evaluation can be obtained by performance metrics provided by the system while the qualitative evaluation can be obtained by effective visualization of the discovered models.

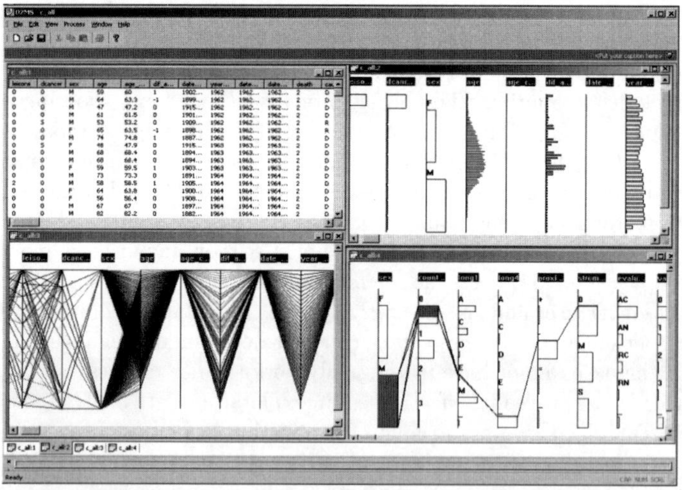

Figure 1: Data visualization in D2MS: the top-left window shows the dataset, the bottom-left window shows the original data view, the top-right window shows summarizing data view, and the bottom-right window shows the querying data view.

2 Visualization support for user-centered model selection

The visualization module is linked to most other modules in D2MS in particular those directly concerned with model selection. It currently consists of a data visualizer, a rule visualizer, and a tree visualizer (for hierarchical structures). These visualizers are integrated to most methods mentioned above in preprocessing, data mining, and postprocessing. We describe each visualizer with focus on its techniques and how it is linked to the steps in the KDD process.

2.1 Data Visualization

We have chosen the parallel technique for visualizing 2D tabular datasets defined by n rows and p columns. D2MS improves parallel coordinates in several ways to adapt them to the knowledge discovery context. It is because when viewing a large dataset with many attributes, particularly categorical attributes, the advantage of parallel coordinates may lose as many polylines or their parts are partially overlapped, and certain kinds of summarization might be needed. Also, the user often needs to see subsets of the dataset in terms of cases and/or attributes.

2.1.1 Viewing original data

The basic idea of viewing a p-dimensional dataset by parallel coordinates is to use p equally spaced axes—which are parallel to one of the screen axes and correspond to attributes and the ends of the axes correspond to minimum and maximum values for each dimension—to represent each data instance as a polyline that crosses each axis at a position proportional to its value for that dimension. This view gives the user a rough idea about the distribution of data on values of each attribute; in particular colors of classes can show clearly how classes are distributed. The original stomach cancer data (top-left windows in Figure 1) is visualized in the bottom-left window.

2.1.2 Summarizing data

This view is significant as the dataset may be very large. The key idea is not to view original data points but to view their summaries on parallel attributes. As WinViz, D2MS uses bar charts in the place of attribute values on each axis. The bar charts in each axis have the same height (depending on the number of possible attribute values) and different widths that signify the frequencies of attribute values. D2MS also provides interactively common statistics on each attribute as mean or mode, median, variance, boxplots, etc. The top-right window in Figure 1 shows the summaries of the stomach cancer data.

2.1.3 Querying data

This view serves the hypothesis generation and hypothesis testing by the user. It allows the user to view subsets of the dataset determined by queries. There are three types of queries: (i) based on a value of lthe class attribute where the query determines the subset of all instances belonging to the indicated class; (ii) based on a value of a descriptive attribute where the query determines the subset of all instances having this value, (iii) based on a conjunction of attribute-values pairs where the query determines the subset of all instances satisfied this conjunction. The queries can be determined by just using

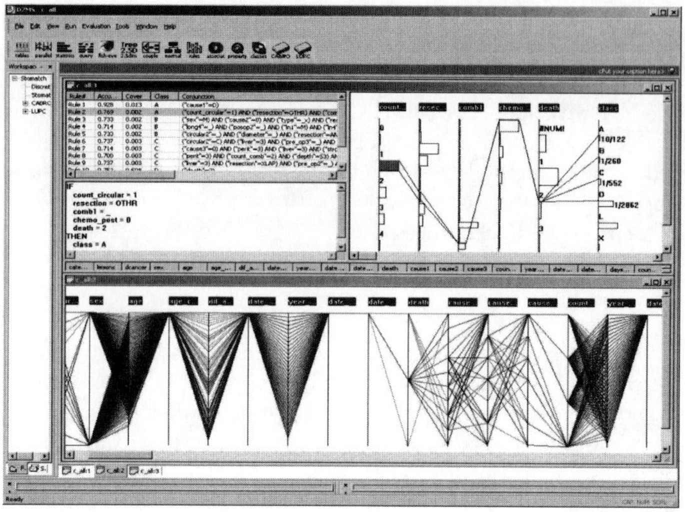

Figure 2: Rule Visualization in D2MS: the top-left window shows the list of discovered rules, the bottom-left window visualizes the rule by parallel coordinates, the top-right window shows the set of instances covered by the rule, and the bottom-right window visualizes this set.

point-and-click. The subset of instances matched the query is visualized in viewing data mode and in summarizing data mode. The gray regions on each axis show the proportions of specified instances on values of this attribute as shown in bottom-right window in Figure 2.

2.2 Rule Visualization

A rule is a pattern related to several attribute-values and a subset of instances. The importance in visualizing a rule is how this local structure is viewed in its relation to the whole dataset, and how the view support the user's evaluation on the rule interestingness. D2MS's rule visualizer allows the user to visualize rules in the form *antecedent* → *consequent* where *antecedent* is a conjunction of attribute-value pairs, *consequent* is a conjunction of attribute-value pairs in case of association rules, and is a value of the class attribute in case of prediction rules. A rule is simply displayed by a subset of parallel coordinates included in *antecedent* and *consequent*. The D2MS's rule visualizer has the following functions:

2.2.1 Viewing rules

Each rule is displayed by polyline that goes through the axes containing attribute-values occurred on the antecedent part of the rule leading to the consequent part of the rule that are displayed with different color. In the case of prediction rules, the ratio associated with each class in the class attribute corresponds to the number of instances of the class covered by the rule over the total number of instances in the class. This view gives a first observation of the rule quality.

2.2.2 Viewing rules and data

The subset of instances covered by a rule is visualized together with the rule by parallel coordinates or by summaries on parallel coordinates. From this subset of instances, the user can see the set of rules each of them cover some of these instances, or the user can smoothly change the values of an attribute in the rule to see other related possible rules. These possible operations facilitate the user in evaluating the quality of this rule: a rule is good if instances covered by it are not recognized by other rules, and vice-versa. The rules for a class can be displayed together, and instances of the class as well of other classes covered by these rule are displayed.

2.2.3 Rule visualization in model selection

There are several ways that support the user in evaluating the quality of the rule together with other measure such as coverage and accuracy of the rule. For example, two rules predicting a target class having the same support and confidence but the one wrongly covered more instances belonging to classes different from the target class would be considered worse. Figure 3 illustrates rule visualization in D2MS where the top-left and bottom left windows display a discovered rule, and the top-right and bottom right windows show the instances covered by that rule.

2.3 Tree Visualization

D2MS provides several visualization techniques that allow the user to visualize effectively large hierarchical structures. The *tightly coupled views* display simultaneously a hierarchy in normal size and tiny size that allows the user to determine quickly the field-of-view and to pan to the region of interest. The *fish-eye view* distorts the magnified image so that the center of interest is displayed at high magnification, and the rest of the image is progressively compressed. Also, the new technique *T2.5D* is implemented in D2MS for visualizing very large hierarchical structures.

2.3.1 Different modes of viewing hierarchical structures

D2MS tree visualizer provides multiple-views of trees or hierarchical structures.

- *Tightly coupled views*: The global view (on the left) shows the tree structure with nodes in same small size without labels and therefore it can display a tree fully or a large part of it, depending on the tree size. The detailed view (on the right) shows the tree structure and nodes with their labels associated with operations to display node information. The global view is associated with a field-of-view or panner (a wire-frame box) that corresponds to the detailed view. These two views are tightly coupled as the field-of-view can be moved around in the global view in order to pan the detailed view. Also, when the detailed view is scrolled the position of the field-of-view will be updated accordingly. The windows for these two views can be resized by the user, and the field-of-view shape and size will be automatically changed. The top-left and top-right windows in Figure 3 show the tightly coupled views of D2MS for the stomach cancer data.

- *Customizing views*: Initially, according to the user's choice, the tree is either displayed fully or with only the root node and its direct sub-nodes. The tree

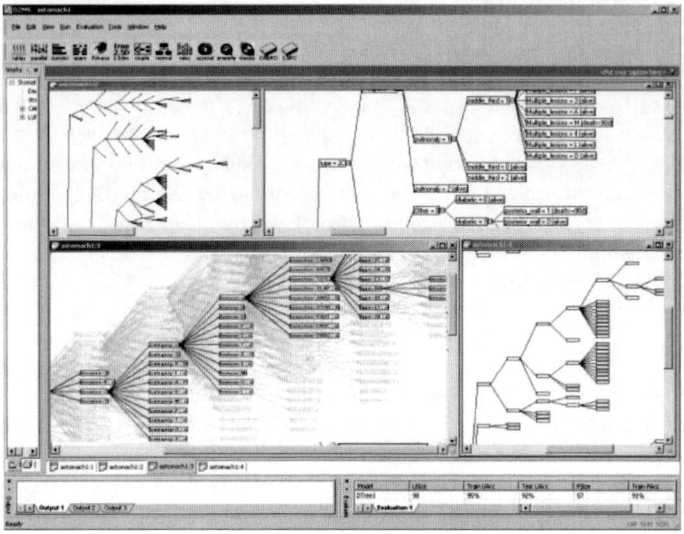

Figure 3: Multiple views of trees: tightly coupled, fish-eye, overview, and T2.5D.

then can be collapsed or expanded partially or fully from the root or from any intermediate node.

- *Tiny mode with fish-eye view*: Note that no current visualization technique allows us to display efficiently the entire tree when it has, says, ten thousands nodes. The tightly coupled views are extended with three viewing modes according to the user's choice: normal size, small size and tiny size. The tiny mode uses much more efficiently the space to visualize the tree structure, on which the user can determine quickly the field-of-view and pan to the region of interest. It allows the user to be able to see the tree structure while focusing on any particular part so that the relationship of parts to the whole can be seen and the focus can be moved to other parts in a smooth and continuous way.

2.3.2 Trees 2.5 Dimensions

The user might find it difficult to navigate a very large hierarchy, even with tightly coupled and fish-eye views. To overcome this difficulty, we have been developing a new technique called T2.5D (stands for Trees 2.5 Dimensions).

Different from tightly-coupled and fish-eye views that can be seen as location-based views, T2.5D can be seen as a relation-based view in the sense that highlighted parts of a tree are relations determined by queries. The starting point of T2.5D is the observation that a large tree consists many subtrees that are not usually and necessarily viewed simultaneously. The key idea of T2.5D is to represent a large tree in a virtual 3D space (subtrees are overlapped to reduce occupied space) while each subtree of interest is displayed in a 2D space. To this end, T2.5D determines the fixed position of each subtree (its root node) in two axes X and Y, and in addition, it computes dynamically a Z-order for this subtree in an imaginary axis Z. A subtree with a given Z-order is displayed "above" its siblings those have higher Z-orders. When visualizing and navigating a tree, at each moment the Z-order of all nodes on the path from the root to a node in focus in the tree is set to zero by T2.5D. The *active wide path* to a node in focus, which contains all nodes on the path from the root to this node in focus and their siblings, is displayed in the front of the screen with highlighted colors to give the user a clear view. Other parts of the tree remain in the background to provide an image of the overall structure. With Z-order, T2.5D can give the user an impression that trees are drawn in a 3D space. The user can easily change the active wide path by choosing another node in focus.

We have experimented T2.5D with various real and artificial datasets. In an experiment, T2.5D can handle well trees with more than 20,000 nodes, and more than 1,000 nodes can be displayed together on the screen. Figure 3 illustrates a pruned tree of 1795 nodes learned from stomach cancer data and drawn by T2.5D (note that the original screen with colors gives a better view than this black-white screen).

2.3.3 Tree Visualization in Model Selection

In D2MS, visualization is integrated with the steps of the KDD process and closely associated with the plan management module in support for model selection. The user can have either views in executing a plan or comparative views of discovered models.

If the user is interested in following the execution of one plan, he/she can view, for example, the input data, the derived data after preprocessing, the generated models with chosen settings, the exported results. Thus, the user can follow and verify the process of discovery by each plan, and change settings to reach alternative results.

With the three modes of viewing of data, D2MS integrates data visualization into different KDD steps by displaying and interactively changing these views of data at any time. Data visualization supports doing data preprocessing and examining the relation between data and discovered knowledge. In the first step of collecting data and formulating the problem, the user can and often need to view the original dataset and its summarization. The visual analysis of collected data may help the user to identify important or redundant attributes or new attributes to be added. The data visualization has shown to be significant in the data preprocessing step that consists of functions on data cleaning, integration, transformation and reduction. For example, many discretization algorithms provide alternative solution of dividing a numerical attribute into intervals, and the visual data query on the discretized attribute and the class attribute can give insights for decision. The data visualization is also very significant in data mining step with data query mode, and particularly in the evaluation step in its synergistic combination with rule and tree visualization.

If the user is interested in comparative evaluation of competing models generated by different plans, he/she can have multiple views on these models. The user can compare performance metrics of all activated plans that are always available in the summary table. Whenever the user highlights a row in the table, the associated model will be automatically displayed. Several windows can be opened simultaneously to display competing models in forms of trees, concept hierarchies, or rule sets. For example, two rules predicting a target class having the same support and confidence but the one wrongly covered more instances belonging to classes different from the target class would be considered worse.

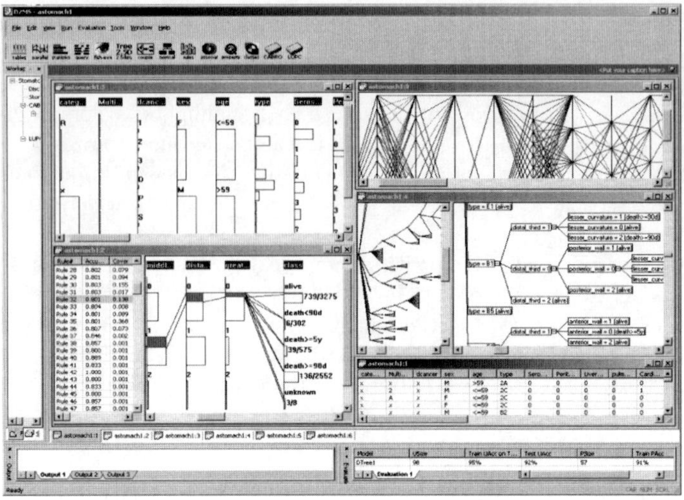

Figure 4: Visualization in studying stomach cancer data.

3 Case-Study: Stomach Cancer Dataset

The stomach cancer dataset collected at the National Cancer Center in Tokyo during the period 1962-1991 is a very important resource for research. It contains data of 7,520 patients described originally by 83 numeric and categorical attributes. These include information on patients, symptoms, type of cancer, longitudinal and circular location, serosal invasion, liver metastasis, pre-operative complication, post-operative complication, etc. One problem to be investigated is to use of attributes containing patient information before operation to predict the patient status after the operation. The domain experts are particularly interested in finding predictive and descriptive rules for the class of patients who "dead within 90 days" after operation among totally 5 classes.

Several well-known systems were applied to this dataset such as See5 and CBA[7], but the obtained results are far from expectation. For example, See5.0 induces rules with an average error of 30.5% on testing data, but with a very high false positive rate of 98.9%. Similarly, CBA also gives poor results on the class "dead within 90 days" even when they produce a large number of rules with small thresholds. We have used visual interactive LUPC to investigate the stomach cancer data, and found significant results some of them are presented here, including preliminary analysis of data by visualization tool, unusual findings in two extreme classes "dead within 90 days" and "alive".

3.1 Preliminary analysis of data with visualization tools

The visualization tools in LUPC allow us to examine the data and to gain better insight into complex data before learning. While the viewing mode of original data offers an intuition about the distribution of individual attributes and instances, the summarizing and querying modes can suggest patterns to be investigated, or to guide which bias could be used to narrow the huge search space.

An illustration of identifying patterns is shown in Figure 4. It is well-known that patients who have symptoms of "liver_metastasis" of all levels 1, 2, or 3 will certainly not survive. Also, "serosal_invasion = 3" is a typical symptom of the class "dead within 90 days". With the visualization tools, we found several unusual events such as among 2329 patients in class "alive", 5 of them have heavy metastasis of level 3, 1 and 8 of them are of metastasis level 2 and 1, respectively. Moreover, the querying data allow us to verify some significant combination of symptoms such as "liver_metastasis = 3" and "serosal_invasion = 3" as shown in Figure 4.

3.2 Finding irregular rules in class "dead within 90 days"

It is common known that the patients will be died when liver metastasis occurs aggressively. Other learning methods when applied to this datasets often yield rules for class "dead within 90 days" containing "liver_metastasis" that are considered acceptable but not useful by domain experts. Also, these discovered rules usually cover only a subset of patients of this class. It means that there are patients of the class who are not concerned with "liver_metastasis" and they are difficult to be detected.

Using visual interactive LUPC, we did different trials and specified parameters and constraints to find only rules that do not contain the characterized attribute "liver_metastasis" and/or its combination with other two typical attributes "Peritoneal_metastasis", "Serosal_invasion". Below is a rule with accuracy 100% discovered by LUPC that can be seen as rare and irregular event in the class.

Rule 8 accuracy = 1.0 (4/4), cover = 0.001 (4/6712)
IF category = R AND sex = F AND proximal_third = 3
AND middle_third = 1
THEN class = death within 90 days

3.3 Finding rare events in class "alive"

In KDD the prediction of rare events is coming to be of particular interest. Thanks to the support for human interaction with data in LUPC, when supposing that some attribute-value pairs may characterize some rare and/or significant events, LUPC allow us examine effectively the hypothesis space and identify rare rules with any small given support or confidence. An example is to find rules in class "alive" that contain the symptom "liver_metastasis". Such events are certainly rare and influence the human decision making. We found events in the class "alive" such as male patients getting "liver_metastasis" at serious level 3 who can survive with the accuracy of 50%.

Rule 1 accuracy = 0.500 (2/4); cover = 0.001(4/6712)
IF sex = M AND type = B1 AND liver_metastasis = 3
AND middle_third = 1
THEN class = alive

4 Conclusion

We have presented the visualization techniques in the knowledge discovery system D2MS for supporting model selection. We emphasize the crucial role of the user's participation and visualization in the model selection process of knowledge discovery and have designed D2MS to support such participation. Our basic idea is to provide the user with the ability of trying various alternatives of algorithm combinations and their settings, and to provide the user with performance metrics as well effective visualization so that the user can get insight into the discovered models before making his/her final selection. D2MS with its visualization support in model selection has been used and shown advantages in extracting knowledge from a real-world application on stomach cancer data.

Acknowledgement

This work was partially supported by the grant-in-aid for scientific research on priority area "Active Mining" funded by the Japanese Ministry of Education, Culture, Sport, Science and Technology.

References

[1] Brunk, C., Kelly, J. and Kohavi, R., MineSet: An Integrated System for Data Mining, Proc. Third Int. Conf. on Knowledge Discovery and Data Mining KDD'97, 135-138, 1997.

[2] Card, S. K., Mackinlay, J. D., Shneiderman, B., Readings in Information Visualization, Morgan Kaufmann, 1999.

[3] Fayyad, U.M., Grinstein. G.G., and Wierse, A. Information Visualization in Data Mining and Knowledge Discovery, Morgan Kaufmann, 2002.

[4] Han, J. and Cercone, N., Visualizing the Process of Knowledge Discovery, J. of Electronic Imaging, No. 4, 404-420, 2000,.

[5] Keim D. A. and Kriegel H.P., Visualization Techniques for Mining Large Databases: A Comparison, J. IEEE Transactions on Knowledge and Data Engineering, 923-938, 1996.

[6] Lee, H.Y., Ong, H.L., and Quek, L.H., Exploiting Visualization in Knowledge Discovery, Proc. of First Inter. Conf. on Knowledge Discovery and Data Mining, 198-203, 1995.

[7] Liu, B., Hsu, W., and Ma, Y., Integrating Classification and Association Rule Mining, Fourth Int. Conf. on Knowledge Discovery and Data Mining KDD'98, 80-86, 1998.

[8] Liu, H. and Motoda, H., Feature Selection for Knowledge Discovery and Data Mining, Kluwer Academic Publishers, 1998.

The Future Direction of Active Mining in the Business World

Katsutoshi Yada
yada@ipcku.kansai-u.ac.jp
Faculty of Commerce,
Kansai University
Yamate, Suita, Osaka 564-8680, JAPAN

Abstract. In this article, we discuss how the active mining system is applied to the data in the real business world and point out the direction of research activities in future. First, we introduce a case of the consumer behavior analysis in the market of nursing care goods. The framework of the knowledge discovery process in database is then reviewed. Finally, we discuss the future prospect and direction of the active mining research.

1 Introduction

In recent years, a great number of studies have been performed on data mining in various fields such as mechanical learning. Active mining is a study to positively support the creation of new knowledge in the entire process of knowledge discovery from acquisition of data to the request of the user based on the results of basic research. However, for the purpose of accomplishing the active mining, it is necessary to solve many technical problems. In the present article, we have taken an example in the process of knowledge discovery in the business field and have tried to clarify and identify the direction of the research in future as it may be required in the business field.

The present article comprises the introduction of the case, the review of the research relating to the process of knowledge discovery, case analysis, and discussion on the research field. First, the results of the study are introduced, which was performed in cooperation with some enterprises in the year 2000. An example is taken in consumer behavior analysis in the market of nursing care goods. Next, the framework relating to the process of knowledge discovery is reviewed from the conventional database. The case is generally reviewed along these processes, and discussion is made on the study, which is especially important in the business fields.

2 The Case of the Consumer Behavior Analysis in the Market of Nursing Care Goods

In the spring, 2000, a common research was initiated by a project under cooperation of a toiletry manufacturer A, a drug store chain B, Kansai University, Kyoto University, and Osaka Industrial University, and regular meetings were held once per month. In the first meeting, the brand manager of the company A introduced the categories to be analyzed and pointed out problems and questions. From the firm B, POS data with ID on 1,200 stores all over the country were provided, and the conditions were given for the analysis of time series data of the customers.

First, the members participating in the analysis, mostly university students, performed detailed and complete research on the matters relating to the market. This included: investigation on retail stores such as pharmacies and drugstores, investigation on hospitals and old people's homes, and survey on users. At the same time, the students tried to learn basic knowledge of computer to acquire computer skill suitable for the analysis.

The first analysis is a basic market analysis. Normally, the first analysis is generally completed by simply providing the results in common sense unless the research staffs have substantial knowledge in the specific fields. Based on these results, critical points on the problems are elucidated through frequent communication with the people engaging in the business activities at each site. In the present case, it was recommended to share the information at regular meetings among the participants, and we acquired professional knowledge of the persons in charge of marketing. Particularly important was the knowledge relating to the combination purchase of the nursing care goods by the consumers. Through qualitative investigation, they already had a hypothesis that combination purchase is different for each individual consumer depending on the intended purpose, the symptoms of the patients, the place of the stores, etc. Analysis was carried out to demonstrate whether this hypothesis is true or not. The reason why we have selected this market is that this market is expected to be one of the most promising market with incessant growth in Japan where elderly population is rapidly increasing.

The paper diapers for adults selected as the object for analysis in the present study are roughly divided to four types: "pad", "flat", "pants", and "tape-fixed diaper". As the result of analysis, it was found that most of the consumers were buying two or more different types of goods. However, it was not possible to find a definite pattern of combination purchase. In this respect, we tried to follow the data in detail. As a result, it was made clear that there was a certain definite pattern in combination purchase in the changes of time series pattern of purchase.

Before finding the pattern, it is essential to clearly define the combination purchase. The combination purchase was divided to 3 categories depending on the symptoms of the patients: "single use", "concomitant combined use", and "multiple use". Then, we were able to find 7 typical patterns. These 7 patterns accounted for more than 60% of all, and the other patterns were also more or less similar to one of these patterns. The pattern thus found was characterized in that the purchase combination was changed only once or less, and that "pad" was used concomitantly in all cases. Also, regardless of the changes in the purchase pattern, it was made clear that "tape-fixed diaper" and "flat type diaper" were purchased only in the cases of concomitant use with "pad". Among the specialists including the developer of the goods, the pattern of these changes, i.e. how the consumers change the purchase from one combination to another, was not clearly recognized. Our discovery of new knowledge was started from this point.

When we analyzed in detail, it was estimated that the change of purchase pattern was probably caused by aggravation of the symptoms. Regrettably, most of the patients changed the purchase pattern as the symptoms were aggravated. That is, it was found that "pad" and "tape-fixed diaper" to be used by the patients with severe symptoms were purchased more often as the concomitantly used goods. The timing of such change was also concentrated within one year from the first purchase. Also, it was discovered that the timing of sales promotion was important. The analysis was continued further, and when we analyzed how the customers with such purchase pattern contribute to each store, it was found that, in the cases of single use, the customers were continuously visiting the store regardless of the type of goods they select. On the contrary, when the customers changed the purchase to the goods of concomitant use and having higher gross return, the customers did not visit the store any more. This may be attributed to the facts that the symptoms of the patients were aggravated and there was no need any more to visit the store or that the cost-consciousness of the customers became better due to learning and the customers may have switched over to the other stores. Among these rules, there was a rule that the customer ceased to visit the store as soon as they began to buy paper diaper for infants, which are typical shopping goods.

This was initially started from the need to acquire information on the types of combination purchase among the persons in charge of marketing, while unexpected and wide variety of knowledge was obtained such as time series pattern changes, customer characteristics, contribution of the customers to the store, etc. These types of knowledge were not necessarily the knowledge requested by the firms during the process of analysis. Many of these types of knowledge are derived from subjective view and personal interest of the analyzer and from steady and sober analysis. This means that a specialist or a professional does not necessarily discover important knowledge. These types of analysis were made possible only by frequent and close communication between the marketing staffs and the analyzers. This is only achieved by such tremendous and frantic efforts as we often see, for instance, when our houses are ablaze in a fire, i.e. the efforts to try to comply with the deadline of the presentation at the final meeting.

3 Knowledge Discovery Process in Database

Here, we wish to review the study of knowledge discovery process from database for the purpose of discussing the above case. Metheus et al. [6] expressed the knowledge discovery process in several steps and explained the entire model and the elements of the knowledge discovery system. Major elements given by them are acquisition and processing of data, extraction of pattern, formation of knowledge, evaluation, and database to support these factors. In the framework of the knowledge discovery process, emphasis is put on the interaction between the analyzer and the system [7, 8, 2]. In fact, in the knowledge discovery, it is important how human factor intervenes this process [5]. They assert that it is necessary to clearly become conscious of the introduction of professional knowledge.

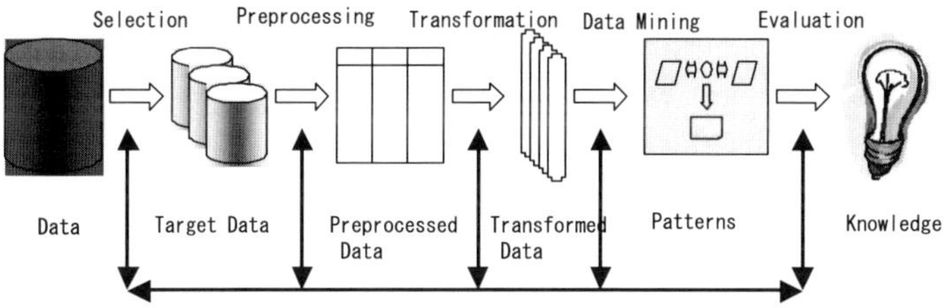

Figure 1: Knowledge discovery process in database

Now, we wish to discuss the knowledge discovery process (Fig. 1) based on the study by Fayyad et al. [3]. The knowledge discovery is initiated from the efforts to clearly identify the ultimate purpose and to completely understand the field of application and the related professional knowledge. Based on these data, the data sets necessary for the discovery are extracted, and various attribute groups are prepared (preparation of target data). Normally, many noises are included in this type of data, and preprocessing is performed by deletion and correction. Important attributes are estimated, selected or prepared, and the adjusted and re-arranged data sets are prepared. By taking the purpose of analysis and the characteristics of data sets into account, the most adequate method is selected, and a really interesting patterns are extracted. Efforts are made to comprehend the pattern discovered from the above steps and to go back to the step in the middle of the way in some cases. While repeating trial-and-error, beneficial pattern is extracted. This collection of patterns is converted to a meaningful new knowledge through feedback of evaluation by the related persons in charge and by check-up with the existing knowledge.

However, these series of conventional research have the focus on the study of the knowledge discovery process with higher efficiency. This is because the studies in the past were based on the assumption that, as seen in the study of human genome, due compensation is given only to those who discovered at the earliest. In contrast, in the field of management, the developed technique or knowledge is imitated within very short time or the dominant position in competition is usurped by an alternative technique in many cases [1]. Therefore, in order to continuously maintain dominant position in competition, it is necessary not only to attain the discovery process with higher efficiency but also to continuously discover the knowledge further [9].

Particularly important in the field of management is the fact that the knowledge discovery process is realized and continued only in the business system and cannot be firmly established without the business system. The business system can be roughly divided to information system, organization system, and inter-company system, and it is a system to accomplish the creation of values. The knowledge in the management is to be created from interaction of those engaged in the business system, and sufficient effects cannot be provided simply by grasping only one side of information technology. In other words, an interesting knowledge is different from the knowledge beneficial to the business. Unless there is a scheme to convert the discovered rules and patterns to actual business action to execute these and to feedback to the system, these rules and patterns are in no way beneficial.

4 Problems in the Study of Active Mining

In this Chapter, we try to elucidate important problems in the study of active mining in future through the cases along the discussion of the knowledge discovery process from the conventional database.

4.1 Setting of the Purpose

In the textbook on data mining, it is pointed out as important to clearly define the purpose to utilize the data mining. However, it is very difficult to set up the themes, and it can be defined only by enormous series of basic analysis. For this purpose, useful opinions and findings must be acquired by fully understanding the rules of the duty and by obtaining the know-how in the field of application in order to avoid misunderstanding on the movement in the entire market. Tremendous series of trial-and-error in this stage may provide a clue for the unexpected and new finding.

In the case under analysis, the initial setting of the themes was limited only to the discovery of combination purchase. The setting of themes was frequently changed in the course of analysis and it was brought to a position closer to the knowledge useful for the practice. Also, it seems to be important for the discovery of new knowledge to flexibly change the setting of the themes through steady communication with the situation at site and through reaction from the users without persisting on the initial setting of the themes.

4.2 Manipulation of Gigantic Data

To perform basic market analysis after the purpose has been clearly defined, the data are processed in order to analyze the initial data from various angles. Ideally, it is desirable that the persons who have both practical knowledge useful at site and skillful computer technique carry out the analysis. In fact, however, this is often executed by an analyzer, who does not have sufficient computer skill, and it is very difficult to handle gigantic mass of initial data.

Even for the skillful person with sufficient computer knowledge and experience in analysis, the data handling task such as data cleaning may impose considerable burden on the analyzer, and it is an important problem in the business field to reduce and to turn this work to the one with higher efficiency. Even for the beginners of data mining, it is important to develop analytical environment for easier handling of data. We have developed a system architecture called "MUSASHI" and have successfully reduced the burden of work by sharing the processing work between the analyzers as components [4].

The study of data mining is a study of effectiveness and efficiency of rule extraction in most cases. The process of basic processing up to this goal is closely related with the effectiveness. There have appeared several studies dealing with these problems in recent years, while further development are required, and it is important to advance toward standardization and to carry out development and analysis with higher efficiency.

4.3 Generation of Attributes

Normally, for the purpose of generating the adjusted and re-arranged data which are to become the object of the data mining, various attributes are newly incorporated from the initial data. The generation of these new attributes and their effectiveness depend much on the factors such as human knowledge at site to process the data and experience of analysis in the past. In particular, when we analyze complicated causal relation such as consumer behavior, knowledge of goods and experience of analysis in the past make up an important basis. Therefore, in order to carry out the process more efficiently, it is essential to have a scheme to share the knowledge and the experience in the past among the analyzers.

The attributes newly incorporated such as expression of knowledge in the case are made up with the components generated by the system architecture "MUSASHI" as described above. The newly generated attributes are designed in such manner that these can be shared through WEB by MUSASHI and that all of the analyzers can utilize them for higher efficiency. For instance, the attribute of the customer such as repeated purchases of a certain commodity can be readily utilized in the new analysis. In future, it would be necessary to introduce the results of the study such as automatic generation of new attributes and to carry out the scheme with high efficiency.

4.4 Evaluation Criteria of Rules

There are also theoretical problems relating to evaluation criteria of the rules. Normally, actual data contains noise and it is far from clean state. Also, distorted or deviated data may be present in many cases. For instance, the customers who easily respond to the direct mail usually account for 5-6 % (at the highest) of all, and positive-to-negative ratio shows an extreme value. The distribution of the population is not known in almost all cases. If a certain premise (normal distribution) is assumed, unexpected error may occur in the analysis. Also, the problem of marketing is really diverse, and it is difficult to adopt a fixed criterion for evaluation. In the past, we also used Gini's index for the evaluation criterion, but it seems to be necessary to develop an optimal criterion suitable for each case.

4.5 Creation of Business Action

Among the studies of data mining, there are relatively many, which aim the improvement in accuracy and speed. These are naturally important factors, but the improvement in the interpretation of the rules is also important. This is because, for the purpose of incorporating the rules into practical business action and of carrying out the rules, the people at working site must understand the importance. In the present case, analysis has been performed mostly by the students and it was not possible to use complicated mathematical formulae and rules. Rather, the results derived from such analysis were easily understandable for the people at the working site. Ironically, the results obtained through high-grade method were often neglected. In the study of data mining, it appears that the studies on visualization of knowledge and expression may contribute to the solution of these problems.

4.6 Framework of Knowledge Discovery Process

In the frameworks of knowledge discovery process by Fayyad et al. [3], typical process of data processing and its cycle are basically expressed. However, it is difficult to obtain useful suggestion for practical analysis from these frameworks. As the problems of conventional frameworks, there are two important points: The first is the process of knowledge conversion, and its details are not clearly expressed. The knowledge discovery process is a process where various types of information and knowledge are integrated and unified and are converted to new knowledge, while it is not possible to understand this from the conventional frameworks. It is necessary to develop a new framework to facilitate the understanding of more concrete details of knowledge conversion.

The second point is the problem relating to the procedure to introduce professional knowledge. In the conventional framework, it is suggested that professional knowledge is needed, but concrete suggestion on how to introduce such knowledge is lacking. It is practically very difficult to adopt the evaluation of professional experts in all processes. A strategy must be developed to introduce valuable professional knowledge into the knowledge discovery process with high efficiency.

5 Conclusion

In the present article, active mining in the business knowledge discovery was discussed through the case with special weight on the themes of future research. The case was discussed along the framework of the conventional knowledge discovery process, and association with the research currently executed in the study of data mining and its importance were emphasized. In future, we will try to solve these problems one by one and will make every effort so that the study of active mining would extensively contribute to the society.

References

[1] D. A. Aaker. Managing Assets and Skills: The Key to Sustainable Competitive Advantage. *California Management Review*, No. 4, pages 91–106, 1989.

[2] R. Brachman and T. Anand. The Process of Human-Centered Apporoach. *In Advances in Knowledge Discovery and Data Mining*, U. M. Fayyad, G. Piatetsky-Shapiro, P. Smyth, and R. Uthurusamy. (eds.),pages 37–58, AAAI Press, 1996.

[3] U. M. Fayyad, G. Piatetsky-Shapiro and P.Smyth. From Data Mining to Knowledge Discovery: An Overview. *In Advances in Knowledge Discovery and Data Mining*, U. M. Fayyad, G. Piatetsky-Shapiro, P. Smyth and R. Uthurusamy. (eds.), pages 1–34, AAAI Press, 1996.

[4] Y. Hamuro, N. Katoh and K. Yada. Data Mining Oriented System for Business Application. *Proc. 1st International Conference on Discovery Science*, (Lecture Notes in Artificial Intelligence 1532), pages 441–442, Springer-Verlag, 1998.

[5] P. Langley. The Computer-aided Discovery of Scientific Knowledge. *Proc. 1st International Conference on Discovery Science*, (Lecture Notes in Artificial Intelligence 1532), pages 25–39, Springer-Verlag, 1998.

[6] C. J. Matheus, P. K. Chan and G. Piatetsky-Shapiro. Systems for Knowledge Discovery in Databases. *IEEE Transaction on Knowledge and Data Engineering*, Vol. 5, pages 903–913, 1993.

[7] R. E. Valdes-Perez. Machine Discovery in Chemistry: New Results. *Artificial Intelligence*, 74, pages 191–201, 1995.

[8] R. E. Valdes-Perez. Principles of Human Computer Collaboration for Knowledge Discovery. *Artificial Intelligence*, 107, pages 335–346, 1999.

[9] K. Yada, N. Katoh, Y. Hamuro and Y. Matsuda. Customer Profiling Makes Profits: How did a Japanese firm achieve competitive advantage through the knowledge creation? *Proceedings of The Practical Application of Knowledge Management 98*, pages 57-66, The Practical Application, 1998.

Active Mining
H. Motoda (Ed.)
IOS Press, 2002

Topographical Expression of a Rule for Active Mining

Takashi Okada
okada@kwansei.ac.jp
Center for Information & Media Studies,
Kwansei Gakuin University
1-1-155 Uegahara, Nishinomiya, Hyogo 662-8501, JAPAN

Abstract. Datascape surveys, introduced in a recent paper, provide a way to overview data by organizing rules into a few principal rules and their relative rules. When a rule illustrates a characteristic that is of interest, the user naturally wishes to examine it in detail. This paper develops a way of expressing rules topographically to guide surveys of micro-datascapes. The cascade model was developed previously from association rule mining, and has several advantages that allow it to lay the foundation for a better expression of rules. One is that a rule denotes local correlations explicitly, and the strength of a rule is given by the numerical value of the *BSS* (between-groups sum of squares). This paper briefly overviews the cascade model, and proposes a new method that expresses the "ridges" of a rule; a ridge indicates the location of a sharp decrease in the *BSS* value. Ridge information is useful in interpreting the data distribution surrounding the supporting instances of a rule. Application to a real medical dataset is also discussed.

1 Introduction

Datascape surveying is a new concept that was proposed in a recent paper [9]; it is very helpful for understanding data using characteristic rules. The datascape refers to the image of a dataset from the perspective of the analyst. The article proposed four conditions for a datascape survey: it (1) generates rules from concise to detailed, (2) quantifies a problem and a rule, (3) identifies the dependencies among the various variables in the supporting instances of a rule, and (4) generates information related to the rule. The cascade model was developed from association rule mining by the author to provide a solution to some of these conditions [4, 5, 7]. I subsequently used the model and developed a new set of rules suitable for datascape surveys [9]. That is, numerous rules are first optimized to a smaller number of rules, which are organized into a few principal rules and their associated relatives. The resulting rules meet the above-mentioned requirements and proved useful in surveying a practical datascape.

In the cascade model, the LHS conditions of a rule are divided into main and precondition clauses. The strength of a rule is measured by its *BSS* value, which is computed from the distribution of dependent variables before and after application of the main condition. This raises the question of whether this precondition is essential, or does the rule show a fairly strong pattern without the precondition? This paper answers this question by developing topographical expression of rule strength. By recognizing and expressing the ridges that are caused by changes in the *BSS* around a rule, information can be mined from the data.

The next section briefly reviews the cascade model. I propose the ridge expression of a rule to give a topographical view in Section 3, and Section 4 applies the scheme to a medical diagnosis problem.

2 The Cascade Model

2.1 Cascades and the Sum of Squares Criterion

The cascade model was originally proposed by the author [4]. It can be considered an extension of association rule mining. The method creates an itemset lattice in which an [attribute: value] pairs are employed as an item to form itemsets.

Let us consider the trivial dataset shown in the sample data of Fig. 1, which discriminates the Y value using two attributes, A and B. When we construct a lattice, the nodes and links can be viewed as lakes and connecting waterfalls, respectively, as shown in Fig. 1. The height of a lake is assumed to denote the purity of its class features, and its area approximates the number of instances that support the itemset.

Sample data

A	B	Y
a1	b1	p
a2	b1	p
a2	b1	p
a1	b2	n
a1	b2	n
a1	b2	p
a2	b2	n
a2	b2	n

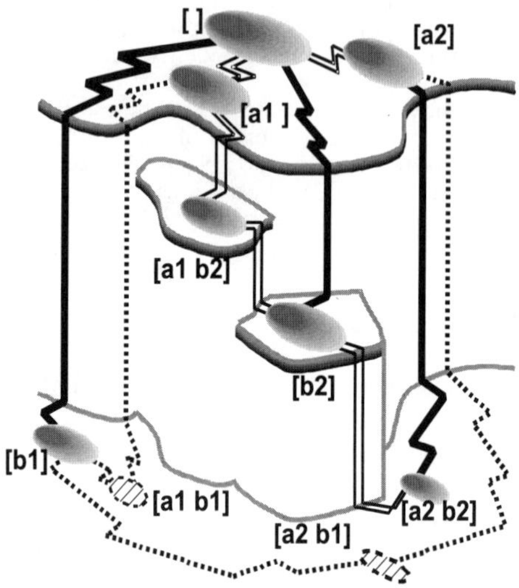

Figure 1: The cascades expression of a lattice

Since the concept behind the cascade model is to select the most powerful waterfalls and to use them as rules, we need to define the power of a waterfall. Gini's definition of SS (sum of squares) for categorical data in Eq. 1 provides a framework for the power of a waterfall [2]. Imagine that the instances are divided into G subgroups. Then, TSS (total sum of squares) can be decomposed into the sum of WSS_g (within-group sum of squares) and BSS_g (between-groups sum of squares) using Eq. 2, if we define BSS_g as in Eq. 3 [6]. I proposed using BSS_g as a measure of rule strength. The BSS value per instance is called *dpot*, as defined in Eq. 4, and this can be used as a measure of the potential difference of a waterfall.

$$SS = \frac{1}{2}(1 - \sum_a p(a)^2) \tag{1}$$

$$TSS = \sum_{g=1}^{G} (WSS_g + BSS_g) \tag{2}$$

$$BSS_g = \frac{n^L}{2} \sum_a \left(p_a^L(g) - p_a^U(g)\right)^2 \tag{3}$$

$$dpot(L, U) = \frac{1}{2} \sum_a \left(p_a^L(g) - p_a^U(g)\right)^2 \tag{4}$$

In these equations, g designates a subgroup; the superscripts U and L indicate the upper and lower nodes, respectively; n is the number of cases supporting a node; and $p(a)$ is the probability of obtaining the value a for the objecive attribute.

Figure 2 illustrates a sample of *SS* decomposition. Here, 1000 instances (p: 800, n: 200) in the top node are split into two lower groups containing 800 (p: 760, n: 40) and 200 (p: 40, n: 160) instances. We see that the sum of *BSS* and *WSS* for the lower nodes equals *TSS* for the top node. The *BSS* value for the lower right node is much larger than that for the lower left node. This means that this measure of rule strength emphasizes the link to the subgroup showing a distribution opposite that in the upper node.

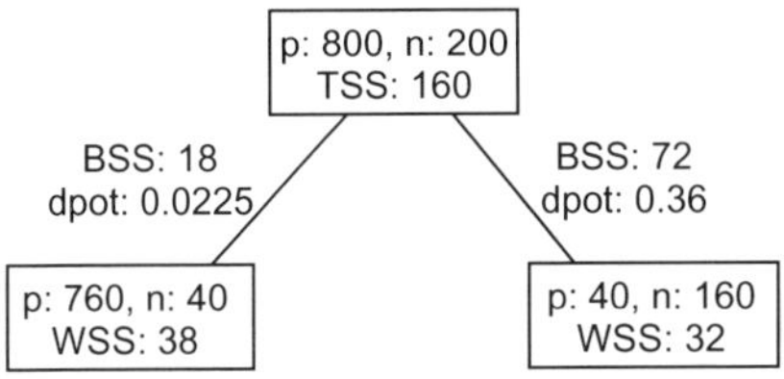

Figure 2: Sample decomposition of the sum of squares

2.2 Rule Link in the Lattice

Powerful links in the lattice are selected and expressed as rules [5]. Figure 3 shows a typical example of a link and its rule expression. Here, the problem contains four explanatory attributes, *A*–*D*, and an objective attribute *Z*, which take (y, n) values. The itemset at the upper end of the link contains item $[A\!:\!y]$, and another item, $[B\!:\!y]$, is added along the link. The items of the other attributes are called veiled items. The three small tables at the center show the frequencies of the veiled items in the upper node. The corresponding *WSS* and *BSS* values are also shown.

The textbox to the right in Fig. 3 shows the derived rule. The large $BSS(Z)$ value is evidence of a strong interaction between the added item and attribute Z, and its distribution change is placed on the RHS of the rule. The added item $[B\!:\!y]$ appears as the main condition on the LHS, while the items in the upper node are placed at the end of the LHS as preconditions. When an explanatory attribute has a large *BSS* value, its distribution change is also denoted on the RHS to show the additional dependency. This information is useful for detecting colinearity among variables in the supporting instances in the lower node.

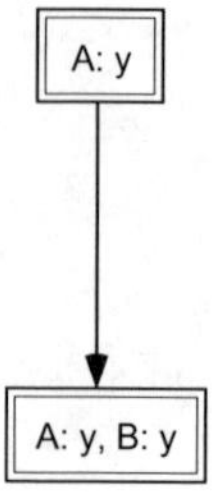

	y	n	WSS
B	60	40	24.0
C	50	50	25.0
D	60	40	24.0
Z	40	60	24.0

	BSS	dpot
B	9.60	0.16
C	0.00	0.00
D	6.67	0.11
Z	5.40	0.90

IF [B: y] added on [A: y]
THEN [Z] BSS=5.40 (.40 .60) ==> (.10 .90)
THEN [D] BSS=6.67 (.60 .40) ==> (.93 .07)

	y	n	WSS
B	60	0	0.0
C	30	30	15.0
D	56	4	3.73
Z	6	54	5.40

Figure 3: A sample link, its rule expression, and the distributions of the veiled items

It is not necessary for items on the RHS of a rule to reside in the lattice. We need only itemsets $[A\!:\!y]$ and $[A\!:\!y,\ B\!:\!y]$ to detect the rule shown in Fig. 3, although we have to count the frequencies of the veiled items. This is in sharp contrast to association rule mining, which requires the itemset $[A:y,\ B:y,\ D:y,\ Z:y]$ to derive the rule in Fig. 3. This property makes it possible to detect powerful links dynamically before constructing the entire lattice.

Combinatorial explosion in the number of nodes is always a problem in lattice-based machine-learning methods. Since an item is expressed in the form [attribute: value], the item distribution is very dense in the cascade model, making the problem more serious. However, the above-mentioned property makes it possible to prune the lattice expansion, allowing us to derive a rule from a link while avoiding the construction of the entire lattice [7]. In fact, by deciding an inequality constraint for the $BSS(Z)$ value, and using this inequality as the pruning criterion, valuable rules can be found, even when the number of attributes reaches several thousand.

2.3 Datascape and Organization of Rules

I introduced the new concept of datascape survey in a previous paper [9]. A datascape is a perspective view of data from a user's viewpoint. I used the analogy of a variable focus lens to point out the importance of datascape surveying. That is, we need to obtain a global view of the data using a few rules, and then proceed to inspect details of the datascape, guided by supplementary rules. However, typical mining systems, such as association rule miners, generate numerous, unorganized rules, which do not allow users to inspect their details.

My solution to this problem is to optimize the rules in the first step. That is, to look for a stronger rule by a greedy search of conditions, using powerful links in the lattice as starting seeds. As a result, a rule takes a local maximum *BSS* value, using disjunction of features in the conditions. Moreover, several seed links converge to a single rule, decreasing the number of resulting rules.

In the second step, I organize the rules into principal rules and their related rules. For example, the difference between a pair of rules is sometimes the addition or deletion of a precondition clause, or a pair of rules may be expressed by completely different main and preconditions, but share most of their supporting instances. Since such pairs of rules can be considered different aspects of a single phenomenon, the organization of these rules into a principal rule and its related rules helps users to examine the data.

The rules are organized using the relevance values computed from the overlap of supporting instances among rules. Relevant rules are classified using the relevance of the supporting instances at the upper and lower nodes. They are (1) *ULrelative*: relevant at both nodes, (2) *Lrelative*: relevant only at the lower node, and (3) *Urelative*: relevant only at the upper node.

The resulting expression of rules consists of a few principal rules, allowing a user to grasp the global characteristics of the data quickly. If some interesting pattern is found, its related rules guide a detailed survey of the datascape. This expression was proved to be useful in an application to real-world data involving medical diagnosis.

3 Topographic Expression

3.1 Analysis of BSS Changes

A rule derived from the cascade model extracts a group of instances specified by its preconditions. There is a strong correlation between the feature in the main condition and the distribution of the objective attribute in these instances. This raises the question of whether this correlation is strictly limited to the instances in the precondition region, or is roughly applicable to all the data.

Let us imagine a sample dataset and a rule derived from it, as shown in Fig. 4. Here, attributes A and B have ordered categorical values ranging from 1 to 5, and the Y values are *pos* or *neg*, of which counts are shown in the contingency table in Fig. 4. This rule states that if A takes a lower-middle value range $(2-3)$, then the probability that the objective attribute Y is *pos* increases from 0.62 to 0.9, if the precondition $[B: 2-3]$ is satisfied. Is Y correlated with A in a similar way, if B takes the value 1? What happens if the value of B changes to 4 or 5?

Sample data (Y: pos/neg)

A	B: 1	2	3	4	5
1	10/10	14/6	14/6	10/10	10/10
2	14/6	18/2	18/2	10/10	10/10
3	14/6	18/2	18/2	10/10	10/10
4	10/10	6/14	6/14	10/10	10/10
5	10/10	6/14	6/14	10/10	10/10

IF [A: 2-3] added on [B: 2-3]
THEN Y: (0.62 0.38) => (0.9 0.1)

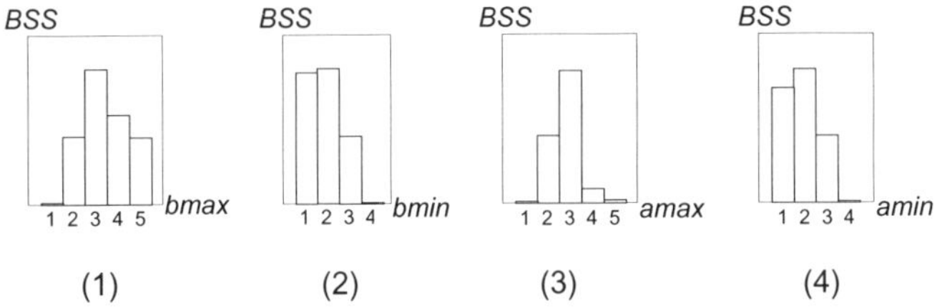

Figure 4: Sample *BSS* changes of a rule

Bar graph (1) in Fig. 4 shows the changes in the *BSS* value as the upper edge *bmax* of B changes from 1 to 5, while the lower edge is fixed at 2; note that the number of instances is equally distributed for all categories of B. The highest *BSS* value specified by the rule is at $bmax = 3$. The *BSS* value decreases sharply as *bmax* moves to 4.

If the correlation is approximately the same as that in the rule region, the resulting *BSS* should be high, because it is proportional to the number of supporting instances. This result indicates that the correlation between A and Y is much weaker than in the rule region, when the value of B is higher. *BSS* also decreases when *bmax* decreases. However, as this is simply because of the decrease in the number of instances, we can say that the level of correlation in the rule region is constant.

Bar graph (2) shows the *BSS* changes when the lower edge of the B value, *bmin*, is altered. The *BSS* decrease for higher *bmin* is again explained by the decrease in instances, but it remains roughly the same when *bmin* moves to 1. From these two graphs, we can conclude that a fairly strong correlation between A and Y holds for $B=1$, and that the correlation is lost if the value of B increases to 4.

Bar graphs (3) and (4) show the *BSS* changes when we alter the edges of the main condition, *amax* and *amin*, respectively. The decrease in *BSS* is gentle at $amin = 1$, but it is very steep at $amax=4$. The latter indicates that the Y value changes to *neg* at this ridge.

The characteristic decrease in *BSS* values obtained here is useful for understanding the datascape near the rule region. The next problem is how to extract and express these characteristics.

3.2 Ridge Detection and Expression

Let us introduce a variable X to denote the value range specified by the preconditions. Then *BSS* is a function of X that takes a local maximum value BSS_0 at precondition X_0 of a rule. Suppose that n is a function of X that returns the number of instances in X. If we consider X_0 as a mountain peak on the map of X, we can draw contour lines of *BSS* values. The interesting characteristics have steep slopes around the peak. However, there is no simple way of expressing the contour of *BSS* on the map of X, and some steep slopes are not interesting because they reflect only a decrease in the number of supporting instances.

Let us call an interesting X region of decreasing *BSS* a ridge ΔX. This may be an addition to or a deletion from region X, and $n(\Delta X)$ can take plus or minus values.

The problem is to extract ridges from the many steep slopes around a *BSS* peak. When the precondition region changes from X to $X+\Delta X$, the difference in the *BSS* values is $n(\Delta X) \cdot BSS(X)/n(X)$ if the probability distribution of Y is the same in ΔX as in X.

To omit such an effect from the change in *BSS*, we employ the $\Delta BSSrate$ defined by Eq. 5, normalized by $n(\Delta X)$ and by $BSS(X_0)/n(X_0)$, as the criterion to judge whether ΔX is a ridge. This expresses a kind of normalized gradient from X to $X+\Delta X$.

$$\Delta BSSrate = \frac{BSS(X+\Delta X) - (n(X+\Delta X)/n(X)) \cdot BSS(X)}{|n(\Delta X)|} \Big/ \frac{BSS(X_0)}{n(X_0)} \tag{5}$$

$$\Delta BSSrate < \textit{thres-ridge-rate} \tag{6}$$

$$|n(\Delta X)| > \max(\textit{min-ins-ratio} \cdot n(X), \textit{min-instances}) \tag{7}$$

If $\Delta BSSrate$ is less than a threshold value, *thres-ridge-rate* (default value: -0.3), as shown in Eq. 6, ΔX is considered a ridge, as there is a sudden change in the probability distribution of Y values when the instance range moves from X to ΔX. However, ΔX might contain so few instances that the data fluctuation results in large negative $\Delta BSSrate$ accidentally. Therefore, to recognize a ridge, another condition is added,

Eq. 7, using 0.1 and 5 as the default values of *min-ins-ratio* and *min-instances*, respectively.

The rule preconditions in the cascade model suggest the existence of adjacent ridges by nature. However, from a rule itself we cannot know whether these ridges are very steep or gentle. Once we recognize a sharp ridge, its region ΔX and the distribution of Y values in ΔX are expressed, as well as its $\Delta BSSrate$. The ridge information provided is expected to contribute to the user's understanding of the datascape.

The next problem that we must consider is the search range of ΔX around the precondition region. As shown in Fig. 4, changes in the upper and lower edges of all ordered categorical precondition clauses should be included. We examine a value range from the inside edge of a precondition to the end of its attribute value. Then, we select ΔX as the ridge with the largest $\Delta BSSrate$ of the ridges satisfying Eq. 7. When a precondition uses a nominal attribute, we need to examine $\Delta BSSrate$ by deleting every item that is present in the precondition and by adding every item that is absent from it. Then all the ΔXs satisfying Eqs. 6 and 7 are described as ridges.

Neighboring regions of rule region X_0 are found by introducing a new precondition. That is, the edge of a new precondition can be cut in a manner similar to that of a normal precondition. Detecting ridges that change with changes in the main condition is also interesting, as discussed in the previous subsection. The above discussion applies to this problem, if variable X is interpreted as denoting the value range at the lower node specified by the main condition. We can expect a very sharp edge if there is an interchange in the value of the objective attribute.

4 Application to Meningoencephalitis Diagnosis

I used the test dataset for meningoencephalitis diagnosis provided at the JSAI KDD Challenge 2001 workshop [11] to detect ridges of a rule using the method proposed in the previous section. The aim of the analysis is to determine whether disease is bacterial or viral meningitis. It is already known that a diagnosis can be obtained by comparing the numbers of polynuclear and mononuclear cells, but there should be additional information related to the diagnosis. The cascade model has already been used to analyze this data [8]. That analysis produced strong rules based on the number of cells that contributed to the diagnosis, but the results were not easy to interpret. Recently, I reanalyzed this data set and organized the resulting rules into principal and related rules in order to facilitate the survey of the datascape [9]. This section shows the results of ridge detection, and their interpretation, for those rules. I use the same categories and parameters as used in [8].

We show three rules and their ridges in Fig. 5. They are selected from 17 principal and relative rules in the previous results [9], because they have steep and interesting ridges. The steepest ridge was found for *Rule* 7 at the top in Fig. 5. The rule employs *FEVER, LOC* and *BT* as precondition attributes, and denotes the percentage of *bacteria* as rising from 27 to 85% if its main condition [*CT_FIND* : *abnormal*] is satisfied. It also denotes that the main condition correlates with *LOC_DAT, SEX, FOCAL* and *CSF_PRO* to varying degrees. For example, applying the main condition increases the percentage of [*LOC_DAT* : +] from 20 to 77%.

I found five interesting ridges for this rule. The ridge description first denotes its location around the rule. *Pre, New* and *Low* show that the ridge is in a precondition, a newly introduced precondition and the main condition, respectively. I attach *inside* or *outside* depending on whether a ridge ΔX is inside of the rule region X_0 or not. When

```
Rule 7:      Cases: 74 -> 13; BSS=  4.311
  IF [CT_FIND: abnormal]
    added on     [FEVER=<6] [LOC=<1] [BT=<39]
  THEN Diag2:         BSS:4.31           0.27 0.73 ==>  0.85 0.15
  THEN SEX:           BSS:2.00           0.61 0.39 ==>  1.00 0.00
  THEN LOC_DAT:       BSS:4.17           0.80 0.20 ==>  0.23 0.77
  THEN FOCAL:         BSS:1.50           0.88 0.12 ==>  0.54 0.46
  THEN CT_FIND:       BSS:8.83           0.18 0.82 ==>  1.00 0.00
  THEN CSF_PRO:       BSS:1.29           0.09 0.31 0.32 0.16 0.11 ==>  0.23 0.00 0.31 0.08 0.38
Ridge information                         dBSSrate    NewBSS    Ridge Upper      -->    Ridge Lower
 New:inside          [FOCAL: +]            -3.69      (1.85)    0.78 0.22 / 9    -->  1.00 0.00 / 6
 New:inside          [LOC_DAT: +]          -1.57      (2.07)    0.67 0.33 / 15   -->  0.80 0.20 / 10
 Pre:right-outside   [BT>39]               -1.43      (3.96)    0.71 0.29 / 14   -->  0.75 0.25 / 4
 Pre:right-outside   [FEVER>10]            -.706      (3.44)    0.18 0.82 / 17   -->  0.40 0.60 / 5
 New:inside          [SEX: M]              -.644      (0.00)    0.40 0.60 / 45   -->  0.85 0.15 / 13

Rule 5:      Cases: 63 -> 7; BSS=  4.587
  IF [CT_FIND: abnormal]
    added on    [NAUSEA=<3] [STIFF=<3] [LOC_DAT: -] [CSF_GLU>40]
  THEN Diag2:         BSS:4.59           0.19 0.81 ==>  1.00 0.00
  THEN HEADACHE:      BSS:.887           0.13 0.30 0.25 0.13 0.19 ==>  0.00 0.14 0.00 0.43 0.43
  THEN NAUSEA:        BSS:.705           0.68 0.32 0.00 ==>  1.00 0.00 0.00
  THEN ONSET:         BSS:.750           0.08 0.87 0.02 0.03 ==>  0.43 0.57 0.00 0.00
  THEN BT:            BSS:.552           0.22 0.17 0.37 0.13 0.11 ==>  0.14 0.14 0.14 0.14 0.43
  THEN FOCAL:         BSS:2.04           0.83 0.17 ==>  0.29 0.71
  THEN CRP:           BSS:1.19           0.60 0.17 0.10 0.13 ==>  0.14 0.14 0.29 0.43
  THEN CT_FIND:       BSS:5.53           0.11 0.89 ==>  1.00 0.00
  THEN CSF_CELL:      BSS:.734           0.11 0.16 0.22 0.24 0.27 ==>  0.29 0.00 0.14 0.00 0.57
  THEN Cell_Poly:     BSS:.935           0.22 0.14 0.25 0.27 0.11 ==>  0.14 0.00 0.14 0.14 0.57
Ridge information                         dBSSrate    NewBSS    Ridge Upper      -->    Ridge Lower
 New:inside          [FOCAL: +]            -2.86      (1.50)    0.45 0.55 / 11   -->  1.00 0.00 / 5
 Pre:outside         [LOC_DAT: +]          -2.78      (2.51)    0.31 0.69 / 16   -->  0.33 0.67 / 12
 Pre:right-outside   [NAUSEA>3]            -1.71      (3.76)    0.19 0.81 / 16   -->  0.40 0.60 / 5
 Pre:left-outside    [CSF_GLU=<40]         -1.40      (4.35)    0.37 0.62 / 8    -->  0.00 0.00 / 0
 New:inside          [EEG_FOCUS: +]        -.528      (2.92)    0.33 0.67 / 15   -->  1.00 0.00 / 3

Rule 1-UL2:  Cases: 99 -> 15; BSS=  8.381
  IF [CSF_CELL>750]
    added on    [Cell_Mono=<750] [CSF_PRO>0]
  THEN Diag2:         BSS:8.38           0.25 0.75 ==>  1.00 0.00
  THEN SEX:           BSS:1.92           0.58 0.42 ==>  0.93 0.07
  THEN KERNIG:        BSS:1.51           0.28 0.72 ==>  0.60 0.40
  THEN LOC_DAT:       BSS:1.51           0.72 0.28 ==>  0.40 0.60
  THEN WBC:           BSS:1.25           0.06 0.30 0.20 0.19 0.24 ==>  0.00 0.27 0.07 0.07 0.60
  THEN CT_FIND:       BSS:1.81           0.25 0.75 ==>  0.60 0.40
  THEN CSF_CELL:      BSS:6.85           0.14 0.20 0.20 0.30 0.15 ==>  0.00 0.00 0.00 0.00 1.00
  THEN Cell_Poly:     BSS:6.21           0.29 0.17 0.18 0.16 0.19 ==>  0.00 0.00 0.00 0.00 1.00

Ridge information                         dBSSrate    NewBSS    Ridge Upper      -->    Ridge Lower
 Pre:right-outside   [Cell_Mono>750]       -2.74      (4.70)    0.40 0.60 / 25   -->  0.40 0.60 / 25
 Pre:left-outside    [CSF_PRO=<0]          -1.69      (7.56)    0.43 0.57 / 14   -->  0.50 0.50 / 2
 New:inside          [CT_FIND: abnormal]-.839      (4.49)    0.60 0.40 / 25   -->  1.00 0.00 / 9
```

Figure 5: Sample rules with ridge information

the ridge attribute is ordered categorical, *right* or *left* is shown to indicate whether the ridge is at the upper or lower edge of the region, respectively. Then, the ridge is described by an $[attribute : value]$ pair, followed by its $\Delta BSSrate$ and $BSS(X+\Delta X)$ values. Shown at the end of a line are the distributions of the objective attribute values and the numbers of supporting instances in the ridge regions of the upper and lower nodes.

First, I interpret the third ridge $[BT \geq 39]$ for the sake of simplicity of explanation, although the first ridge is the steepest. For the third ridge, $\Delta BSSrate$ is -1.43 and adding 14 instances with higher BT values decreases the BSS from 4.31 to 3.96. The distribution of data selected by the preconditions $[FEVER \leq 6,\ LOC \leq 1]$ is illustrated in Fig. 6.(1); the x- and y-axes denote ridge attribute BT and main condition attribute *CT_FIND*, respectively, and *bacteria/virus* regions are shown as black/white areas in each bar, respectively. Applying the main condition involves selecting the lower bars, and two right-most bars in the figure are the detected ridge. The correlation between the main condition $[CT\_FIND : abnormal]$ and the *bacteria/virus* ratio is seen by the increase in the black *bacteria* region in the 4 bars at the lower left. The ridge information tells us that the rise in the percentage of *bacteria* on applying the main condition diminishes sharply in this ridge region (71% → 75%). We can reconfirm the existence of the ridge by inspecting the *bacteria/virus* ratio near the ridge in the Figure. Similar ridges exist around the other two preconditions as expected, but the $[LOC \geq 2]$ area contains only 4 instances and is not recognized as a ridge.

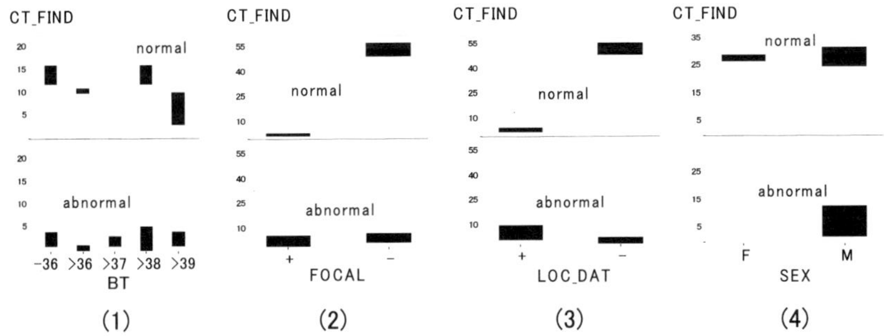

Figure 6: Distribution of *bacteria*(black)/*virus*(white) around the ridges of *Rule* 7

The steepest ridge shown in the top row is the introduction of a new precondition $[FOCAL:+]$; where $\Delta BSSrate$ is -3.69 and BSS decreases from 4.31 to 1.85. Unlike the ridges already interpreted, this ridge is contained in the region defined by the preconditions. The omission of 9 instances in this ridge lowers the BSS value greatly. The distribution of data is illustrated in Fig. 6.(2) using *FOCAL* and *CT_FIND* as the x- and y-axes, respectively. Application of the main condition means selecting the lower two bars, where the ratio of the black *bacteria* region increases greatly. A weaker correlation between the main condition *CT_FIND* and the ridge attribute *FOCAL* is understood by the relative heights of the 4 bars, neglecting their colors.

The ridge region is depicted by the two left bars in the Fig. 6.(2). The *bacteria* percentage in the ridge rises from 78 to 100%; this is much gentler than that of the entire region of the rule. This ridge indicates that excluding the instances in the two left bars greatly decreases the discrimination power of the main condition. Unlike ridge $[BT \geq 39]$, omission of this ridge from the rule region lowers BSS. Therefore, we should interpret this ridge as containing the data that most powerfully discriminate *bacteria/virus* by the main condition. The importance of this ridge is indicated by the fact that 6 out of 11 *bacteria* instances exist in the lower left bar of the two lower bars. In conclusion, this ridge indicates that the distribution of the *bacteria/virus* ratio is not uniform along the attribute *FOCAL*, and that we must be careful in applying this rule to instances with $[FOCAL:-]$. The distributions of the other ridges, $[LOC\_DAT:+]$ and $[SEX:M]$, are also shown in the Fig. 6, and can be interpreted in a similar way.

Why doesn't the complementary region $[FOCAL:-]$ appear as a ridge? Employing this region as a ridge results in a very low BSS (0.30). However, $\Delta BSSrate$ is also lowered to -0.06 because of numerous instances in the ridge region, and hence it is not shown as a ridge. We can say that an instance with $[FOCAL:+]$ has a larger effect on the BSS value than one with $[FOCAL:-]$.

We can find another typical example of a ridge in *Rule* 5 in Fig. 5. This rule has the same main condition as *Rule* 7, but it shares no supporting instances at the lower node. The second ridge, $[LOC\_DAT:+]$, is outside the rule region for a nominal attribute. It adds 16 instances, but their *bacteria/virus* ratio changes little, leading to a sharp ridge.

Its distribution is shown in Fig. 7.(1).

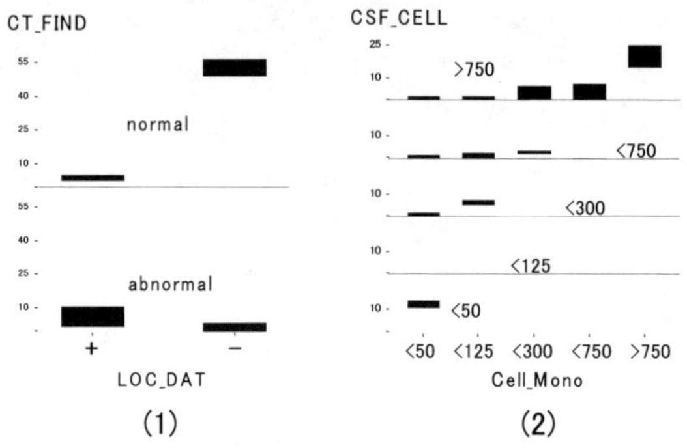

Figure 7: Distribution of *bacteria*(black)/*virus*(white) ratio around the ridges of *Rule* 5 and *Rule* 1-UL2

One of the *ULrelative* rules of *Rule* 1, shown at the bottom of Fig. 5, has another interesting ridge: [*Cell_Mono* > *750*]. Fig. 7.(2) shows the distribution using the attributes in the main condition: [*CSF_CELL* > *750*] and a precondition [*Cell_Mono* > *750*]. There are no instances in the lower right region, as *CSF_CELL* is defined by (*Cell_Mono* + *Cell_Poly*). Black *bacteria* boxes appear only in the upper left part of the figure. The top right-most bar is the ridge region, where the distribution is the opposite of that in the rule region, leading to a sharp ridge.

There were many other ridges in addition to those shown here, but a more detailed interpretation is beyond the scope of this paper. We can conclude that ridge information is effective for mining valuable knowledge from rules efficiently.

5 Conclusion

This paper defines ridges of *BSS* values based on the cascade model. The omission or addition of data in a ridge has a large effect on the *BSS* values of a rule, and hence is of importance in a detailed analysis of the data. The results of application to a real medical data set showed that sharp ridges can be extracted by this method. Combined with the visualization technique, the details of the datascape in and around a rule can be recognized. These results cannot be reached by using visualization alone, as the search space is too large to be inspected by a human expert. The method proposed in this paper can pinpoint a place in a high-dimensional data space, making detailed analysis of data possible. This will surely invoke an active reaction from users and will lead to the redesign of the overall framework of the mining task.

Let us consider a high-dimensional data space. A conventional rule indicates the existence of some characteristic pattern in the space resulting from the conjunction of the condition clauses. This knowledge corresponds to the position of the mean for a group of data with the characteristic pattern. The separation of the main condition from the rest of the clauses helps to carve the sharp relief of the data against the distributions specified by the preconditions. This step corresponds to recognizing the most important axis in the space. In this sense, a ridge as defined in this paper can be considered as an attempt to detect some kind of higher order bias in the data space.

The field of association rule mining has explored several directions, aside from speeding up the mining process. One aim is to reduce the number of rules, thereby reducing the load that a user faces in examining numerous rules [10, 12]. Another direction is to incorporate new semantics to the link in the lattice, leading to novel knowledge [1, 3]. However, no works have facilitated the microscopic analysis of data. I expect that the direction indicated in this paper will lead the way to a fruitful research area in active mining that invokes active user reactions.

References

[1] R. Agrawal and R. Srikant. Mining sequential patterns. In *Proc. International Conf. Data Engineering*, pages 3–14, 1995. IEEE computer society.

[2] C.W. Gini. Variability and mutability, contribution to the study of statistical distributions and relations. *Studi Economico-Giuridici della R. Universita de Cagliari*, 1912. Reviewed in R.J. Light and B.H. Margolin. An analysis of variance for categorical data. *J. Amer. Stat. Assoc.*, 66:534–544, 1971.

[3] A. Inokuchi, T. Washio, and H. Motoda. An apriori-based algorithm for mining frequent substructures from graph. In *Principles of Data Mining and Knowledge Discovery, PKDD2000*, pages 13–23, 2000. LNAI 1910, Springer-Verlag.

[4] T. Okada. Finding discrimination rules using the cascade model. *J. Jpn. Soc. Artificial Intelligence*, 15:321–330, 2000.

[5] T. Okada. Rule induction in cascade model based on sum of squares decomposition. In *Principles of Data Mining and Knowledge Discovery, PKDD'99*, pages 468–475, 1999. LNAI 1704, Springer-Verlag.

[6] T. Okada. Sum of squares decomposition for categorical data. *Kwansei Gakuin Studies in Computer Science*, 14:1–6, 1999. http://www.media.kwansei.ac.jp/home/kiyou/kiyou99/kiyou99.html.

[7] T. Okada. Efficient detection of local interactions in the cascade model. In *Knowledge Discovery and Data Mining, PAKDD-2000*, pages 193–203, 2000. LNAI 1805, Springer-Verlag.

[8] T. Okada. Medical knowledge discovery on the meningoencephalitis diagnosis studied by the cascade model. In *New Frontiers in Artificial Intelligence, Joint JSAI 2001 Workshop Post-Proceedings*, pages 533–540, 2001. LNCS 2253, Springer-Verlag.

[9] T. Okada. Datascape survey using the cascade model. submitted to KDD 2002.

[10] N. Pasquier, Y. Bastide, R. Taouil and L. Lakhal. Discovering frequent closed itemsets for association rules. In *Database Theory, Proc. ICDT '99*, pages 398–416, 1999. LNCS 1540, Springer-verlag.

[11] T. Washio. JSAI KDD Challenge, 2001. http://wwwada.ar.sanken.osaka-u.ac.jp/pub/washio/jkdd/jkddcfp.html.

[12] M.J. Zaki. Generating non-redundant association rules. In *Proc. KDD 2000*, pages 34–43, Boston, 2000. ACM.

Active Mining
H. Motoda (Ed.)
IOS Press, 2002

The Effect of Spatial Representation of Information on Decision Making in Purchase

Hiroko Shoji[†] [‡] and Koichi Hori[†]
{hiroko, hori}@ai.rcast.u-tokyo.ac.jp
[†]Department of Advanced Interdisciplinary Studies, University of Tokyo
4-6-1 Komaba, Meguro-ku, Tokyo 153-8904, JAPAN
[‡]Department of Informatics Education, Kawamura Gakuen Women's University
1133 Segedo, Abiko-shi, Chiba 270-1138, JAPAN

Abstract. The process of concept formation changes as strategy changes. For creatively forming concepts, a strategy specially tailored for that purpose should be adopted. The authors have been claiming that a change in representation is as effective as a change in strategy. This study performed a microscopic analysis of cognitive process, especially with a focus on how the strategy changes as representation changes. To be more specific, we have built an experimental online-shopping system called S-Conart (Concept Articulator for Shoppers) and conducted an experiment to compare two ways of representing information, that is, spatial and listing representations. Especially, we have made a detailed analysis of cognitive process of users doing their shopping. The results of this study justify our hypothesis that a change in representation has similar effects as a change in strategy. This study can be a basis for successful support of daily activities such as online-shopping.

1 Introduction

When observing human behavior in the actual purchase activities, the underlying mental process may be roughly categorized into the following two types: *problem-solving* and *concept-formation*. When customers follow the problem-solving type, they have clear image and functional requirements on desired products, and perform problem-solving in a way that they look for the products which meet their requirements. When they follow the concept-formation type, on the other hand, they only have vague requirements on their needs, and try to make a gradual clarification and/or refinement of their requirements through the interaction with salesclerks.

Most of existing online shopping sites assume that customers' requirements have been already determined [6]. That is, they only target the problem-solving type of purchase. This study aims at developing online shopping systems which can help the customers make a concept-formation type of purchase. Its purpose is specifically to establish information presentation methods to effectively support the customer's concept formation process, and to build the design methodology for Human-Computer Interaction (HCI) to realize them.

This study started with observing human behavior in the actual purchase activities. Then, the protocol analysis of actual conversation between the customer and salesclerk revealed that appropriate information given by the clerk in a timely manner often causes the customer's focus to be changed to lead the change of their search goal itself in their decision-making process when shopping. It also found that this interaction is effective in decision-making for the concept-formation type of purchase [7].

Based on these knowledge acquired from the analysis of human behavior in the actual purchase activities, this study has created a system, called S-Conart(CONcept ARTiculator for Shoppers), to support the concept-formation type of purchase. The authors are developing a system which puts special emphasis on the appropriate information presentation to support the customer's concept formation instead of replacing human communication with HCI as is [8].

This paper describes the system overview of S-Conart, and introduces the result of the experiment conducted with S-Conart. We conducted an experiment to compare two ways of representing information, that is, *spatial* and *listing* representations. Especially, we have made a detailed analysis of cognitive process of users doing their shopping. Through this experiment and its analysis, the authors argue that changing the content and/or presentation method of information provided by the system can bring an equivalent change to the human mental world, although it is in the different form from the human-human interaction. The results of this study justify our hypothesis that a change in representation has similar effects as a change in strategy.

This paper also includes related studies and mentions what should be addressed in the future.

2 System Overview

The goal of this study is to investigate the effect of online-shopping system's interface on the consumers' purchase behavior, especially decision-making process for item selection. Comparative control experiments will be performed from various aspects on what information and how it should be presented to consumers. Currently, an experiment is being conducted to compare two styles of interfaces, i.e. *spatial arrangement* and *listing*. Each of Figure 1 and Figure 2 shows one of screens of our experimental online-shopping system called S-Conart (CONcept ARTiculator for Shoppers). At this moment, the experiment only deals with Japanese sake as product items and creates for use a database consisting of sake data with 12 attributes and of 193 kinds.

Previous researches have verified that the spatial-arrangement style of presentation is useful for creativity support [4, 9]. Therefore, also in online shopping, information presented through the spatial arrangement may be expected to promote customer's decision-making during shopping. Especially if there are many kinds of items available, representing their interrelationship as a spatial image may cause underlying information to be a trigger to help the customer's mental leap.

Accordingly, this study applies the spatial arrangement to our experimental online-shopping system called S-Conart [8], which is often used in the studies of creativity support, and takes an approach that places items on a two-dimensional space using multi-dimensional scaling method (MDS) to indirectly show the relationship between them. This approach calculates distances between individual items based on the similarities between them and represents degrees of similarity between their attributes as spatial geometric relationship (i.e., distance) between them. By comparing this spatial arrangement style of representation with the selecting-from-list style, this study will verify the usefulness of the spatial arrangement for customer's mental leap in decision-making during shopping.

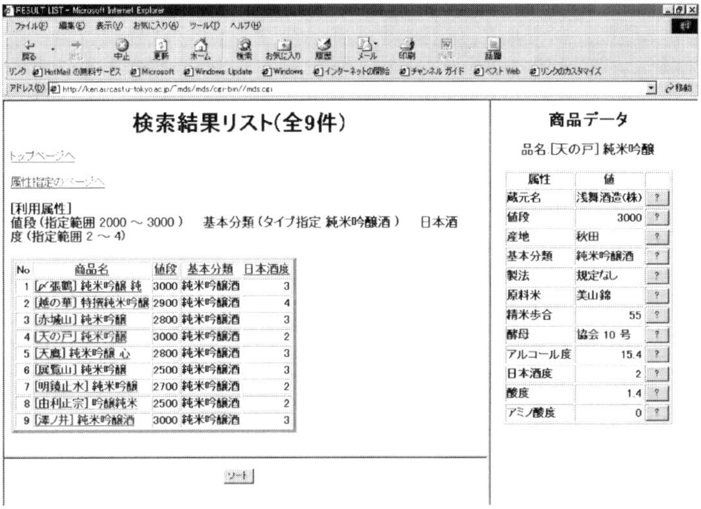

Figure 1: Listing style interface of S-Conart

3 Experiment

Through the experiment of choosing sake as an item, the difference in human cognitive process between using spatial arrangement and listing as a presentation style will be examined.

3.1 Subjects

The content of the experiment was designed based on a preliminary experiment on students in our lab as subjects and then the main experiment was performed for eight subjects. These subjects are grouped as shown below based on the familiarity with computer/interface and the interest in and knowledge of sake as a target item. These criteria were categorized from responses to the pre-experiment questionnaire.

- group 1 (subject 1, 2): much familiar with computer/interface; much interested in and knowledgeable about sake as a target item.
- group 2 (subject 3, 4): little familiar with computer/interface; much interested in and knowledgeable about sake as a target item.
- group 3 (subject 5, 6): much familiar with computer/interface; little interested in and knowledgeable about sake as a target item.
- group 4 (subject 7, 8): little familiar with computer/interface; little interested in and knowledgeable about sake as a target item.

3.2 Procedure

Each subject was given a document describing the content of the experiment and assignment and then followed the procedure shown below to perform their respective experiment.

1. Fill in the questionnaire before starting the experiment. (15 min. at the maximum)

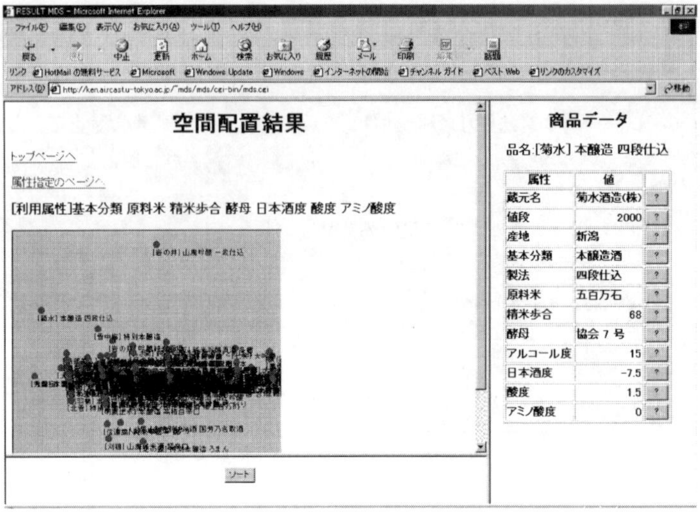

Figure 2: Spatial-arrangement style interface of S-Conart

2. Use listing or spatial arrangement style of representation to do the assignment #1. (30 min. at the maximum)

3. Have an interview and respond to a questionnaire regarding assignment #1. (60 min. at the maximum)

4. Break. (10 min.)

5. Use a different style from the one used in step 2 to do assignment #2. (30 min. at the maximum)

6. Have an interview and respond to a questionnaire regarding assignment #2. (60 min. at the maximum)

Two kinds of assignments were used in the experiment. The subjects were given the instructions describing the purpose, procedure, notes and so on of the experiment as well as the assignment descriptions shown below.

Assignment #1 Choose appropriate Japanese sake for your home party with your friends of your generation (about six persons including you, both men and women). Suppose it is in winter and strongly seasoned Ishikari-nabe is scheduled for a meal. Choose three bottles of sake in total such that the amount is within ten thousands yen.

Assignment #2 Supposing that a welcoming party will be held in your office (your lab or seminar if you are a student) in April, choose appropriate Japanese sake for the occasion. About 15 persons will participate in the party. Suppose that there will be a wide range of participants in terms of age and taste, ranging from those who are sake drinkers to those who don't drink it at all. The budget is fifteen thousands yen. The number of bottles is not specified, however, buy at least three bottles for such many participants. You may buy more than one bottle of same kind of sake.

Among two subjects in each group, one used the listing style to do assignment #1 and used spatial arrangement style to do assignment #2. The other used spatial arrangement style to do assignment #1 and used listing style to do assignment #2. What was happening during doing assignments was shot with a video camera. When each subject has an interview regarding their assignments, viewing this video and their

Table 1: Grouping of the unit cognitive process used for our protocol analysis

category	name	description	example
conceptual	plan	plans actions	PlanHowToSelect
	decide	determines actions	DecideToSelect
functional	compare	compares objects	CompareDatas
	investigate	evaluates and examines objects	InvestigateData
	remember	remembers objects	RememberAttribute
	confirm	confirms objects	ConfirmAttributes
perceptual	look	looks at objects	LookAtList
	read	reads objects	ReadData
physical	select	Selects objects	SelectItem
	set	sets values	SetAttribute
	display	displays the result	DisplaySpace
	explore	performs operations for exploration	ExploreSpace

operation history stored in the system, they were asked for as detailed explanation as possible about why they performed each operation and what they had in their mind at that time. What they answered was recorded and used for the protocol analysis.

4 Result of analysis

4.1 Method of Analysis

Referring to a protocol analysis technique used by Suwa for the cognitive process in the architectural design domain [10], this study divided the cognitive process during item selection (purchase) using our experimental system into the four levels of conceptual, functional, perceptional, and physical to define the unit cognitive processes as shown in the Table 1. Following these unit cognitive processes, the behavior of subjects when they used the spatial arrangement and listing styles of interface was analyzed microscopically to draw the transition diagram of their cognitive process [1]. Then, the difference in the process between when using the two styles was examined through the diagram.

4.2 An Example with Listing-Style Interface

Figure 3 is a transition diagram to show a part of the cognitive process when a subject did his/her assignment using the listing style interface.

For selecting items, a subject must first plan what items to select according to what policy (PlanHowToSelect). After reading and considering the description in the assignment, the subject shown in this figure came to the conclusion that "a sake from Hokkaido suits Ishikari-nabe as a main dish" and "inexpensive and popular Ginjo-shu [1] would be better for the casual party." Once he/she has determined the plan for selection, he/she first selects attributes according to the plan. The cycle (1) in the figure represents the process where viewing the attribute selection screen (LookAtAttributes), the subject repeats several times his/her tasks to set required attributes and their ranges (SetAttribute). After setting the attributes, he/she confirms them (ConfirmAttributes) and has the result displayed (DisplayList).

[1]Ginjo-shu is a quality sake brewed from the finest rice.

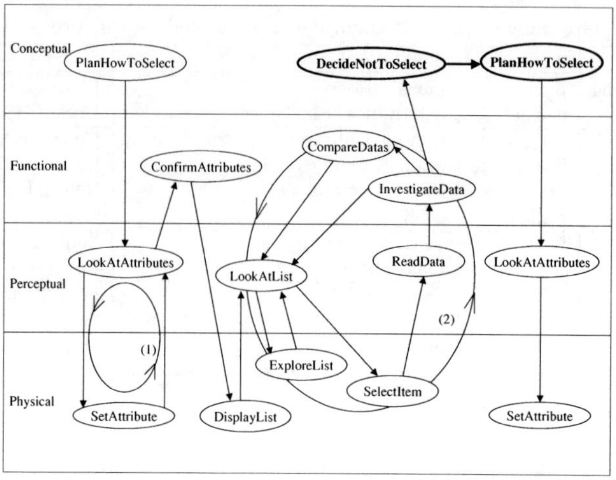

Figure 3: An example of the cognitive process when using list style

Next, viewing the resultant list (LookAtList) – the listing region is often larger than the screen, in which case, moving up and down (ExploreList) to perform this action –, the subject clicks the names of items of his/her interest to browse these item data and make a comparative examination of them. The cycle (2) in the figure represents the process where viewing several item data (SelectItem, ReadData), the subject compares them (CompareDatas) to examine for promising item data (InvestigateData).

In the case shown in the Figure 3, the subject chose to buy nothing yet (DecideNot-ToSelect) because "nothing favorite was found in spite of viewing several item data." And, he/she thought "Ginjo-shu is expensive and hard to meet the criteria," decided to "loosen the criteria and search for other sake than Ginjo-shu as well" (PlanHowToSelect) and then returned to the attribute selection screen to redo the task (LookAtAttributes, SetAttribute).

4.3 An Example with Spatial-Arrangement Style Interface

Figure 4 is a transition diagram to show a part of the cognitive process when a subject did his/her assignment using the spatial-arrangement style interface.

The subject in this figure thought that "the space to be displayed should reflect preferably the difference in taste" during the prior process, decided to "construct the space with attributes likely to have a strong relationship with the taste," selected those attributes, and then displayed and examined the result.

As a result, finding that "the space is not easy to view and the price and kind should be specified as well," the subject chose to use the focusing functionality and decided that "reasonable Ginjo-shu from northern regions should be focused" (PlanHowToSelect).

Once the policy has been established, attributes are first selected accordingly. The cycle (1) in the figure represents the process where viewing the attribute selection screen (LookAtAttributes), the subject repeats several times his/her tasks to set required attributes and their ranges (SetAttribute). After setting the attributes, he/she confirms them (ConfirmAttributes) and has the result displayed (DisplaySpace).

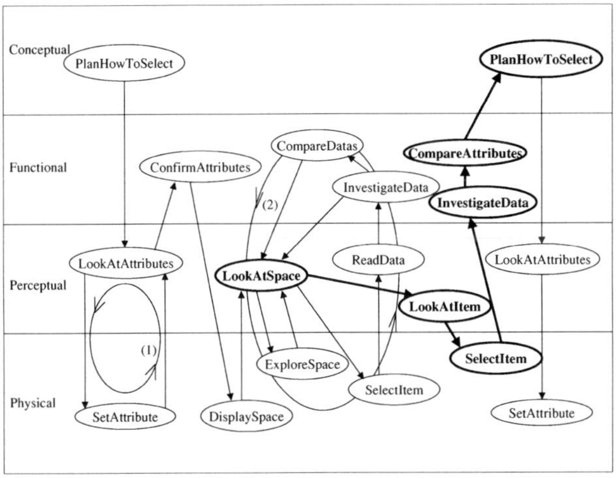

Figure 4: An example of the cognitive process when using spatial arrangement style

Next, viewing the space displayed (LookAtSpace) – the space is often larger than the screen, in which case, moving up and down (ExploreSpace) to perform this action –, the subject clicks the names of items of his/her interest to browse these item data and make a comparative examination of them. The cycle (2) in the figure represents the process where viewing several item data (SelectItem, ReadData), the subject compares them (CompareDatas) to examine for promising item data (InvestigateData).

In the case shown in the Figure 4, the subject decided to try to display with a focus on Junmai-shu [2] (PlanHowToSelect), because while looking at items colored orange in the focused view, an item (colored blue) not corresponding to the current view caught his/her attention (LookAtItem) and then clicking it to view the data (SelectItem) showed that it is Junmai-shu, which inclined him/her toward Junmai-shu. Thus, he/she returned to the attribute selection screen to redo the task (LookAtAttributes, SetAttribute).

4.4 The Effect of Spatial Arrangement

The example with listing style interface in the Figure 3 shows that the process make a sequential transition from planning to action, evaluation & examination, determination, plan reexamination, and so on. To the contrary, the example with spatial-arrangement style interface in the Figure 4 shows that something that catches the subject's attention before coming to the conclusion, i.e., before making some judgement at the conceptual level, affects the subject's mental process to cause a change in the plan.

Thus, the cases which have a direct transition from lower level cognition to the next plan before coming to the conclusion at the conceptual level were extracted and counted from protocol data for all of eight subjects. Table 2 shows the result. This shows that if spatial arrangement is used, *another item which happens to catch the subject's attention often triggers a shift to a different plan* in all subject groups. If listing representation is used, it could be a matter of course that such an effect is not frequent because only items matching the criteria are presented, however, it may be of significance that the effect of

[2]Junmai-shu is a sake brewed from the pure rice.

Table 2: Number of the direct transitions from lower level cognition to the next plan

(a) List style interface

Triggered by	LookAtAttributes	LookAtList	ReadData	ReadHelp	ReadLog	Total
Group 1	0	0	2	1	0	3
Group 2	0	0	1	2	0	3
Group 3	0	0	0	2	0	2
Group 4	0	0	0	1	0	1

(b) Spatial arrangement interface

Triggered by	LookAtAttributes	LookAtList	ReadData	ReadHelp	ReadLog	Total
Group 1	0	5	1	0	0	6
Group 2	0	6	0	2	0	8
Group 3	0	2	1	1	0	4
Group 4	0	2	0	1	0	3

the spatial arrangement could be recognized as a difference in transition pattern in the unit cognitive process in terms of whether or not the determination at the conceptual level is undergone.

The Table 2 also shows that the effect of spatial arrangement as mentioned above is more frequently observed in subjects belonging to group 1 and 2. This may suggest that in case of subjects with much knowledge or interest in target items, *something which happens to catch their attention* is quite likely to trigger a call for the attention causing a transition to the next plan. To the contrary, in case of subjects with little knowledge or interest, "item which happens to catch their attention" may be less likely to be selected, and even selecting it to view its data may less frequently trigger an inspiration. The difference in the familiarity with interface such as spatial arrangement style may have no effect.

Other than spatial arrangement, what is described in the help information also has proved to have a great deal of effect as a trigger. An example of the trigger by help information might be the case where reading a help because of no knowledge about the degree of sake causes a user to notice its importance and add it to the criteria.

5 Related Work

It has become an important issue in the field of cognitive science to analyze the effect of the difference in representation of information on human cognitive process. For example, Zhang [11] conducted an experiment to study the effect of the form of external representation on problem solving process, and proposed the *Representational Determinism*, in which perceivable structures are determined by the form of information representation.

On the other hand, some studies in the field of creativity support show that the representation of information affects articulation, and that articulation or concept formation can be supported by adopting effective information representation methods [5, 10].

This study investigated the effect of spatial arrangement as one form of information representation. In the field creativity support, there also has been several studies on information representation in the form of spatial arrangement, which indicate the effectiveness of spatial arrangement [1, 4, 9].

Based on previous studies, we hypothesized that information representation in the form of spatial arrangement is effective in promoting decision-making. To verify the hypothesis, we conducted a comparison study under online shopping setting as a everyday life situation which requires decision-making.

6 Conclusion

We have built an experimental online-shopping system called S-Conart and conducted an experiment to compare two ways of representing information, that is, spatial arrangement and listing representations. Especially, we have made a detailed analysis of the cognitive process of users doing their shopping. This study can be the basis for successful support of daily activities such as shopping by information media.

Our result showed that if spatial arrangement is used, "another item which happens to catch the subject's attention" often triggers a shift to a different plan in all examinee groups. It may be of significance that the effect of the spatial arrangement could be recognized as a difference in transition pattern in the unit cognitive process in terms of whether or not the determination at the conceptual level is undergone.

7 Looking ahead

The authors think that in the future we need to make more detailed analysis of what characteristics of the spatial representation caused the user's mental world to be changed in what way. And, making various devices to the listing representation as well as the spatial representation is expected to cause the user's mental world to be effectively changed. This point also needs to be examined. Enough analysis of how changing the information representation can change human mental world has not yet been made. Knowledge about this problem is being gradually accumulated from the studies by various researchers including us.

The result from the experiment described in this paper may also suggest that the relationship between listing and spatial arrangement styles of user interface is analogous to that between *expected and unexpected reactions* observed in the protocol of actual purchase behavior [7]. This analogy will be further more addressed in our future work.

The goal of our study is not to build the current S-Conart system but to use it examine human mental process and continue to make improvements to the system that reflect the result from the examination. We ourselves would like to explore the interaction design desirable in terms of conceput formation through this iteration.

References

[1] Amitani, S., Supporting Musical Composition by Externalizing the Composer's Mental Space, Master thesis, University of Tokyo, 2001.

[2] Boden, M., *The Creative Mind: Myths and Mechanisms*, Basic Books, 1991.

[3] Gero, J.S., Computational models creative design processes, Artificial Intelligence and Creativity(T. Dartnall, ed), *Studies in Cognitive Systems*, Vol.17, pp.269-281, Kluwer Academic Publishers, 1994.

[4] Hori, K., Concept space connected to knowledge processing for supporting creative design, *Knowledge-Based Systems*, Vol.10, No.1, pp.29-35, 1997.

[5] Nakakoji, K. and Fischer, G., Intertwining Knowledge Delivery, Construction, and Elicitation: A Process Model for Human-Computer Collaboration in Design, *Knowledge-Based Systems Journal*: Special Issue on Human-Computer Collaboration, Vol.8, No.2-3, pp.94-104, 1995.

[6] Pu, P. and Faltings, B.: Enriching buyers' experiences: the SmartClient approach, *Proceedings of ACM CHI2000*, pp.289-296, 2000.

[7] Shoji, H. and Hori, K.: Chance Discovery by Creative Communicators Observed in Real Shopping Behavior, T. Terano et al.(Eds.), *Post Proceedings of JSAI2001 Workshops*, LNAI2253, pp.462-467, 2001.

[8] Shoji, H. and Hori, Strategy Emergence from Human-Computer Interaction, J. S. Gero and K. Hori(eds), *Strategic Knowledge and Concept Formation III*, pp.87-99, 2001.

[9] Sugimoto, M., Hori, K. and Ohsuga, S., A method to assist building and expanding subjective concepts and its application to design problems, *Knowledge-Based Systems*, Vol.7, No.4, pp.233-238, 1994.

[10] Suwa, M., Purcell, T. and Gero, J., Macroscopic analysis of design processes based on a scheme for coding designers' cognitive actions, *Design Studies*, Vol.19, No.4, pp.455-483, 1998.

[11] Zhang, J., The Nature of External Representation in Problem Solving, *Cognitive Science*, Vol.21, No.2, pp.179-217, 1997.

Active Mining
H. Motoda (Ed.)
IOS Press, 2002

A Hybrid Approach of Multiscale Matching and Rough Clustering to Knowledge Discovery in Temporal Medical Databases

Shoji Hirano and Shusaku Tsumoto
hirano@ieee.org, tsumoto@computer.org
Department of Medical Informatics Shimane Medical University, School of Medicine
89-1 Enyacho, Izumo, Shimane 693-8501, Japan

Abstract. Knowledge discovery in time-series medical databases has been receiving considerable attention since it provides a way of revealing hidden relationships between temporal course of examination and onset time of diseases. This paper presents a novel approach to pattern recognition in temporal sequences. The key techniques employed here are multiscale structure matching and rough clustering. Multiscale matching enables us cross-scale comparison of the sequences, namely, it enable us to compare temporal patterns by partially changing observation scales. On the other hand, rough clustering enable us to construct interpretable clusters of the sequences even if their similarities are given as relative similarities. We combine these methods and attempt to cluster the sequences according to multiscale similarity of patterns. First, we apply multiscale stricture matching to all pairs of sequences and obtain similarity for each of them. Next, we apply rough-sets based clustering technique to cluster the sequences based on the obtained similarity. We applied this method to a time-series laboratory examination dataset acquired from a hospital information system. The results show that sequences that have similar patterns are successfully gathered up as an identical cluster.

1 Introduction

As twenty years passed since the hospital information system (HIS) started working at large hospitals, time-series laboratory examination databases, which store results of laboratory examinations (blood exam, biochemical exam etc.) become available for analysis. Laboratory examination data is collected by continuously tracking a patient's record through the duration of one week to several years. Such time-series medical databases have been attracting much interests because they contain important information that can be used to reveal underlying relationships between temporal course of examination and onset of diseases. Long-term laboratory examination databases may also enable us to validate hypothesis about temporal course of chronic diseases that has not been evaluated yet on large samples. However, despite their importance, time-series medical databases have not widely been considered as the subject of analysis. This is primarily due to inhomogeneity of the data. Basically, the data were collected without considering further use in automated analysis. Therefore it involves the following problems. (1) Missing values: Examinations are not performed on every day when a patient comes to the hospital. It depends on the needs for examination. (2) Irregular interval of data acquisition: A patient consults a doctor in different interval of date depending on his/her condition, hospital's vacancy, and so on. The intervals can vary from a few days to several months. (3) Noise: The data can be distorted due to contingent change of patient's condition.

In this paper, we present a hybrid approach to the analysis of such inhomogeneous time-series medical databases. The techniques employed here are multiscale structure matching [1] and rough-sets based clustering technique [2]. The first one, multiscale structure matching, is a method to effectively compare two objects from various scales of view. We apply this method to the time-series data, and examine similarity of two sequences in both long-term and short-term points of view. It has an advantage that connectivity of segments is preserved in the matching results even when the partial segments are obtained from different scales. The second technique, rough-sets based clustering, classifies sequences based on their indiscernibility defined in the context of rough set theory [3]. The method can produce interpretable clusters even under the condition that similarity of objects is defined only as a relative similarity. Our method attempts to cluster the time sequences according to their long- and short-term similarity by combining the two techniques. First, we apply multiscale stricture matching to all pairs of sequences and obtain similarity for each of them. Next, we apply rough-sets based clustering technique to cluster the sequences based on the obtained similarity. After then, features of the clustered patterns can be compared to the class of diagnosis to understand relationships between them.

The remaining part of this paper is organized as follows. In Section 2 we introduce the related work. In Section 3 we describe the procedure of our method including explanation of each process such as pre-processing of data, multiscale structure matching and rough sets-based clustering. Then we show some experimental results in Section 4 and finally conclude the technical results.

2 Related Work

Data mining in time-series data has received much interests in both theoretical and applicational areas. A widely used approach in time-series data mining is to cluster sequences based on the similarity of their primary coefficients. Agrawal et al. [4] utilize discrete Fourier transformation (DFT) coefficients to evaluate similarity of sequences. Chan et al. [5] obtain the similarity based on the frequency components derived by the discrete wavelet transformation (DWT). Korn et al. [6] use singular value decomposition (SVD) to reduce complexity of sequences and compare the sequences according to the similarity of their eugenwaves. Another approach includes comparison of sequences based on the similarity of forms of partial segments. Morinaka et al. [7] propose the L-index, which performs piecewise comparison of linearly approximated subsequences. Keogh et al. [8] propose a method called piecewise aggregate approximation (PAA), which performs fast comparison of subsequences by approximating each subsequence with simple box waves having constant length.

These methods can compare the sequences in various scales of view by choosing proper set of frequency components, or by simply changing size of the window that is used to translate a sequence into a set of simple waves or symbols. However, they are not designed to perform cross-scale comparison. In cross-scale comparison, connectivity of subsequences should be preserved across all levels of discrete scales. Such connectivity is not guaranteed in the existing methods because they do not trace hierarchical structure of partial segments. Therefore, similarity of subsequences obtained on different scales can not be directly merged into the resultant sequences. In other words, one can not capture similarity of sequences by partially changing scales of observation.

Table 1: An example of laboratory examination data.

PATID	Date	GOT	GPT	LDH	ALP	TP	ALB	UA	UN
0001	860419								
0001	860430	25	12	162	76	7.9	4.6	4.7	18
0001	860502	22	8	144	68	7	4.2	5	18
0001	860506								
0001	860507								
0001	860508	22	13	156	66	7.6	4.6	4.4	15
0001	860512	22	9	167	64	8	4.8	4.5	14
0001	860519	28	13	185	60	7.5	4.5	4	13
0001	860526	21	12	134	56	7.2	4.4		
0001	860527	23	10	165	55	7.1	4.2		
0001	860528								
0001	860630	23	10	137	66	7.6	4.4	3.2	12

On the other hand, clustering has a rich history and a lot of methods have been proposed. They include, for example, k-means [9], EM algorithm [10], CLIQUE [11], CURE [12] and BIRCH [13]. However, the similarity provided by multiscale matching is relative and not guaranteed to satisfy triangular inequality. Therefore, the methods based on the center, gravity or other types of topographic measures can not be applied to this task. Although classical agglomerative hierarchical clustering [14] can treat such relative similarity, in some case it has a problem that the clustering result depends on the order of handling objects.

3 Methods

3.1 Overview

The overall procedure is summarized as follows. First, we apply pre-processing to all the input sequences and obtain the interpolated sequences resampled in a regular interval. This procedure rearranges all data on the same time-scale and is required to compare long- and short-term variance using their length of trajectory. Next, we apply multiscale structure matching to all possible combinations of two sequences and obtain their similarity as a matching score. We here restricted combinations of pairs to which have the same attributes such as GOT-GOT, because our interest is not on the cross-attributes relationships. After obtaining similarity of the sequences, we cluster the sequences by using rough-set based clustering. Consequently, the similar sequences are clustered into the same clusters and their features are visualized.

3.2 Feature of Labo-Exam Data and Pre-Processing

Table 1 shows an example of the clinical laboratory examination database. Each line in the data corresponds to each date of consultation. Typically, such database include information about patient id, date of consultation (in the form of YYMMDD) and examination results (GOT, GPT etc). In this example, interval of the consultation dates irregularly varies from one day to one month. There are some days that no examination was performed, as seen in 860506 and 860507. Types of examination could also change depending on the needs.

In order to equally compare these data over different patients, we rearrange the data to fit the regular interval. Because continuous dates can rarely be observed in the actual data, and data themselves may contain missing values, the re-sampling process

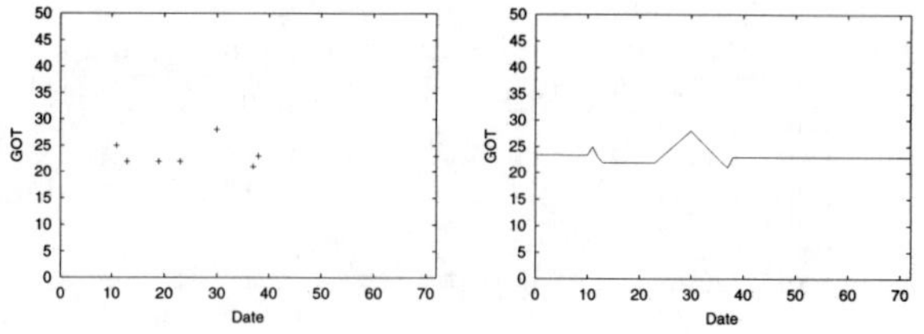

Figure 1: Plot of the GOT data before/after interpolation.

essentially requires interpolation. There are a lot of interpolation/estimation techniques such as mean of sequences, mean of adjacent values, linear interpolation, trend, auto-regression, and so on [15]. Considering reasons that might cause the irregular acquisition interval, we used combination of linear interpolation of adjacent values and trend. We classified the reasons as follows:

(c1) Patient's condition is stable and no rapid examination is required: In this case, the first examination is important because it could be performed after the doctor had suspected some abnormality on the patient. Therefore we interpolate the missing value before the first examination using linear trend of the available data and emphasize changes in the early phase.

(c2) Examination is postponed because the date is close to the previous examination: In this case, we interpolate the missing value between two adjacent examinations using linear interpolation of two close examination results.

(c3) Examination is performed in large interval to follow patient's prognosis: In this case, we consider the variance is small and interpolate the missing values on the late phase by using linear trend of the data.

Figure 1 shows original and interpolated plots of the GOT data in Table 1. In this example, the missing data between two adjacent available data were interpolated using linear interpolation if the two available data were taken within two weeks. Otherwise, data were interpolated using linear trend of the original data. Dates in the horizontal axis correspond to relative dates from the first exam date, 860419. Dates 0-10, 11-37 and 33-end correspond to the above cases c1, c2 and c3, respectively.

The resampling interval can be chosen arbitrary to fit the nature of the diseases. For example, one day may be good for acute diseases, whereas one month may be enough for chronic diseases.

3.3 Multiscale Structure Matching

Multiscale structure matching, proposed by Mokhtarian [1], is a method to describe and compare objects in various scales of view. Its matching criterion is similarity between partial contours. It seeks the best pair of partial contours throughout all scales, not only in the same scale. This enables matching of object not only from local similarity but also from global similarity. The method required much computation time because it should continuously change the scale, however, Ueda et al. [16] solved this problem by introducing a segment-based matching method which enabled the use of discrete scales. We use Ueda's method to perform matching of time sequences between patients. We associate a convex/concave structure in the time-sequence as a convex/concave

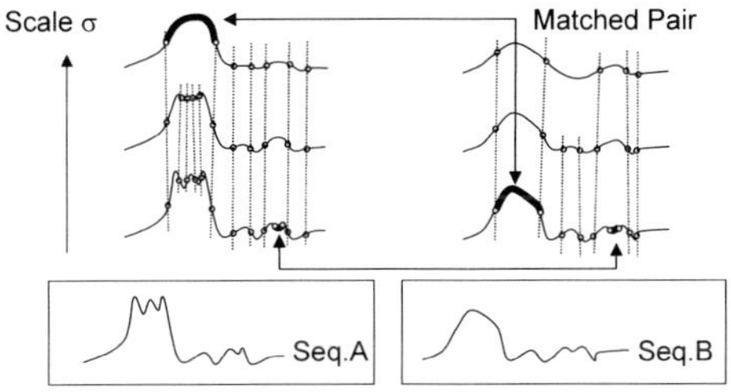

Figure 2: Multiscale matching.

structure of partial contour. Such a structure can be generated by increase/decrease of examination values. Then we can compare the sequences from different terms of observation.

Now let $x(t)$ denote a time sequence where t denotes time of examination. The sequence at scale σ, $X(t,\sigma)$, can be represented as a convolution of $x(t)$ and a Gauss function with scale factor σ, $g(t,\sigma)$, as follows:

$$\begin{aligned} X(t,\sigma) &= x(t) \otimes g(t,\sigma) \\ &= \int_{-\infty}^{+\infty} x(u) \frac{1}{\sigma\sqrt{2\pi}} e^{-(t-u)^2/2\sigma^2} du. \end{aligned}$$

Figure 2 shows an example of sequences in various scales. It shows that the sequence will be smoothed at higher scale and the number of inflection points is also reduced at higher scale. Curvature of the sequence can be calculated as

$$K(t,\sigma) = \frac{X''}{(1+X'^2)^{3/2}},$$

where X' and X'' denotes the first- and second-order derivative of $X(t,\sigma)$, respectively. The m-th derivative of $X(t,\sigma)$, $X^{(m)}(t,\sigma)$, is derived as a convolution of $x(t)$ and the m-th order derivative of $g(t,\sigma)$, $g^{(m)}(t,\sigma)$, as

$$X^{(m)}(t,\sigma) = \frac{\partial^m X(t,\sigma)}{\partial t^m} = x(t) \otimes g^{(m)}(t,\sigma).$$

The next step is to find inflection points according to change of the sign of the curvature and to construct segments. A segment is a partial contour whose ends correspond to the adjacent inflection points. Let $\mathbf{A}^{(k)}$ be a set of N segments that represents the sequence at scale $\sigma^{(k)}$ as

$$\mathbf{A}^{(k)} = \left\{ a_i^{(k)} \mid i = 1, 2, \cdots, N^{(k)} \right\}.$$

Then, difference between segments $a_i^{(k)}$ and $b_i^{(h)}$, $d(a_i^{(k)}, b_j^{(h)})$ is defined as follows:

$$d(a_i^{(k)}, b_j^{(h)}) = \frac{\mid \theta_{a_i}^{(k)} - \theta_{b_j}^{(h)} \mid}{\theta_{a_i}^{(k)} + \theta_{b_j}^{(h)}} \left| \frac{l_{a_i}^{(k)}}{L_A^{(k)}} - \frac{l_{b_j}^{(h)}}{L_B^{(h)}} \right|,$$

where $\theta_{a_i}^{(k)}$ and $\theta_{b_j}^{(h)}$ denote rotation angles of tangent vectors along the contours, $l_{a_i}^{(k)}$ and $l_{b_j}^{(h)}$ denote length of the contours, $L_A^{(k)}$ and $L_B^{(h)}$ denote total segment length of

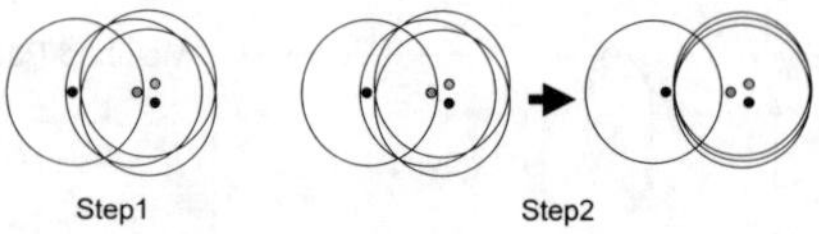

Figure 3: Rough clustering.

the sequences A and B at scales $\sigma^{(k)}$ and $\sigma^{(h)}$. According to the above definition, large differences can be assigned when difference of rotation angle or relative length is large. Continuous $2n-1$ segments can be integrated into one segment at higher scale. Difference between the replaced segments and another segment can be defined analogously, with additive replacement cost that suppresses excessive replacement.

The above similarity measure can absorb shift of time and difference of sampling duration. However, we should suppress excessive back-shift of sequences in order to correctly distinguish the early-phase events from late-phase events. Therefore, we extend the definition of similarity as follows.

$$d(a_i^{(k)}, b_j^{(h)}) = \frac{1}{3}\left(\left| \frac{d_{a_i}^{(k)}}{D_A^{(k)}} - \frac{d_{b_j}^{(h)}}{D_B^{(h)}} \right| + \frac{\mid \theta_{a_i}^{(k)} - \theta_{b_j}^{(h)} \mid}{\theta_{a_i}^{(k)} + \theta_{b_j}^{(h)}} + \left| \frac{l_{a_i}^{(k)}}{L_A^{(k)}} - \frac{l_{b_j}^{(h)}}{L_B^{(h)}} \right| \right),$$

where $d_{a_i}^{(k)}$ and $d_{b_j}^{(h)}$ denote dates from first examinations, $D_A^{(k)}$ and $D_B^{(h)}$ denote durations of examinations. By this extension, we simultaneously evaluate the following three similarities: (1)dates of events (2) velocity of increase/decrease (3) duration of each event.

The remaining procedure of multiscale structure matching is to find the best pair of segments that minimizes the total difference. Figure 2 illustrates the process. For example, in the upper part of Figure 2, five contiguous segments at the lowest scale of Sequence A are integrated into one segment at the highest scale, and this segment is well matched to one segment in Sequence B at the lowest scale. While, another pair of segments is matched at the lowest scale. In this way, matching is performed throughout all scales. The matching process can be fasten by implementing dynamic programming scheme. For more details, see ref [16]. After matching process is completed, we calculate the remaining difference and use it as a measure of similarity between sequences.

3.4 Rough Clustering

Generally, if similarity of objects is represented only as a relative similarity, it is not an easy task to construct interpretable clusters because some of important measures such as inter- and intra-cluster variances are hard to be defined. The rough-set based clustering method is a clustering method that clusters objects according to the indiscernibility of objects. It represents denseness of objects according to the *indiscernibility degree*, and produces interpretable clusters even for the objects mentioned above. Since similarity of sequences obtained through multiscale structure matching is relative, we use this clustering method to classify the sequences.

The clustering method lines its basis on the *indiscernibility* of objects, which forms basic property of knowledge in rough sets. Let us first introduce some fundamental definitions of rough sets related to our work. Let $U \neq \phi$ be a universe of discourse and X be a subset of U. An equivalence relation, R, classifies U into a set of subsets $U/R = \{X_1, X_2, ...X_m\}$ in which following conditions are satisfied:

$$\begin{array}{ll}(1)X_i \subseteq U, X_i \neq \phi & \text{for any } i,\\ (2)X_i \cap X_j = \phi & \text{for any } i,j,\\ (3)\cup_{i=1,2,...n} X_i = U. & \end{array}$$

Any subset X_i, called a category, represents an equivalence class of R. A category in R containing an object $x \in U$ is denoted by $[x]_R$. For a family of equivalence relations $\mathbf{P} \subseteq \mathbf{R}$, an indiscernibility relation over $\mathbf{P}$ is denoted by $IND(\mathbf{P})$ and defined as follows

$$IND(\mathbf{P}) = \bigcap_{R \in \mathbf{P}} IND(R).$$

The clustering method consists of two steps: (1)assignment of initial equivalence relations and (2)iterative refinement of initial equivalence relations. Figure 3 illustrates each step. In the first step, we assign an initial equivalence relation to every object. An initial equivalence relation classifies the objects into two sets: one is a set of objects similar to the corresponding objects and another is a set of dissimilar objects. Let $U = \{x_1, x_2, ..., x_n\}$ be the entire set of n objects. An initial equivalence relation R_i for object x_i is defined as

$$R_i = \{\{P_i\}, \{U - P_i\}\},$$

$$P_i = \{x_j |\ s(x_i, x_j) \geq S_i\}, \quad \forall x_j \in U.$$

where P_i denotes a set of objects similar to x_i. Namely, P_i is a set of objects whose similarity to x_i, s, is larger than a threshold value S_i. Here, s corresponds to the inverse of the output of multiscale structure matching, and S_i is determined automatically at a place where s largely decreases. A set of indiscernible objects obtained using all sets of equivalence relations corresponds to a cluster. In other words, a cluster corresponds to a category X_i of $U/IND(\mathbf{R})$.

In the second step, we refine the initial equivalence relations according to their global relationships. First, we define an indiscernibility degree, γ, which represents how many equivalence relations commonly regards two objects as indiscernible objects, as follows:

$$\gamma(x_i, x_j) = \frac{1}{|U|} \sum_{k=1}^{|U|} \delta_k(x_i, x_j),$$

$$\delta_k(x_i, x_j) = \begin{cases} 1, & \text{if } [x_k]_{R_k} \cap ([x_i]_{R_k} \cap [x_j]_{R_k}) \neq \phi \\ 0, & \text{otherwise.} \end{cases}$$

Objects with high indiscernibility degree can be interpreted as similar objects. Therefore, they should be classified into the same cluster. Thus we modify an equivalence relation if it has ability to discern objects with high γ as follows:

$$R'_i = \{\{P'_i\}, \{U - P'_i\}\}$$

$$P'_i = \{x_j | \gamma(x_i, x_j) \geq T_h\}, \quad \forall x_j \in U.$$

This prevents generation of small clusters formed due to the too fine classification knowledge. T_h is a threshold value that determines indiscernibility of objects. Therefore, we associate T_h with roughness of knowledge and perform iterative refinement of equivalence relations by constantly decreasing Th. Consequently, coarsely classified set of sequences are obtained as $U/IND(\mathbf{R}')$.

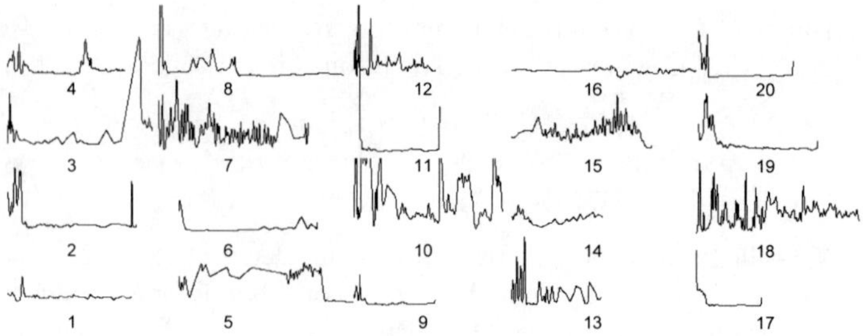

Figure 4: Test patterns.

Table 2: Similarity of the sequences

	1	2	3	4	5	6	7	8	9	10	11	12	13	14	15	16	17	18	19	20
1	1.00	0.70	0.68	0.78	0.00	0.63	0.48	0.71	0.72	0.61	0.73	0.66	0.64	0.72	0.50	0.00	0.53	0.00	0.74	0.45
2		1.00	0.61	0.73	0.00	0.68	0.22	0.46	0.68	0.67	0.72	0.73	0.72	0.68	0.54	0.00	0.68	0.00	0.77	0.41
3			1.00	0.75	0.45	0.51	0.68	0.47	0.71	0.70	0.69	0.73	0.71	0.81	0.68	0.00	0.62	0.00	0.72	0.55
4				1.00	0.00	0.60	0.52	0.47	0.75	0.71	0.64	0.79	0.75	0.82	0.47	0.00	0.60	0.00	0.75	0.48
5					1.00	0.23	0.62	0.49	0.33	0.53	0.44	0.45	0.50	0.44	0.56	0.01	0.00	0.26	0.53	0.30
6						1.00	0.00	0.00	0.59	0.00	0.58	0.39	0.61	0.65	0.00	0.00	0.47	0.00	0.47	0.48
7							1.00	0.49	0.54	0.80	0.57	0.73	0.73	0.59	0.76	0.00	0.00	0.44	0.62	0.39
8								1.00	0.53	0.47	0.57	0.56	0.51	0.49	0.54	0.00	0.00	0.00	0.66	0.51
9									1.00	0.00	0.68	0.00	0.00	0.00	0.00	0.00	0.82	0.00	0.00	0.00
10										1.00	0.59	0.83	0.76	0.75	0.81	0.00	0.47	0.11	0.59	0.37
11											1.00	0.76	0.54	0.68	0.00	0.00	0.74	0.00	0.76	0.00
12												1.00	0.81	0.78	0.67	0.00	0.70	0.00	0.63	0.40
13													1.00	0.75	0.00	0.00	0.64	0.00	0.67	0.35
14														1.00	0.00	0.00	0.66	0.00	0.71	0.00
15															1.00	0.00	0.43	0.20	0.55	0.39
16																1.00	0.00	0.00	0.43	0.19
17																	1.00	0.00	0.00	0.00
18																		1.00	0.39	0.03
19																			1.00	0.00
20																				1.00

4 Experimental Results

We applied the proposed method to time-series GPT sequences acquired on a running hospital information system. For checking applicability of multiscale matching to time-series data analysis, we randomly constructed a small subset containing 20 sequences. Figure 4 shows all the preprocessed sequences. Each sequence originally has different sampling intervals from one day to one year. From preliminary analysis we found that the most frequently appeared interval was one week; this means that most of the patients took examinations on a constant day of a week. According to this observation, we determined resampling interval to seven days.

Table 2 shows normalized similarity of the sequences derived by multiscale matching. Since consistency of self-similarity ($s(A, B) = s(B, A)$) holds, the lower-left half of the matrix is omitted. We can observe that higher similarity was successfully assigned to intuitively similar pairs of sequences.

Based on this similarity, the rough clustering produced nine clusters: $U/IND(\mathbf{R}) =$ {{1,2,9,11,17,19}, {4,3,8}, {7,14,15}, {10,12,13}, {5}, {6}, {16}, {18}, {20}}. A parameter Th for rough clustering was set to $Th = 0.6$. Refinement was performed up to five times with constantly decreasing Th toward $Th = 0.4$. It can be seen that similar sequences are clustered into the same cluster. Some sequences, for example

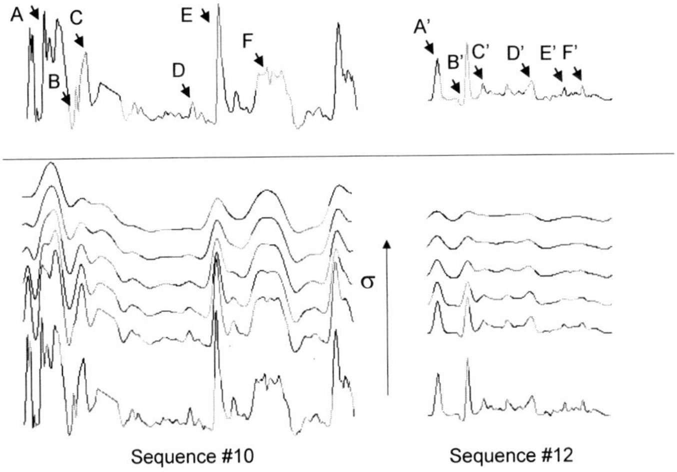

Figure 5: Matching result of sequences #10 and #12.

#16, were clustered into independent clusters due to remarkably small similarity to other sequences. This is because multiscale matching could not find good pairs of subsequences.

Figure 5 shows the result of multiscale matching on sequences #10 and #12, that have high similarity. We changed σ from 1.0 to 13.5, with intervals of 2.5. At the bottom of the figure there are original two sequences at $\sigma = 1.0$. The next five sequences represent sequences at scales σ = 3.5, 6.0, 8.5, 11.0, and 13.5, respectively. Each of colored line corresponds to a segment. The matching result is shown at the top of the figure. Here the lines with same color represent the matched segments, for example, segment A matches segment A' and segment B matches segment B'. We can clearly observe that increase/decrease patterns of sequences are successfully captured; large increase (A and A'), small decrease with instant increase (B and B'), small increase (C and C') and so on. Segments $D - F$ and $D' - F'$ have similar patterns and the feature was also correctly captured. It can also be seen that the well-matched segments were obtained in the sequences with large time difference.

5 Conclusions

In this paper, we have presented an analysis method of time-series clinical laboratory examination databases using multiscale structure matching and a rough sets-based clustering method. Since both of long-term and short-term change can be found in the examination data, multiscale analysis for both terms are essential. Multiscale structure matching is one of the methods that satisfy this requirement and able to effectively compare objects throughout various scales. On the other hand, clustering techniques that can deal with these 'relatively' compared sequences are also required. Rough sets-based clustering technique, which lies its basis on the indiscernibility, is one of such clustering techniques. The proposed method is a hybrid method of these techniques and would be a powerful tool for analysis of time-series medical databases. It remain as a future work to validate the method in the large databases.

Acknowledgments

This work was supported in part by the Grant-in-Aid for Scientific Research on Priority Area (B)(No.759) "Implementation of Active Mining in the Era of Information Flood" by the Ministry of Education, Culture, Science and Technology of Japan.

References

[1] F. Mokhtarian and A. K. Mackworth (1986): Scale-based Description and Recognition of planar Curves and Two Dimensional Shapes. IEEE Transactions on Pattern Analysis and Machine Intelligence, PAMI-8(1): 24-43.

[2] S. Hirano and S. Tsumoto (2001): Indiscernibility Degrees of Objects for Evaluating Simplicity of Knowledge in the Clustering Procedure. Proceedings of the 2001 IEEE International Conference on Data Mining. 211–217.

[3] Z. Pawlak (1991): Rough Sets, Theoretical Aspects of Reasoning about Data. Kluwer Academic Publishers, Dordrecht.

[4] R Agrawal, C. Faloutsos, and A. N. Swami (1993): Efficient Similarity Search in Sequence Databases. Proceedings of the 4th International Conference on Foundations of Data Organization and Algorithms: 69–84.

[5] K. P. Chan and A. W. Fu (1999): Efficient Time Series Matching by Wavelets. Proceedings of the 15th IEEE International Conference on Data Engineering: 126–133.

[6] F. Korn, H. V. Jagadish, and C. Faloutsos (1997): Efficiently Supporting Ad Hoc Queries in Large Datasets of Time Sequences. Proceedings of ACM SIGMOD International Conference on Management of Data: 289–300.

[7] Y. Morinaka, M. Yoshikawa, T. Amagasa and S.Uemura (2001): The L-index: An Indexing Structure for Efficient Subsequence Matching in Time Sequence Databases. Proceedings of International Workshop on Mining Spatial and Temporal Data, PAKDD-2001: 51-60.

[8] E. J. Keogh, K. Chakrabarti, M. J. Pazzani, and S. Mehrotra (2001): "Dimensionality Reduction for Fast Similarity Search in Large Time Series Databases" Knowledge and Information Systems 3(3): 263-286.

[9] S. Z. Selim and M. A. Ismail (1984): K-means-type Algorithms: A Generalized Convergence Theorem and Characterization of Local Optimality. IEEE Transactions on Pattern Analysis and Machine Intelligence, 6(1): 81–87.

[10] A. P. Dempster, N. M. Laird, and D. B. Rubin (1977): Maximum likelihood from incomplete data via the EM algorithm. J. of Royal Statistical Society Series B, 39: 1–38.

[11] R. Agrawal, J. Gehrke, D. Gunopulos, and P. Raghavan (1998): Automatic Subspace Clustering of High Dimensional Data for Data Mining Applications. Proceedings of ACM SIGMOD International Conference on Management of Data: 94–105.

[12] S. Guha, R. Rastogi, and K. Shim(1998): CURE: An Efficient Clustering Algorithm for Large Databases. Proceedings of ACM SIGMOD International Conference on Management of Data: 73–84.

[13] T. Zhang, R. Ramakrishnan, and M. Livny (1996): BIRCH: An Efficient Data Clustering Method for Very Large Databases. Proceedings of ACM SIGMOD International Conference on Management of Data: 103–114.

[14] M. R. Anderberg (1973): Cluster Analysis for Applications. Academic Press, New York.

[15] R. H. Shumway and D. S. Stoffer (2000): Time Series Analysis and Its Applications. Springer-Verlag, New York.

[16] N. Ueda and S. Suzuki (1990): A Matching Algorithm of Deformed Planar Curves Using Multiscale Convex/Concave Structures. IEICE Transactions on Information and Systems, J73-D-II(7): 992–1000.

Active Mining
H. Motoda (Ed.)
IOS Press, 2002

Meta Analysis for Data Mining

Shusaku Tsumoto
tsumoto@computer.org
Department of Medicine Informatics,
Shimane Medical University, School of Medicine,
Enya-cho Izumo City, Shimane 693-8501 JAPAN

Abstract. Rule induction methods have been introduced to extract simple patterns from datasets and have been found useful to apply to real-world databases. However, the power of rule induction method as hypothesis synthesis and the method for quantitative evaluation of extracted rules have not been discussed in the knowledge discovery and data mining communities. In this paper, first, epidemiological approach is introduced and the quality of evidence of rules is discussed. Then, meta analysis which is frequently used for quantitative evaluation of published medical papers, is introduced as the method of upgrading the quality of rule evidence. Since a set of rule corresponds to the contingency table, meta analysis can be applied to integration of the statistics for rules. We illustrate rule induction using tables obtained by leave one out method.

1 Introduction

Rule induction method, which are originally developed for machine learning, has been applied to data mining since 1980's[3, 6, 8, 9]. While these methods has been reported to discover interesting knowledge in real-world databases, its power for hypothesis generation and the way how to quantatively evaluate and integrate the rules obtained from heterogeneous datasets collected in different institutes have not been discussed in data mining community. Rules represent part of knowledge in a contingency table. For example, the following rule gives several information about a contingency table shown in Table 1 if the number of examples which satisfies [b=0] is equal to 6.

$$[a = 0] \rightarrow [b = 0], accuracy : 0.5, coverage : 0.5, N = 10$$

Since four rules can be extracted from one two-way contingency tables, it is easy to see that a set of rule with the same attributes corresponds to a contingency table. Thus, using several statistics obtained from a given contingency table, we can evaluate the power of the set of rules in a statistical way. This methodology can also apply to evaluation of rule obtained by resampling method[1].

Table 1: Example of Contingency Table

	b=0	b=1	
a=0	3	3	6
a=1	3	1	4
6	4	10	

In this paper, we introduce meta analsis[4], which is well known for systematic review methods in EBM (Evidence-based medicine) and discuss the way how to evaluate a set of rules obtained from different datasets quantitatively. Although meta analysis was originally introduced in psychology to evaluate different conclusions from different papers with statistical tests, it became popular in EBM to integrate and evaluate statistical evidence of the results reported in medical journals[5].

The paper is organised as follows. Section 2 discusses the characteristics of data mining from the viewpoints of data collection and EBM. Data mining is a kind of retrospective study and the type of the evidence of acquired knowledge corresponds to $III_{b,c}$. If meta analysis is applied, then the evidence can be strengthened to III_a. Section 3 presents the methodologies of meta analysis and Section 4 shows how to apply meta analysis to quantative evaluation of rule sets. Finally, Section 5 concludes this paper.

2 Data Mining from the viewpoint of Data Collection

2.1 From Epidemiology

The author argues that data mining method should be used as a method for hypothesis generation in data mining contests, such as discovery challenge[10, 11]. One of the reason is that data mining method mainly deals with observational datasets, which can be viewed as a retrospective study in epidemiology. Studies on Epidemiology and public health have shown that data collection determines the characteristics of analysis of collected data and the quality of evidence obtained from data.

Most of data used in data mining, including medical datasets extracted from hospital information systems, are collected without any hypothesis. Although it has been discussed that data mining is used to extract information from a huge amount of data collected almost automatically (also called observational data), the quality of knowledge from observational data is limited, compared the data collected with a hypothesis, which has been discussed in the epidemiology literature. Epidemiology classifies data analysis into the following two types. The one is called a prospective study in which data are collected with a given hypothesis in a experimental way. The other one is called a retrospective study in which data are collected without any firm hypothesis.

Prospective study is also called a cohort study in which a record is classified into several groups (usually, binary classified, such as factor(+) or (-)) just when the observation (data collection) of this record starts. Then, data required for the analysis of hypothesis, such as occurrence of a disease, are collected/observed. In this study, after the existence of factors in a given hypothesis is fixed, the existence of response will be observed. Thus, the reason why this type of study is called prospective is that the flow of data/information collection follows the causality from a condition to a conclusion.

On the other hand, retrospective study is called case-control study in which a record with response(a disease) is classified as a case (positive example) and one without response is classified as a control (negative example). After classification, examples in case and control groups are compared with respect to the existence of factors. Since th flow of data/information is the reverse of the causality from a conclusion to a condition, this type of study is called retrospective.

Table 2 shows the advantages and disadvantages of these two studies. Since most of the datasets used in data mining belongs to retrospective study, the analysis of these data have problems with noise (variance) and bias. If we want to obtain the results with high reliability and evidence, a prospective study is much better. However, the cost of a prospective study is much higher than that of a retrospective study because the former study includes a follow-up study, sometimes a record has to be followed for several years. Therefore, one of the best way for application of data mining to scientific area is to extract a hypothesis from observational data as a retrospective analysis and collect data based on the hypothesis obtained.

In order to generate a good hypothesis, we should validate and interpret patterns extracted from data using domain specific knowledge. That is, postprocessing of patters and interpretation by domain experts are indispensable to find a good hypothesis.

Table 2: Comparison between Prospective and Retrospective Study

	Prospective Study	Retrospective Study
Alias	Cohort Study	Case-Control Study
Flow of Data Collection	Factor → Response	Response → Factor
Estimation of Statistical Indices	Good	Difficult
Analysis of Rare Factors	Possible	Difficult for Rare Factors
Multiplicity of Analysis	One Factor to Multiple Responses	One Response to Multiple Factors
Effect of Noise	Small	Large
Calculation of Relative Risk	Difficult for a High-Prevalence Response	Possible
Cost	High	Low
Analysis of Rare Response	Difficult	Possible
Follow-up Period	Long	Short
Follow-Up (Censored case)	Observed	No

2.2 EBM

Coupling of the above epidemiological approach and clinical medicine is evidence-based medicine (EBM). According to David L. Sacket, who is one of the core members of EBM, EBM is defined as "the conscientious, explicit and judicious use of current best evidence in making decisions about the care of individual patients." Although EBM is called *clinical epidemiology* for a long time, the name EBM has become popular since the method for acquisition of statistical evidence has been established. Evidence in EBM mainly means evidence derived by scientific data analysis, mainly by statistical technique.

One of the most important points in EBM is to acquire the best scientific evidence. Although the best way is that each clinician can collect and analyse data by using statistical technique, it is very difficult to collect and analyse data in the way recommended in EBM. Thus, in general, clinicians collect all the papers in medical journals, check the validities of these papers with respect to statistical evidence, and extract statistical evidence from the papers.

Figure 1 shows the procedure of EBM. EBM processes will be executed as follows. First, the problem will be fixed. Then, clinicians will look for all the papers in the medical literature by using information retrieval system, such as MEDLINE. If they cannot collect papers, clinicians have to collect data and proceed into decision analysis or cost-effective analysis from the collected data. If they can collect papers, clinicians have to check the quality of evidence. When the conclusions of the papers are different, clinicians should apply meta-analysis to integrate all the results. If the conclusions are stable, clinicians will check the applicability of evidence to the problem.

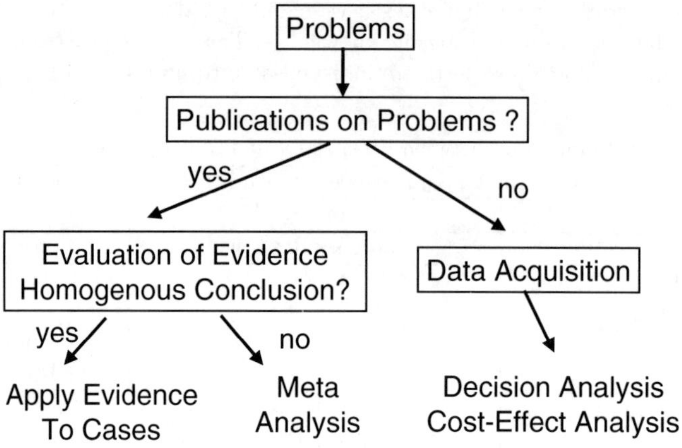

Figure 1: EBM Process

EBM focuses on the data collection method and statistical analysis used in each paper when it evaluates the quality of evidence. Especially, the data collection determines whether the results obtained in the paper can be generalised or not. The method for data collection is called "experimental design" and the assignment of rank on generalisation is called "validation power". Experimental designs are: randomised comparison test, cohort study, case-control study, transversal study and case studies, sorted in descending order with respect to the validation power.

Randomised comparison test, the most powerful experimental design, is a prospective study in which classifies each record applicable to the experiment into case and control randomly. This test performs very well with respect to statistical comparison between case and control groups. Furthermore, the sampling theory gives the number of samples needed for a statistical test with a given significant level α and power of test β. For example, when α and β are set to 0.01 and 0.95, respectively, if the frequencies of two groups are 0.4 and 0.45, then 3486 samples are needed for randomised comparison test. If the frequencies are 0.3 and 0.7, then only 52 samples are needed.

Table 3 shows the classification of the quality of evidence.

As shown in this table, the outcome of data mining is classified into III_b since data mining is a kind of retrospective study.

Meta analysis integrates the results obtained by several papers on the same experimental design with a fixed validation of power and generalises the results as a statistical evidence. Thus, when meta analysis is applied to the outcomes of data mining, the type of evidence is upgraded from III_b to III_a.

Table 3: Type of the Quality of Evidence

I_a:	Meta analysis of randomised comparison test
I_b:	At least one randomised comparison test
I_c:	No negative examples (therapy/prognosis: death) in randomised comparison test
II_a:	Meta analysis of at least one well-designed non-randomised comparison test
II_b:	At least one well-designed non-randomised comparison test
II_c:	At least one well-designed semi-experimental study
III_a:	Meta Analysis of well-designed non-experimental descriptive studies, such as case-control or transversal studies
III_b:	Well-designed non-experimental descriptive studies, such as case-control or transversal studies
IV:	Case studies / Not-well designed cohort and case-control studies
V:	Report or proposal from experts' committee or Clinical test by authorities

3 Meta Analysis

Meta analysis is a type of systematic review in which all the results of statistical tests in the papers on a given problem are integrated and generalised by using statistical technique. Systematic review is different from general reviews with respect to the following points:

1. Focus on a fixed hypothesis, while general surveys focus on several topics
2. Detailed description on the way how papers are retrieved from what kind of databases
3. Clear criteria on the selection of papers
4. Evaluation of papers with critiques
5. Summarise the results in the papers in a quantative way.
6. Interpretation and evaluation are based on statistical evidence.

The important point of systematic review is to fix a hypothesis, which is needed for quantative comparison. From the fixed hypothesis, all the results can be compared with respect to data collection and statistical tests used in each paper. Meta analysis plays an important role in the last two process: quantative summarisation and evaluation of the results. Figure 2 shows the process of systematic review.

The most import process in meta analysis is a test for homogeneity. This test is mainly based on chi square test, but test statistics are different for each meta analysis method. Thus, first, we will discuss the main three method of meta analysis in the next subsequent sections, in which the discussions on the corresponding test of homogeneity will be included.

3.1 Mantel-Haenszel Method

Mantel-Haenszel proposes a statistical test of contingency table in which data are collected with stratification[2], called Mantel-Haenzal method. This method can be applied to a ratio statistic, including odd ratio.

Let us illustrate this method with a contingency table T_i given in Table 4.

1. Search exhaustively for papers on a fixed problem (Retrieval)
2. Select papers for analysis from all the retrieved papers (Selection)
3. Summarise each paper
4. Meta Analysis
   a. Test of Homogeneity
   b. If homogeneous: āpply fixed effect model:
      Mantel-Haenzel, Peto or general variance-based
      (odds ratio, relative risk, difference)
   c. Else: random effect model:
      Random effect model: DerSimonian-Laird
      (ratio, difference)
5. Search for the reason on inhomogeneity

Figure 2: Systematic Review Process

Table 4: Example of Contigency Table (T_i)

	$R_1 = 0$	$R_1 = 1$	
$R_2 = 0$	a_i	b_i	g_i
$R_2 = 1$	c_i	d_i	h_i
	e_i	f_i	n_i

The odds ratio in Table T_i is defined as:

$$OR_i = \frac{a_i \times d_i}{b_i \times c_i}.$$

Since the variance of OR_i is given as:

$$var_i = \frac{n_i}{b_i \times c_i},$$

the weight w_i for each table T_i is obtained as:

$$w_i = \frac{1}{var_i} = \frac{b_i \times c_i}{n_i}$$

Therefore, the odd ratio integrated from $T_i (i = 1, 2, \cdots, n)$ is equal to:

$$OR_{mh} = \frac{\sum_{i=1}^{n} w_i \times OR_i}{\sum_{i=1}^{n} w_i},$$

and 95% confidence interval is:

$$95\%C.I. = e^{\ln OR_{mh} \pm 1.96\sqrt{var(OR_{mh})}}.$$

Concerning to calculation of $var(OR_{mh})$, reader may refer to [5].

In the same way, integrated chi-square test statistic can be calculated. In Table T_i, the chi-square test statistic is:

$$\begin{aligned}\chi^2 &= \frac{(a_i - e_i g_i/n_i)^2}{e_i g_i/n_i} + \frac{(b_i - f_i g_i/n_i)^2}{f_i g_i/n_i} + \\ &\quad \frac{(c_i - e_i h_i/n_i)^2}{e_i h_i/n_i} + \frac{(d_i - f_i h_i/n_i)^2}{f_i h_i/n_i} \\ &= \frac{n_i(a_i d_i - b_i c_i)^2}{e_i f_i g_i h_i}.\end{aligned}$$

Since this statistics can be reformulated as:

$$\frac{\dfrac{(a_i d_i - b_i c_i)^2}{n_i^2}}{\dfrac{e_i f_i g_i h_i}{n_i^2 \times n_i}},$$

chi-square test statistics based on Mantel-Haenszal is given as:

$$\chi_{mh} = \frac{\left(\sum_{i=1}^{n} \dfrac{a_i d_i - b_i c_i}{n_i}\right)^2}{\sum_{i=1}^{n} \dfrac{e_i f_i g_i h_i}{(n_i - 1)n_i^2}},$$

where the formula of unbiased variance estimator is applied. The test of homogeneity with respect to the relative odds OR is defined as:

$$Q = \sum_{i=1}^{n} (w_i \times (\ln OR_{mh} - \ln OR_i)^2),$$

which follows chi-square distribution with the freedom of $n - 1$.

3.2 Peto Method

Peto method is an extension of Mantel-Haenszel method, where the estimator is represented as a logarithmic function.

The odds ratio is defined as:

$$\ln OR_p = \frac{\sum_{i=1}^{n}(O_i - E_i)}{\sum_{i=1}^{n} var_i},$$

where O_i denotes each observed frequencies and E_i and var_i are given as:

$$E_i = \frac{e_i \times g_i}{n_i}$$

$$var_i = \frac{E_i \times f_i \times h_i}{n_i(n_i - 1)}.$$

The test statistic of homogeneity with respect to relative odds OR is:

$$Q = \sum_{i=1}^{n} (w_i \times (O_i - E_i)^2) - \frac{\sum_{i=1}^{n}(O_i - E_i)^2}{\sum_{i=1}^{n} var_i}.$$

3.3 General Variance based Method

While Mantel-Haenszel and Peto method are applied to integration of ratio statistic, general variance based method is applied to calculating summarised estimators for differences.

Let RD_i denote the difference between the ratio in the table T_i. The summarised estimator RD_s is defined as:

$$RD_s = \frac{\sum_{i=1}^{n} w_i \times RD_i}{\sum_{i=1}^{n} w_i},$$

where w_i and var_i are given as:

$$w_i = \frac{1}{var_i}, \text{ and } var_i = \frac{g_i \times h_i}{e_i \times f_i \times n_i}.$$

The 95% confidence interval for OR is obtained as:

$$95\%C.I. = RD_s \pm 1.96\sqrt{var_s},$$

where var_s is defined as:

$$var_s = \frac{1}{\sum_{i=1}^{n} w_i}.$$

When the ratio of statistics is that of risk, general variance based method will be applied after logarithm transformation. For example, the integrated estimator RR_s for $RR_i (i = 1, \cdots, n)$ is given as:

$$\ln RR_s = \frac{\sum_{i=1}^{n} w_i \times \ln RR_i}{\sum w_i},$$

where $var_i = 1/w_i$ is defined as:

$$var_i = \frac{h_i \times n_i}{e_i \times f_i \times g_i}$$

Then, the 95% confidence interval is:

$$95\%C.I. = e^{RR_s \pm 1.96\sqrt{var_s}},$$

and test of homogeneity is given as:

$$Q = \sum_{i=1}^{n} w_i \times (\ln OR_s - \ln OR_i)^2,$$

which follows chi-square distribution.

3.4 DerSimonian-Laird Method

DerSimonian-Laird method is based on random-effect model. The summarised estimator for odds ratio is given as:

$$\ln OR_{dl} = \frac{\sum_{i=1}^{n} w_i^* \times \ln OR_i}{\sum_{i=1}^{n} w_i^*},$$

where w_i^* is:

$$w_i^* = \frac{1}{\left[D + \frac{1}{w_i}\right]}.$$

D is given as:

$$D = \frac{\{Q - (S-1)\} \times \sum_{i=1}^{n} w_i}{(\sum_{i=1}^{n} w_i)^2 - \sum_{i=1}^{n} w_i^2},$$

where S is the number of papers and Q is:

$$Q = \sum_{i=1}^{n} w_i(\ln OR_i - \ln OR_{mh}).$$

One of the problems with meta analysis is that it cannot be applied to evaluation of test statistics in binary classification or two-way contingency table.

In the case of contingency table ($m \times n$ ($m \geq 3$ or $n \geq 3$)), the definition of a summarised statistic will become complicated.

4 Rule Induction and Meta Analysis

As shown in Section 3, meta analysis methods integrate the statistics of two-way tables, which suggests that statistics of rules for binary classification, such as odds ratio and risk ratio can be integrated into a summarised test statistic.

For example, let us consider the case in Table 4. This contigency table gives a rule set: { $[R_1 = 0] \to [R_2 = 0]$, $[R_1 = 0] \to [R_2 = 1]$, $[R_1 = 1] \to [R_2 = 0]$, $[R_1 = 1] \to [R_2 = 1]$ } and the odd ratio (OR) and relative risk (RR) for this rule set are given as:

$$OR = \frac{a_i d_i}{b_i c_i}, RR = \frac{a_i c_i}{g_i h_i},$$

Thus, summarised odds ratio for OR is obtained by using Manzel-Haenszel method and the estimator for RR is obtained by general-based variance method.

Since meta analysis can be regarded as method for comparison or integration of several contingency tables, this method can be applied into the following problems: (1) integration of rule sets obtained from the datasets collected at different institutes and (2) evaluation of rule sets obtained by cross-validation and bootstrap method.

In this paper, we focus on the latter case, especially on the integration of rule sets obtained by leave-one out method, using an illustrative example shown in Table 5.

Table 5: Example of Dataset

No.	age	location	nature	prodrome	nausea	M1	class
1	50-59	occular	persistent	no	no	yes	m.c.h.
2	40-49	whole	persistent	no	no	yes	m.c.h.
3	40-49	lateral	throbbing	no	yes	no	migra
4	40-49	whole	throbbing	yes	yes	no	migra
5	40-49	whole	radiating	no	no	yes	m.c.h.
6	50-59	whole	persistent	no	yes	yes	psycho

DEFINITIONS. M1: tenderness of M1, m.c.h.: muscle contraction headache, migra: migraine, psycho: psychological pain

From this table, a rule $[nausea = yes] \to migra$ will be obtained and the corresponding table will be given as Table 6.

The relative risk ratio is:

$$RR = \frac{2/2}{1/4} = 4.$$

Note that odds ratio cannot be defined for Mantel-Haenszel method in this case since zeros are included.

Table 6: Contingency Table for nausea and migra

	nausea		
	yes	*no*	
$migraine = yes$	2	0	2
$migraine = no$	1	3	4
	3	3	6

In the case of leave-one out method, we can make six different contingency tables since we have six examples in the table. For example, the contingency table where the sixth example is eliminated is given as Table 7. Note that the relative risk cannot be defined in Table 7.

Table 7: Contigency Table with the Elimination of Sixth Examples

	nausea		
	yes	*no*	
$migraine = yes$	2	0	2
$migraine = no$	0	3	3
	2	3	5

From the contingency tables obtained from Table 5, a relative risk and its variance for each table are summarised into Table 8. From this table, RR_s is defined as:

$$\begin{aligned}
\ln RR_s &= \frac{\sum_{i=1}^{n} w_i \times \ln RR_i}{\sum w_i} \\
&= \frac{0.8 \ln 3 + 0.8 \ln 3 + 0.3 \ln 4 + 0.3 \ln 4}{0.8 + 0.8 + 0.3 + 0.3} \\
&= \frac{1.6 \ln 3 + 0.6 \ln 4}{2.2} \\
&= 0.958,
\end{aligned}$$

by using the general variance based method. Since var_s is obtained as $1/2.2 = 0.455$, its 95% confidence interval is obtained as:

$$95\%C.I. = e^{0.958 \pm 1.96\sqrt{0.455}} = (0.693, 9.796).$$

Table 8: Relative Risk and Variance obtained by using Leave-one out method

	1	2	3	4	5	6
Relative Risk	3	3	4	4	3	–
Variance	5/4	5/4	10/3	10/3	5/4	5/4
Weight	0.8	0.8	0.3	0.3	0.8	0.8

Table 9: Expected values and variances for Peto Method

	1	2	3	4	5	6
Expected Value	1.2	1.2	1.2	1.2	1.2	1.2
Variance	0.36	0.36	0.72	0.72	0.36	0.54
Weight	0.8	0.8	0.3	0.3	0.8	0.8

On the other hand, Table 9 gives the odds ratio obtained by Peto method, where the expected value E_i and its corresponding variance var_i are summarised.

From Table 9, the relative odds ratio is calculated as:

$$\begin{aligned} \ln OR_p &= \frac{\sum_{i=1}^{n}(O_i - E_i)}{\sum_{i=1}^{n} var_i} \\ &= 0.915 \end{aligned}$$

and 95% confindence interval is:

$$\begin{aligned} 95\%C.I. &= e^{\ln OR_p \pm 1.96\sqrt{\sum_{i=1}^{6} var_i}} \\ &= e^{0.915 \pm 1.96 \times 1.75} \\ &= (0.589, 458) \end{aligned}$$

In this case, Q-statistics of Peto method is calculated as:

$$Q = \sum_{i=1}^{n}(w_i \times (O_i - E_i)^2) - \frac{\sum_{i=1}^{n}(O_i - E_i)^2}{\sum_{i=1}^{n} var_i} = 4.65,$$

whose p-value is 0.03. Since both confidence intervals derived by Peto method and general variance based method are large, we can conclude that the reliability of knowledge acquired from this table is not so high. Moreover, p-value shows that this datasets are not so homogeneous if the threshold α is set to 0.05.

5 Conclusion

This paper consists of the two parts. The first part introduces epidemiological approach which argues that data collection and data analysis determines the quality of analysis. Especially, prospective study, hypothesis-based data collection gives better results for generalisation than retrospective studies where data comes first. Then, the discussions from EBM show the quality of evidence on results obtained by data mining methods. The type of evidence is III_b from the viewpoint of EBM and meta analysis can be upgraded into

The second part shows meta analysis and discusses that this method can be applied to rule induction methods since a rule set corresponds to a contingency table. We illustrate meta analysis with integration of rules from the tables obtained by leave-one out method.

Acknowledgements

This research was supported by the Grant-in-Aid for Scientific Research on Priority Areas(B) (No.759) "Implementation of Active Mining in the Era of Information Flood" by the Ministry of Education, Science, Culture, Science and Technology of Japan.

References

[1] Efron, B. *The Jackknife, the Bootstrap and Other Resampling Plans.* Society for Industrial and Applied Mathematics, Pennsylvania, 1982.

[2] Mantel, N. and Haenszal, W. Statistical aspects of the analysis of data from retrospective studies of disease. *J Natl Cancer Inst*, **22: 719-748**. 1959.

[3] Michalski, R.S. A Theory and Methodology of Machine Learning. Michalski, R.S., Carbonell, J.G. and Mitchell, T.M., *Machine Learning - An Artificial Intelligence Approach*, 83-134, Morgan Kaufmann, CA, 1983.

[4] Mullen, B. *Advanced Basic Meta Analysis.* Lawrence-Erlbaum, 1989.

[5] Petitti, D.B. *Meta-Analysis, Decision Analysis, and Cost-Effectiveness Analysis*, Oxford University Press, Oxford, 1994.

[6] Quinlan JR: *C4.5 - Programs for Machine Learning.* Morgan Kaufmann, Palo Alto CA, 1993.

[7] Skowron, A. and Grzymala-Busse, J. From rough set theory to evidence theory. In: Yager, R., Fedrizzi, M. and Kacprzyk, J.(eds.) *Advances in the Dempster-Shafer Theory of Evidence*, pp.193-236, John Wiley & Sons, New York, 1994.

[8] Tsumoto, S. Knowledge discovery in clinical databases and evaluation of discovered knowledge in outpatient clinic. *Information Sciences*, **124**, 125-137, 2000.

[9] Tsumoto, S. Automated Discovery of Positive and Negative Knowledge in Clinical Databases based on Rough Set Model., *IEEE EMB Magazine*,**19**: 56-62, 2000.

[10] Tsumoto, S. and Takabayashi, K. Comparison and Evaluation of Knowledge KDD methods Journal of Japanese AI society, **15**(5), 790-797, 2000 (Japanese).

[11] Tsumoto, S. Possiblity of Knowledge Discovery in Medicine (Special Issue: Data Mining Contest), Information Processing , vol. 42, pp. 472–477, 2001 (Japanese).

Author Index